# PEARSON EDEXCEL INTERNATIONAL GCSE (9–1)

# CHEMISTRY

## Student Book

Jim Clark
Steve Owen
Rachel Yu

Published by Pearson Education Limited, 80 Strand, London, WC2R 0RL.

www.pearsonglobalschools.com

Copies of official specifications for all Edexcel qualifications may be found on the website: https://qualifications.pearson.com

Text © Pearson Education Limited 2017
Edited by Lesley Montford
Designed by Cobalt id
Typeset by Tech-Set Ltd, Gateshead, UK
Original illustrations © Pearson Education Limited 2017
Illustrated by © Tech-Set Ltd, Gateshead, UK
Cover design by Pearson Education Limited
Cover photo © NASA

The rights of Jim Clark, Steve Owen and Rachel Yu to be identified as authors of this work have been asserted by them in accordance with the Copyright, Designs and Patents Act 1988.

First published 2017

20 19 18
10 9 8 7 6 5 4

**British Library Cataloguing in Publication Data**
A catalogue record for this book is available from the British Library

ISBN 978 0 435 18516 9

Printed by L.E.G.O. S.p.A. - Lavis - TN - Italy

**Endorsement Statement**
In order to ensure that this resource offers high-quality support for the associated Pearson qualification, it has been through a review process by the awarding body. This process confirms that this resource fully covers the teaching and learning content of the specification or part of a specification at which it is aimed. It also confirms that it demonstrates an appropriate balance between the development of subject skills, knowledge and understanding, in addition to preparation for assessment.

Endorsement does not cover any guidance on assessment activities or processes (e.g. practice questions or advice on how to answer assessment questions), included in the resource nor does it prescribe any particular approach to the teaching or delivery of a related course.

While the publishers have made every attempt to ensure that advice on the qualification and its assessment is accurate, the official specification and associated assessment guidance materials are the only authoritative source of information and should always be referred to for definitive guidance.

Pearson examiners have not contributed to any sections in this resource relevant to examination papers for which they have responsibility.

Examiners will not use endorsed resources as a source of material for any assessment set by Pearson. Endorsement of a resource does not mean that the resource is required to achieve this Pearson qualification, nor does it mean that it is the only suitable material available to support the qualification, and any resource lists produced by the awarding body shall include this and other appropriate resources.

**Picture Credits**
The author and publisher would like to thank the following individuals and organisations for permission to reproduce photographs:

(Key: b-bottom; c-centre; l-left; r-right; t-top)

**123RF.com:** 85t, 157tr, 167tr, Gregory Bruev 293c, Scandal 277cl, unlim3d 164c; **Alamy Stock Photo:** Aurora 163tl, 264bl, Bon Appetit 285tl, Trevor Chriss 81b, Ashley Cooper 305tr, Cultura Creative 163br, 167tl, Phil Degginger 9tl, DJC 165cl, Dpa picture alliance archive 288bl, Robert Gilhooly 164tl, LGPL / Ian Cartwright 157cl, Nikos Pavlakis 145tr, David Taylor 193-18.6t, World History Archive 30tl, Zoonar GmbH 306br; **Fotolia.com:** Chungking 98tl, Grinchh 17bl, Arpad Nagy-Bagoly 302c, Satit Srihin 290cl; **Getty Images:** In Pictures / Corbis / Gideon Mendel 223b, Miguel Malo 145tl, NASA / National Geographic 2r, PaulFleet 208tl, TASS / Maxim Grigoryev 122b; **Nature Picture Library:** Nature Production 235; **Pearson Education Ltd:** Oxford Designers & Illustrators Ltd 305tl; **Science Photo Library Ltd:** 130bl, 165c, 180, 193-18.8t, 193-18.10t, 235bl, 241, Andrew Lambert Photography 20c, 54tl, 57b, 130tr, 132r, 141tr, 142cl, 195t, 195br, Caia Image 287t, Carol and Mike Werner 254c, Martyn F Chillmaid 44tl, 70c, 123tr, 124cl, 127tl, 141cr, 155br, 162c, 170cl, 175tr, 177b, 180bl, 182, 184tl, 191bc, 192bl, 193cr, 193br, 194c, 207b, 215cr, 240bl, 284bl, 294bl, 302tl, Crown Copyright / Health and Safety Laboratory 190tc, Phil Degginger 209bl, Fundamental Photos 149bl, GIPhotoStock 168, 169bl, 186b, 193-18.9t, 227-20.2tc, Gusto Images 6tl, Mikkel Juul Jensen 75c, MARTYN F. CHILLMAID 182, 207b, Alfred Pasieka 263cl, Lea Paterson 255tl, PATRICK WALLET / EURELIOS 300t, Volker Steger 302tr, David Taylor 193-18.7t, Trevor Clifford 101tr, UIG / Dorking Kindersley 232tl, Charles D Winters 64tr, 89tr, 131br, 132l, 183tl, 246tl, 306; **Shutterstock.com:** 160cr, 272c, Aivolie 46t, Anastasios71 137tr, Galyna Andrushko 173tr, ArtisticPhoto 98tl, H E Benson 3tl, Bitt24 282tr, Sinisa Botas 271cl, Sherri R. Camp 227tl, Checubus 237, Marcel Clemens 36cl, Daizuoxin 303br, Dashu 268c, Demarcomedia 35cl, Dencg 14cl, Docent 310bl, Dr.OGA 255, Everett Historical 206c, Extender_01 82tl, Mike Flippo 164b, Peter Hermes Furian 82bl, Jiri Hera 94cl, IM photo 68tl, JPC Prod 297t, Anan Kaewkhammul 269tr, KDEdesign 94tl, Jatuporn Khuansuwan 56c, Ktsdesign 38t, Ilya Malov 227-20.3t, Musicman 271br, Mylisa 14tr, Romanova Natali 24br, Oleg1969 24bl, Palmaria 307b, Werayuth Piriyapornprapa 272t, Vadim Ratnikov 272b, Remedios55 94cr, Albert Russ 160cl, Aygul Sarvarova 305tc, Scanrail1 95cl, Jeff Schultes 227tr, Sam Strickler 255tr, Kuttelvaserova Stuchelova 67tl, Mary Terribery 273bc, Dave Turner 125bl, VanHart 3tr, Ventin 36tl, XXLPhoto 273tc, Ron Zmiri 156cr

**Cover images:** *Front:* **NASA**

*Inside front cover:* **Shutterstock.com:** Dmitry Lobanov

All other images © Pearson Education

**Neither Pearson, Edexcel nor the authors take responsibility for the safety of any activity.** Before doing any practical activity you are legally required to carry out your own risk assessment. In particular, any local rules issued by your employer must be obeyed, regardless of what is recommended in this resource. Where students are required to write their own risk assessments they must always be checked by the teacher and revised, as necessary, to cover any issues the students may have overlooked. The teacher should always have the final control as to how the practical is conducted.

# CONTENTS

# UNIT 1

## PRINCIPLES OF CHEMISTRY

# UNIT 2

## INORGANIC CHEMISTRY

# UNIT 3

## PHYSICAL CHEMISTRY

# UNIT 4

## ORGANIC CHEMISTRY

# ABOUT THIS BOOK

This book is written for students following the Pearson Edexcel International GCSE (9–1) Chemistry specification and the Edexcel International GCSE (9–1) Science Double Award specification. You will need to study all of the content in this book for your Chemistry examination. However, you will only need to study some of it if you are taking the Double Award specification. The book clearly indicates which content is in the Chemistry examination and not in the Double Award specification. To complete the Double Award course you will also need to study the Physics and Biology parts of the course.

In each unit of this book, there are concise explanations and worked examples, plus numerous exercises that will help you build up confidence. The book also describes the methods for carrying out all of the required practicals.

The language throughout this textbook is graded for speakers of English as an additional language (EAL), with advanced Chemistry-specific terminology highlighted and defined in the glossary at the back of the book. A list of command words, also at the back of the book, will help you to learn the language you will need in your examinations.

You will also find that questions in this book have Progression icons and Skills tags. The Progression icons refer to Pearson's Progression scale. This scale – from 1 to 12 – tells you what level you have reached in your learning and will help you to see what you need to do to progress to the next level. Furthermore, Edexcel have developed a Skills grid showing the skills you will practise throughout your time on the course. The skills in the grid have been matched to questions in this book to help you see which skills you are developing. Both Skills tags and Progression icons are not repeated where they are same in consecutive questions. You can find Pearson's Progression scale and Edexcel's Skills grid at www.pearsonglobalschools.com/igscienceprogression along with guidelines on how to use them.

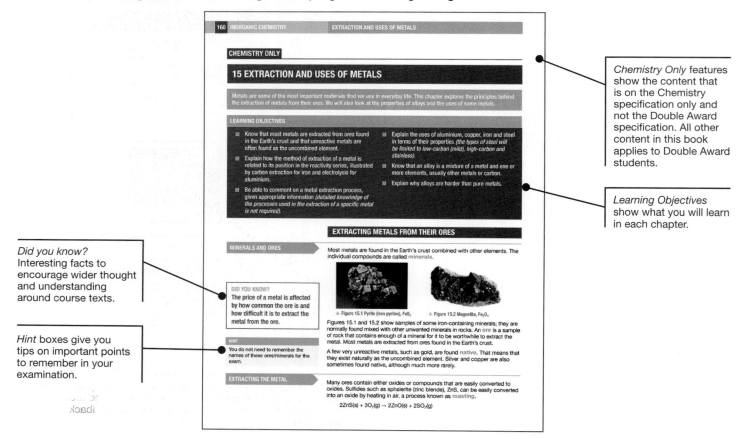

*Did you know?* Interesting facts to encourage wider thought and understanding around course texts.

*Hint* boxes give you tips on important points to remember in your examination.

*Chemistry Only* features show the content that is on the Chemistry specification only and not the Double Award specification. All other content in this book applies to Double Award students.

*Learning Objectives* show what you will learn in each chapter.

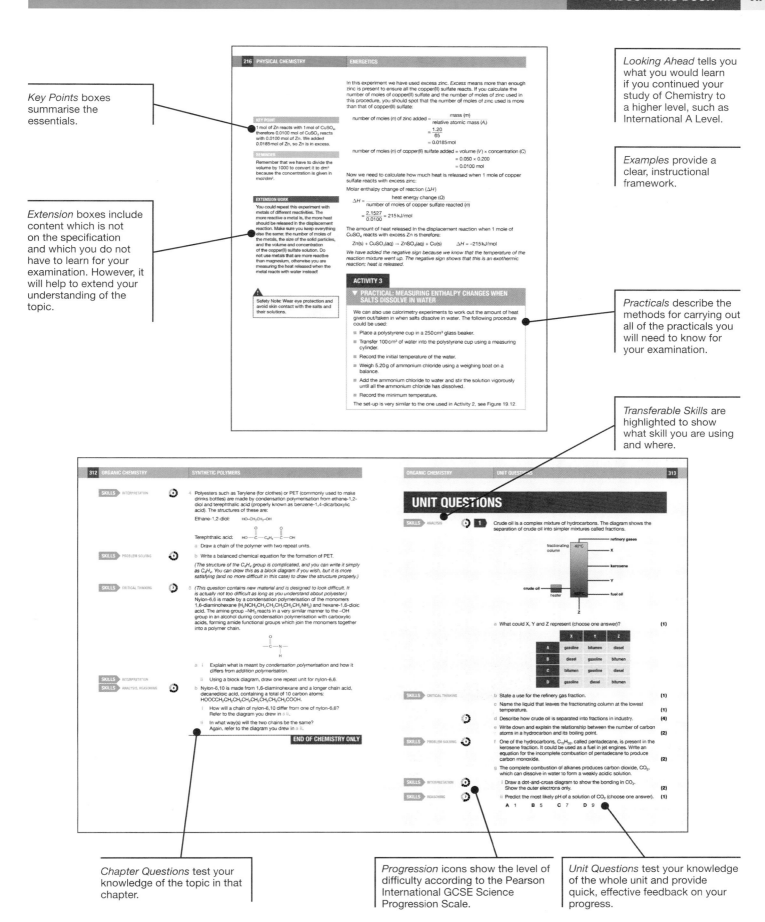

Key Points boxes summarise the essentials.

Extension boxes include content which is not on the specification and which you do not have to learn for your examination. However, it will help to extend your understanding of the topic.

Looking Ahead tells you what you would learn if you continued your study of Chemistry to a higher level, such as International A Level.

Examples provide a clear, instructional framework.

Practicals describe the methods for carrying out all of the practicals you will need to know for your examination.

Transferable Skills are highlighted to show what skill you are using and where.

Chapter Questions test your knowledge of the topic in that chapter.

Progression icons show the level of difficulty according to the Pearson International GCSE Science Progression Scale.

Unit Questions test your knowledge of the whole unit and provide quick, effective feedback on your progress.

# ASSESSMENT OVERVIEW

The following tables give an overview of the assessment for this course.

We recommend that you study this information closely to help ensure that you are fully prepared for this course and know exactly what to expect in the assessment.

| PAPER 1 | SPECIFICATION | PERCENTAGE | MARK | TIME | AVAILABILITY |
|---|---|---|---|---|---|
| Written examination paper<br>Paper code 4CH1/1C and 4SD0/1C<br>Externally set and assessed by Edexcel | Chemistry<br>Science Double Award | 61.1% | 110 | 2 hours | January and June examination series<br>First assessment June 2019 |
| **PAPER 2** | **SPECIFICATION** | **PERCENTAGE** | **MARK** | **TIME** | **AVAILABILITY** |
| Written examination paper<br>Paper code 4CH1/2C<br>Externally set and assessed by Edexcel | Chemistry | 38.9% | 70 | 1 hour 15 mins | January and June examination series<br>First assessment June 2019 |

If you are studying Chemistry then you will take both Papers 1 and 2. If you are studying Science Double Award then you will only need to take Paper 1 (along with Paper 1 for each of the Physics and Biology courses).

## ASSESSMENT OBJECTIVES AND WEIGHTINGS

| ASSESSMENT OBJECTIVE | DESCRIPTION | % IN INTERNATIONAL GCSE |
|---|---|---|
| AO1 | Knowledge and understanding of chemistry | 38%–42% |
| AO2 | Application of knowledge and understanding, analysis and evaluation of chemistry | 38%–42% |
| AO3 | Experimental skills, analysis and evaluation of data and methods in chemistry | 19%–21% |

# EXPERIMENTAL SKILLS

In the assessment of experimental skills, students may be tested on their ability to:

- solve problems set in a practical context

- apply scientific knowledge and understanding in questions with a practical context

- devise and plan investigations, using scientific knowledge and understanding when selecting appropriate techniques

- demonstrate or describe appropriate experimental and investigative methods, including safe and skilful practical techniques

- make observations and measurements with appropriate precision, record these methodically and present them in appropriate ways

- identify independent, dependent and control variables

- use scientific knowledge and understanding to analyse and interpret data to draw conclusions from experimental activities that are consistent with the evidence

- communicate the findings from experimental activities, using appropriate technical language, relevant calculations and graphs

- assess the reliability of an experimental activity

- evaluate data and methods, taking into account factors that affect accuracy and validity.

# CALCULATORS

Students are permitted to take a suitable calculator into the examinations. Calculators with QWERTY keyboards or that can retrieve text or formulae will not be permitted.

# UNIT 1 PRINCIPLES OF CHEMISTRY

The universe is made of three things!

Up to the present day scientists have discovered 118 elements. Most of these have been made naturally in stars but some are made artificially. As far as we know these are the only elements in the universe, so we basically have a model kit containing 118 different atoms. Chemistry can be described as the study of how these different atoms are joined together in various ways to make everything around us, from a tree, to a person, to the tallest skyscraper. Many of these elements are not very common so most of the things we see around us are made up of different combinations of only about a quarter of these elements. What makes this even more amazing is that each atom is made up of just three subatomic particles, which are called protons, neutrons and electrons. So, the world around us is made of only three things arranged in different ways.

▲ Figure 1.1 Southern view of the Milky Way

# 1 STATES OF MATTER

Everything around us is made of particles that we can't see because they are so small. This chapter looks at the arrangement of particles in solids, liquids and gases, and the ways in which the particles can move around. The nature of the different sorts of particles will be explored in Chapter 3.

▲ Figure 1.2 Everything you look at is a solid, a liquid or a gas . . .

▲ Figure 1.3 . . . metals, concrete, water, air, clouds – everything!

## LEARNING OBJECTIVES

■ Understand the three states of matter in terms of the arrangement, movement and energy of the particles

■ Understand the interconversions between the three states of matter in terms of:

   ■ the names of the interconversions

   ■ how they are achieved

   ■ the changes in arrangement, movement and energy of the particles

■ Understand how the results of experiments involving the dilution of coloured solutions and diffusion of gases can be explained

■ Know what is meant by the terms:

   ■ solvent           ■ solute

   ■ solution          ■ saturated solution

### CHEMISTRY ONLY

■ Know what is meant by the term solubility in the units g per 100 g of solvent.

■ Understand how to plot and interpret solubility curves.

■ Practical: Investigate the solubility of a solid in water at a specific temperature.

## STATES OF MATTER

Solids, liquids and gases are known as the three states of matter.

**THE ARRANGEMENT OF THE PARTICLES**

Think about these facts:

■ You can't walk through a brick wall, but you can move (with some resistance – it pushes against you) through water. Moving through air is easy.

■ When you melt most solids their volume increases slightly. Most liquids are less dense than the solid they come from.

■ If you boil about 5 cm³ of water, the steam will fill an average bucket.

The arrangement of the particles in solids, liquids and gases explains these facts.

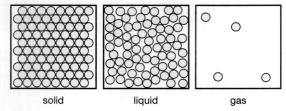

▲ Figure 1.4 The arrangement of particles in different states of matter

In a solid, the particles are usually arranged regularly and packed closely together. The particles are only able to vibrate about fixed positions; they can't move around. The particles have strong forces of attraction between them, which keep them together.

In a liquid, the particles are still mostly touching, but some gaps have appeared. This is why liquids are usually less dense than solids. The forces between the particles are less effective, and the particles can move around each other. The particles in a liquid are arranged randomly.

The particles in a gas are moving randomly at high speed in all directions. In a gas, the particles are much further apart and there are (almost) no forces of attraction between them.

The particles in a solid have less kinetic (movement) energy than the particles in a liquid, which have less kinetic energy than the particles in a gas.

# INTERCONVERSIONS BETWEEN THE THREE STATES OF MATTER

**CHANGING STATE BETWEEN SOLID AND LIQUID**

If you heat a solid, the energy provided by the heat source makes the particles in the solid vibrate faster and faster. Eventually, they vibrate so fast that the forces of attraction between the particles are no longer strong enough to hold them together; the particles are then able to move around each other – the solid melts to form a liquid. The temperature at which the solid melts is called its **melting point**. The particles in the liquid have more kinetic energy than the particles in the solid so energy has to be supplied to convert a solid to a liquid.

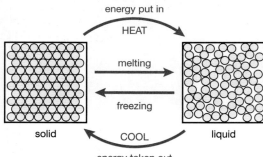

▲ Figure 1.5 Melting to become a liquid – and freezing to become a solid.

If the liquid is cooled again, the liquid particles will move around more and more slowly. Eventually, they are moving so slowly that the forces of attraction between them will hold them in a fixed position and the particles pack more closely together into a solid. The liquid **freezes**, forming a solid. The temperature at which this occurs is called the **freezing point**.

Although they are called different things depending which way you are going, the temperature of the melting point and that of the freezing point of a substance are exactly the same.

## CHANGING STATE BETWEEN LIQUID AND GAS

There are two different ways this can happen, called boiling and evaporation.

### BOILING

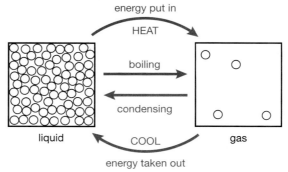

▲ Figure 1.6 Boiling to become a gas – and condensing to become a liquid.

Boiling occurs when a liquid is heated so strongly that the particles are moving fast enough to overcome all the forces of attraction between them. The stronger the forces of attraction between particles, the higher the boiling point of the liquid. This is because more energy is needed to overcome these forces of attraction.

If a gas is cooled, the particles eventually move slowly enough that forces of attraction between them start to form and hold them together as a liquid. The gas condenses.

### EVAPORATION

Evaporation is different. In any liquid or gas, the average speed of the particles varies with the temperature. But at each temperature, some particles will be moving faster than the average and others more slowly.

Some very fast particles at the surface of the liquid will have enough energy to overcome the forces of attraction between the particles – they will break away to form a gas. This is evaporation. You don't see any bubbling; the liquid just slowly disappears if it is open to the air. If the liquid is in a closed container, particles in the gas will also be colliding with particles at the surface of the liquid. If they are moving slowly enough they will be held by the attractive forces and become part of the liquid. In a closed container evaporation and condensation will both be occurring at the same time.

**KEY POINT**

Evaporation occurs at any temperature, but boiling only occurs at one temperature – the boiling point of the liquid. Puddles of water disappear quite quickly despite the outside temperature often being below 5 °C in the winter in the UK. The water in the puddles certainly does not boil at this temperature; the water evaporates. So water will evaporate at, for example, 5 °C but only boil at 100 °C.

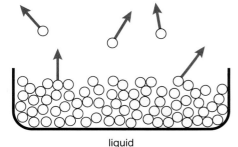

faster moving particles escaping from the surface to form a gas

liquid

▲ Figure 1.7 Evaporation.

## CHANGING STATE BETWEEN SOLID AND GAS: SUBLIMATION

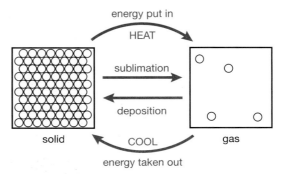

▲ Figure 1.8 This change of state goes directly from a solid to a gas and from a gas to a solid.

A small number of substances can change directly from a solid to a gas, or from a gas to a solid, at normal pressure without involving any liquid in the process. The conversion of a solid into a gas is known as sublimation and the reverse process is usually called deposition.

**KEY POINT**

The process of a gas changing into a solid is given various names. Some people call it 'de-sublimation' or 'deposition' and others just use the word 'sublimation' again.

▲ Figure 1.9 Dry ice subliming. Notice the white solid carbon dioxide in the beaker. The white cloud is because the carbon dioxide gas produced is so cold that it causes water vapour in the air to condense. Carbon dioxide gas itself is invisible.

An example of a substance that sublimes is carbon dioxide. At ordinary pressures, there is no such thing as liquid carbon dioxide – it turns directly from a solid to a gas at –78.5 °C. Solid carbon dioxide is known as dry ice.

## WORKING OUT THE PHYSICAL STATE OF A SUBSTANCE AT A PARTICULAR TEMPERATURE

A substance is a solid at temperatures below its melting point, between its melting point and its boiling point it is a liquid, and above its boiling point it is a gas.

In science we can decide whether a substance is a solid, a liquid or a gas at room temperature by looking at where its melting and boiling points are in relation to room temperature.

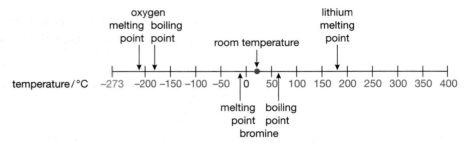

▲ Figure 1.10 A temperature line can be used to work out whether substances are solids, liquids or gases.

If we look at the temperature line in Figure 1.10 we can see that room temperature is above the boiling point of oxygen; this means that oxygen is a gas at room temperature.

Let's look at what happens when we heat bromine from –100 °C to 100 °C. As –100 °C is below bromine's melting point, bromine is a solid at –100 °C. As it is heated to –7 °C (its melting point) it becomes a liquid and it remains as a liquid until its temperature reaches the boiling point at 59 °C. Room temperature is between the melting point and the boiling point, which means that bromine is a liquid at room temperature. Above 59 °C bromine is a gas.

Lithium's melting point is above room temperature and so it is a solid at room temperature.

### KEY POINT

Room temperature is different in different places but in science it is usually taken to mean a temperature between 20 and 25 °C. Because there is not just one fixed value, for changes of state that occur near room temperature we must be careful when making comparisons and make clear what value is being used as room temperature.

## DIFFUSION

### DIFFUSION IN GASES

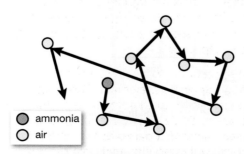

○ ammonia
○ air

▲ Figure 1.11 An ammonia particle bouncing off air particles.

Suppose someone accidentally releases some smelly gas in the lab, ammonia for example. Within a minute or so, everybody in the lab will be able to smell it. That isn't surprising – particles in the gas are free to move around. What does need explaining, though, is why it takes so long.

At room temperature, ammonia particles travel at speeds of about 600 metres per second so they should be able to travel from one end of a lab to the other in less than 1/100 s (0.01 s). This would be the case if they travelled in a straight line without bumping into anything else. However, each particle is bouncing off air particles on its way. In the time that it takes for the smell to reach all corners of the lab, each ammonia particle may have travelled 30 or more kilometres!

The spreading out of particles in a gas or liquid is known as **diffusion**. We say that ammonia particles *diffuse* through the air. A formal definition of diffusion is:

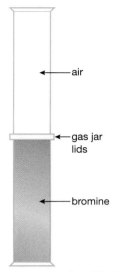

▲ Figure 1.13 Demonstrating diffusion in gases

Safety Note: The teacher demonstration must be prepared in a working fume cupboard wearing eye protection and chemical-resistant gloves. Inhalation of bromine by anyone with breathing difficulties may produce a reaction, possibly delayed, requiring urgent medical attention.

### SHOWING THAT PARTICLES OF DIFFERENT GASES TRAVEL AT DIFFERENT SPEEDS

#### HINT

Don't worry if you don't know how to write symbol equations. This one is included here so that you can refer to it again in later revision.

Safety Note: The teacher demonstration requires eye protection and the avoidance of skin contact and inhalation of any fumes. The apparatus has to be cleaned up in a working fume cupboard.

#### KEY POINT

You will learn about relative molecular mass in Chapter 5. The relative molecular mass of ammonia is 17 and that of hydrogen chloride is 36.5.

*Diffusion is the spreading out of particles from where they are at a high concentration (there are lots of them in a certain volume) to where they are at a low concentration (there are fewer of them in a certain volume).*

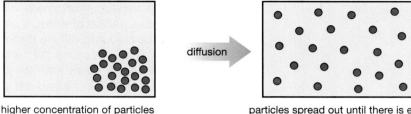

higher concentration of particles in a certain region

particles spread out until there is equal concentration throughout the container

▲ Figure 1.12 Diffusion involves the spreading out of particles.

You can show diffusion in gases very easily by using the apparatus in Figure 1.13. The lower gas jar contains bromine gas; the top one contains air. If the lids are removed, the brown colour of the bromine diffuses upwards until both gas jars are uniformly brown (the air particles also diffuse downwards). The bromine particles and air particles move around at random to give an even mixture – both gas jars contain air and bromine particles.

You can carry out the same experiment with hydrogen and air, but in this example you have to put a lighted splint in at the end to find out where the gases have gone. People often expect that the much less dense hydrogen will all go to the top gas jar. In fact, you will get identical explosions from both jars.

This experiment relies on the reaction between ammonia ($NH_3$) and hydrogen chloride (HCl) gases to give white solid ammonium chloride ($NH_4Cl$):

$$NH_3(g) + HCl(g) \rightarrow NH_4Cl(s)$$

cotton wool soaked in concentrated ammonia solution

cotton wool soaked in concentrated hydrochloric acid

white ring forms closer to the hydrochloric acid end

▲ Figure 1.14 Demonstrating that particles in ammonia and hydrogen chloride travel at different speeds.

Pieces of cotton wool are soaked in concentrated ammonia solution (as a source of ammonia gas) and concentrated hydrochloric acid (as a source of hydrogen chloride gas). These are placed in the ends of a long glass tube with rubber bungs to stop the poisonous gases escaping.

Ammonia particles and hydrogen chloride particles diffuse along the tube. A white ring of solid ammonium chloride forms where they meet. The white ring of ammonium chloride takes time to form (as it takes some time for the particles of ammonia and hydrogen chloride to diffuse along the tube), and appears *closer to the hydrochloric acid end*. Ammonia particles are lighter than hydrogen chloride particles and therefore move faster. The ammonia particles travel further in the same amount of time, which means that the ring forms further away from the ammonia end.

## DIFFUSION IN LIQUIDS

Diffusion through a liquid is very slow if the liquid is completely still. For example, if a small jar of strongly coloured solution (such as potassium manganate(VII) solution) is placed in a gas jar of water, it can take days for the colour to diffuse throughout all the water. This is because *the particles in a liquid move more slowly than the particles in a gas*. The particles in a liquid are also much closer together than those in a gas and so there is less space for particles to move into without colliding with another one.

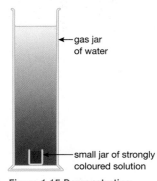

gas jar of water

small jar of strongly coloured solution

▲ Figure 1.15 Demonstrating diffusion in liquids

## THE DILUTION OF COLOURED SOLUTIONS

Imagine you dissolve 0.01 g of potassium manganate(VII) in 1 cm³ of water to make a deep purple solution. If we take the volume of 1 drop as 0.05 cm³ we can work out that there are 20 drops in 1 cm³ and each drop will contain 0.0005 g of potassium manganate(VII).

If you dilute this solution by adding water until the total volume is 10 000 cm³, you should still just be able to see the purple colour.

There are now 200 000 drops in the solution. In order to see the colour each drop must contain at least one 'particle' of potassium manganate(VII), so there must be at least 200 000 'particles' in 0.01 g of potassium manganate(VII). This means that each 'particle' can't weigh more than 50 billionths of a gram (0.00000005 g).

This answer is not even close to the real answer. A potassium manganate(VII) 'particle' actually weighs about 0.00000000000000000000026 g and there are about 38 000 000 000 000 000 000 particles in 0.01 g! In reality, you need very large numbers of particles in each drop in order to see the colour.

### REMINDER

Why the inverted commas around 'particle'? Potassium manganate(VII) is an *ionic compound* and contains more than one sort of particle. You will find out more about ionic compounds in Chapter 7.

# THE SOLUBILITY OF SOLIDS

## SOLUTES, SOLVENTS AND SOLUTIONS

When a solid dissolves in a liquid:
- the substance that dissolves is called the **solute**
- the liquid it dissolves in is called the **solvent**
- the liquid formed is a **solution**.

When you make a solution, the attractive forces between the particles in the solute (the solid) are being broken. At the same time, new attractive forces are being formed between the solvent particles and the solute particles. Whether a particular solid is soluble in any solvent depends on whether the new attractive forces are strong enough to overcome the old ones.

Only a certain amount of solute will dissolve in a fixed amount of solute at a particular temperature. When the maximum amount is dissolved a saturated solution is obtained. A **saturated solution** is a solution which contains as much dissolved solid as possible at a particular temperature. There must be some undissolved solute present.

## MEASURING SOLUBILITY

# CHEMISTRY ONLY

The **solubility** of a solid in a solvent at a particular temperature is usually defined as *'the mass of solute which must dissolve in 100 g of solvent at that temperature to form a saturated solution'*.

▲ Figure 1.16 A saturated solution

! 

Safety Note: Wear eye protection and heat gently to avoid burns from hot solid 'spitting' out of the basin.

In other words, it is the maximum mass of solute that dissolves in 100 g of solvent at a particular temperature.

For example, the solubility of sodium chloride (common salt) in water at 25 °C is about 36 g per 100 g of water.

## ACTIVITY 1

### ▼ PRACTICAL: INVESTIGATING THE SOLUBILITY OF A SOLID IN WATER

A procedure we can use to measure the solubility of potassium nitrate in water at 40 °C is as follows:

1. Weigh an evaporating basin.
2. Heat a boiling tube of water to just above 40 °C.
3. Add potassium nitrate to the water in the boiling tube and stir rapidly until no more of it will dissolve and there is undissolved solid left over.
4. Allow the solution to cool to exactly 40 °C.
5. Pour off some of the solution into the evaporating basin (it is important that you only pour off solution and no solid). You do not have to pour off all the solution.
6. Weigh the evaporating basin and contents.
7. Heat the evaporating basin and contents gently to evaporate off all the water.
8. When it looks as if all the water has evaporated weigh the evaporating basin and contents.
9. Heat the evaporating basin and contents again and then re-weigh. This is to make sure that all the water has, indeed, evaporated and is called *heating to constant mass*.

This procedure is summarised in Figure 1.17.

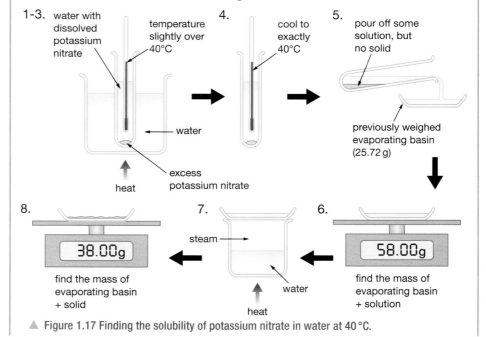

▲ Figure 1.17 Finding the solubility of potassium nitrate in water at 40 °C.

We heat the solution gently to make sure that none spits out. If some did spit out we would record a lower mass of solid and the solubility would appear to be lower than the actual value.

The results for this experiment could be:

| Mass of evaporating basin/g | 25.72 |
| --- | --- |
| Mass of evaporating basin + solution/g | 58.00 |
| Mass of evaporating basin + dry crystals/g | 38.00 |

We need to calculate the mass of the solid and also the mass of water evaporated from the solution:

mass of crystals = 38.00 – 25.72 = 12.28 g

mass of water = 58.00 – 38.00 g = 20.00 g

12.28 g of solid is the maximum mass that dissolves in 20.00 g of water, therefore 5 times as much would dissolve in 100 g of water. That works out at 61.4 g. The solubility of potassium nitrate at 40 °C is therefore 61.4 g per 100 g of water.

More generally, we can calculate the solubility of a substance in 100 g of solvent using the equation:

$$\text{solubility (g/100 g)} = \frac{\text{mass of solute}}{\text{mass of solvent}} \times 100$$

## SOLUBILITY CURVES

The solubility of solids changes with temperature and you can plot this on a solubility curve. Most solids have solubility curves like those for the salts shown in Figure 1.18. Their solubility increases with temperature – either dramatically or just a little.

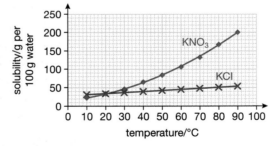

▲ Figure 1.18 Solubility curves for potassium nitrate and potassium chloride

You can use solubility curves to work out what mass of crystals you would get if you cooled a saturated solution.

Consider the solubility curve for potassium nitrate ($KNO_3$) in Figure 1.18. At 90 °C 200 g of potassium nitrate dissolves in 100 g water. At 30 °C only 50 g will dissolve. Therefore, if we have a solution containing 200 g of potassium nitrate dissolved in 100 g of water and let it cool down from 90 °C to 30 °C, 150 g of potassium nitrate must be released from the solution, which it does as crystals. We say that potassium nitrate *crystallises out of the solution* or *precipitates out of the solution*.

The table shows the solubility of potassium chloride at various temperatures.

| Temperature/°C | 10 | 30 | 40 | 70 | 90 |
|---|---|---|---|---|---|
| Solubility/g per 100 g of water | 31.2 | 37.2 | 40.0 | 48.5 | 53.9 |

a  Plot a solubility curve for potassium chloride.

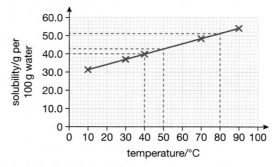

### KEY POINT

The dashed lines marked on the graph come from answers to the next part of the question. In this case the solubility curve is virtually a straight line – this will not always be the case. If you are asked to draw a line of best fit this can be either a straight line or a curve.

▲ Figure 1.19 Solubility curve for potassium chloride

b  Use the solubility curve to find:
   i   the solubility of potassium chloride at 50 °C
   ii  the maximum mass of potassium chloride that would dissolve in 50 g of water at 40 °C.
   iii the temperature at which crystals will first appear if you cooled a hot solution containing 51.0 g of potassium chloride in 100 g of water.

Use the graph to find the information you want.

Part i: The graph shows that at 50 °C the solubility is 42.5 g per 100 g of water.

Part ii: From the graph we can see that the solubility at 40 °C is 40 g per 100 g of water. From this we can deduce that half as much, that is 20 g, will dissolve in 50 g (half the mass) of water.

The numbers we have used here are quite simple, but if you were, for instance, asked to work out the maximum mass that would dissolve in 34.6 g of water at 40 °C you could use the equation:

$$\frac{\text{mass of water (g)}}{100\text{ (g)}} \times \text{solubility (g per 100 g)} = \text{maximum mass that dissolves (g)}$$

So, with 34.6 g of water we would get:

$$\frac{34.6}{100} \times 40 = 13.84\text{ g of potassium chloride dissolves}$$

Part iii: Solubility measures the maximum mass of potassium chloride which will dissolve in 100 g of water at a particular temperature. Crystals will start to appear as soon as the solution becomes saturated. From the graph, it can be seen that more than 51.0 g of solid will be soluble at 90 °C, but as the solution is cooled the solubility decreases. Drawing a line across at 51.0 g shows that this is the maximum mass that will dissolve at 80 °C and therefore crystals will first appear at temperatures below this.

c What mass of potassium chloride would crystallise from the solution in biii if the temperature fell to 10 °C?

You can use the data in the table to find the solubility at 10 °C. This value is 31.2 g per 100 g of water. That means that 31.2 g of potassium chloride will stay in solution at 10 °C. Since you started with 51.0 g, the rest of it must have formed crystals:

mass of crystals = 51.0 – 31.2 g = 19.8 g

19.8 g of potassium chloride will crystallise out.

**END OF CHEMISTRY ONLY**

**HINT**

In different circumstances, you might have to find the 10 °C figure from the graph as well.

### CHAPTER QUESTIONS

**SKILLS** CRITICAL THINKING

1 What name is given to each of the following changes of state?
  a Solid to liquid
  b Liquid to solid
  c Solid to gas
  d Gas to solid

**SKILLS** INTERPRETATION

2 a Draw diagrams to show the arrangement of the particles in a solid, a liquid and a gas.

  b Describe the difference between the movement of particles in a solid and a liquid.

  c The change of state from a liquid to a gas can be either evaporation or boiling. Explain the difference between evaporation and boiling.

**SKILLS** ANALYSIS

3 The questions refer to the substances in the table.

|   | Melting point/°C | Boiling point/°C |
|---|---|---|
| **A** | −259 | −253 |
| **B** | 0 | 100 |
| **C** | 3700 (sublimes) | |
| **D** | −116 | 34.5 |
| **E** | 801 | 1413 |

  a Write down the physical states of each compound at
    i 30 °C.
    ii −100 °C
    iii 80 °C

**SKILLS** PROBLEM SOLVING

  b Which substance has the greatest distance between its particles at 25 °C? Explain your answer.

**SKILLS** REASONING

  c Why is no boiling point given for substance **C**?

  d Which liquid substance would evaporate most quickly in the open air at 25 °C? Explain your answer.

4 Refer to Figure 1.14 on page 7 showing the diffusion experiment.

a Explain why the ring takes a few minutes to form.

b i If you heat a gas, what effect will this have on the movement of the particles?

ii In the light of your answer to i, what difference would you find if you did this experiment outside on a day when the temperature was 2 °C instead of in a warm lab at 25 °C? Explain your answer.

c Explain why the ring was formed nearer the hydrochloric acid end of the tube.

d Suppose you replaced the concentrated hydrochloric acid with concentrated hydrobromic acid. This releases the gas hydrogen bromide (HBr). Hydrogen bromide also reacts with ammonia to form a white ring.

i Suggest a name for the white ring in this case.

ii Hydrogen bromide particles are about twice as heavy as hydrogen chloride particles. What effects do you think this would have on the experiment?

5 Use the words given below to complete the following paragraph. Each word may be used once, more than once or not at all.

Sodium chloride dissolves in water to form a _____. The water is called the _____ and the sodium chloride is the _____. If the solution is heated to 50 °C some of the water _____ until the solution becomes _____ and sodium chloride crystals start to form.

**boils   solution   solute   saturated   evaporates   solvent   condenses**

## CHEMISTRY ONLY

6 The solubility of sodium chlorate in water was measured at a number of different temperatures.

| Temperature/°C | 0 | 20 | 40 | 60 | 80 | 100 |
|---|---|---|---|---|---|---|
| Solubility/g per 100 g of water | 3 | 8 | 14 | 23 | 38 | 55 |

a Use these figures to plot a solubility curve, with the temperature on the horizontal axis and the solubility on the vertical one.

b Use your graph to find the solubility of sodium chlorate at 50 °C.

c Determine the maximum mass of sodium chlorate that will dissolve in 40 g of water at 30 °C.

d 20 g of sodium chlorate was added to 100 g of water and the mixture heated to about 70 °C. It was then left to cool with the thermometer in the solution. Use your graph to answer the following questions.

i At what temperature would crystals first appear in the solution?

ii If the solution was cooled to 17 °C, calculate the total mass of crystals formed.

**END OF CHEMISTRY ONLY**

# 2 ELEMENTS, COMPOUNDS AND MIXTURES

Most of the substances that we are familiar with from everyday life are mixtures. For example, the air that we breathe is a mixture containing elements such as nitrogen and oxygen, and compounds such as carbon dioxide and nitrogen oxides. The food that we eat and the drinks that we drink are mixtures. This chapter looks at the properties of elements, compounds and mixtures, and also how to separate the components of a mixture. Separation of mixtures is very important in the analysis of substances, such as in forensics.

▲ Figure 2.1 Gold is an element, but a gold ring made from 18-carat gold only contains 75% gold. The metal is a mixture of gold and, usually, copper.

▲ Figure 2.2 Pure water is a compound, but the water we drink is a mixture of water and other dissolved substances.

## LEARNING OBJECTIVES

- ■ Understand how to classify a substance as an element, compound or mixture

- ■ Understand that a pure substance has a fixed melting and boiling point, but that a mixture may melt or boil over a range of temperatures

- ■ Describe these experimental techniques for the separation of mixtures:
  - ■ simple distillation
  - ■ filtration
  - ■ paper chromatography
  - ■ fractional distillation
  - ■ crystallisation

- ■ Understand how a chromatogram provides information about the composition of a mixture

- ■ Understand how to use the calculation of $R_f$ values to identify the components of a mixture

- ■ Practical: Investigate paper chromatography using inks/food colourings

**REMINDER**

You might want to look at Chapter 3 if you do not already know the term 'atom'.

**KEY POINT**

It isn't completely true to say that elements consist of only one type of atom. A better way of saying it would be that *all the atoms in an element have the same atomic number*. Most elements consist of mixtures of isotopes, which have the same atomic number, but different mass numbers (due to different numbers of neutrons). When we draw diagrams or make models, we aren't usually interested in the differences between the isotopes. Isotopes will be discussed in Chapter 3.

## ELEMENTS

**Elements** are *substances that can't be split into anything simpler by chemical means*. An element contains only one type of atom (but see the key point in the margin). In models or diagrams they are shown as atoms of a single colour or size.

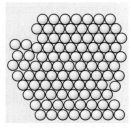

a pure metal such as magnesium

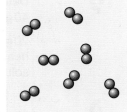

oxygen gas

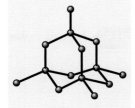

diamond (a form of carbon)

▲ Figure 2.3 Elements contain only one type of atom.

There are 118 elements and these are shown in the Periodic Table. Most of the elements occur naturally, such as hydrogen, helium and sulfur. Some others have to be made artificially, such as einsteinium.

## COMPOUNDS

**Compounds** are formed when *two or more elements chemically combine*. The elements always combine in fixed proportions. For example, hydrogen and fluorine always combine to form hydrogen fluoride, with formula HF, whereas magnesium and fluorine always combine to form magnesium fluoride, with formula $MgF_2$ – the elements must combine in these ratios. Examples of other compounds are carbon dioxide ($CO_2$) and methane ($CH_4$). Diagrams of compounds show more than one type of atom bonded together.

water     silicon dioxide     sodium chloride

▲ Figure 2.4 Some compounds

## MIXTURES

In a **mixture**, the various substances are mixed together and no chemical reaction occurs. Mixtures can be made from elements and/or compounds. The various components can be in any proportion, for example you can put any amount of sugar into your cup of tea or coffee (until it becomes saturated).

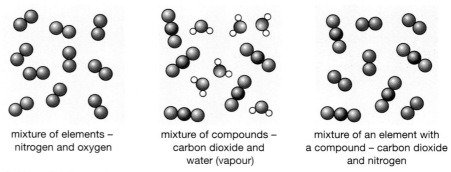

mixture of elements – nitrogen and oxygen     mixture of compounds – carbon dioxide and water (vapour)     mixture of an element with a compound – carbon dioxide and nitrogen

▲ Figure 2.5 Some mixtures

## SIMPLE DIFFERENCES BETWEEN MIXTURES AND COMPOUNDS

**PROPORTIONS**

In water (a compound), every single water molecule has two hydrogen atoms combined with one oxygen atom. It never varies. In a mixture of hydrogen and oxygen gases, the two could be mixed together in any proportion.

If you had some iron metal and some sulfur, you could mix them in any proportion you wanted to. In iron sulfide (FeS), a compound, the proportion of iron to sulfur is always exactly the same.

## PROPERTIES

### REMINDER

You can find out about the reactions of metals with dilute acids on pages 174–175. The reaction between iron sulfide and acids isn't needed for exam purposes at International GCSE.

In a mixture of elements, each element keeps its own properties, but the properties of the compound are quite different. For example, in a mixture of iron and sulfur, the iron is grey and the sulfur is yellow. The iron reacts with dilute acids such as hydrochloric acid to produce hydrogen; the sulfur doesn't react with the acid. However, the compound iron sulfide (FeS) reacts quite differently with acids to produce poisonous hydrogen sulfide gas, which smells of bad eggs.

A mixture of hydrogen and oxygen is a colourless gas which explodes when you put a flame to it. The compound, water, is a colourless liquid which just puts out a flame.

## EASE OF SEPARATION

Mixtures can be separated by **physical means**. Physical means are things like changing the temperature or dissolving part of the mixture in a solvent such as water; in other words, methods that don't involve any chemical reactions.

For example, a mixture of iron and sulfur is quite easy to separate into the two elements using a magnet. The iron sticks to the magnet and the sulfur doesn't. The elements in a compound cannot be separated by physical means. To convert iron sulfide into separate samples of iron and sulfur requires chemical reactions.

You can cool a mixture of hydrogen and oxygen gases to separate it by a physical process. Oxygen condenses into a liquid at a much higher temperature than hydrogen (–183 °C as opposed to –253 °C). This would leave you with liquid oxygen and hydrogen gas, which are easy to separate. But to separate water into hydrogen and oxygen, you have to change it chemically using electrolysis. Electrolysis is explained in Chapter 10.

## MELTING POINT AND BOILING POINT

Pure substances, such as elements and pure compounds, melt and boil at fixed temperatures. For example, the melting point of water is 0 °C and the boiling point 100 °C. However, mixtures usually melt or boil over a *range of temperatures*.

### KEY POINT

A mixture is not a pure substance. If a sample contains only a small amount of an unwanted substance, the unwanted substance might be called an *impurity*.

The presence of impurities lowers the melting point of a substance and raises the boiling point. For instance, dissolving 10 g of common (table) salt (sodium chloride) in 1 litre of water lowers the melting point to about –0.6 °C and raises the boiling point to about 100.2 °C.

The melting point can be very useful in determining whether or not a substance is pure. If you continue to study chemistry you might carry out a practical experiment to make some aspirin. In order to determine whether your sample is pure or not you can measure the melting point. You would record the temperature at which your sample starts to melt, and then you would record the temperature at which it has fully melted to completely form a liquid. Aspirin is a white powder that melts at 138 °C. If the melting point of the sample you made is 128–134 °C you can see that it is quite impure because it melts over a wide range of temperature (below the melting point of pure aspirin).

## SEPARATION OF MIXTURES

Separating mixtures is extremely important in chemistry. For example, we can see this in the processing of crude oil, in producing fresh water from salt water and in the enrichment of uranium. In forensic science, the components of a mixture usually have to be separated before they can be analysed.

## FILTRATION

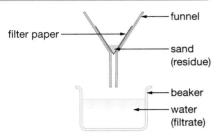

Figure 2.6 Filtration can be used to separate a mixture of sand and water.

Filtration can be used to separate a solid from a liquid.

For example, sand can be separated from water by filtration. The apparatus for filtration is shown in Figure 2.6.

The substance left in the filter paper is called the residue and the liquid that comes through is called the filtrate.

Filtration can also be used to separate two solids from each other if only one of them is soluble in water (see below – rock salt).

## CRYSTALLISATION

Crystallisation can be used to separate a solute from a solution. For example, it could be used to separate sodium chloride from a sodium chloride solution. The solution is heated in an evaporating basin to boil off some of the water until an almost saturated solution is formed. This can be tested by dipping a glass rod into the solution and seeing if crystals form quickly on its surface when it is removed. The Bunsen burner is then turned off and the crystals allowed to form as more water evaporates and the solution cools. The crystals can now be removed from the mixture by filtration.

The apparatus for crystallisation is shown in Figure 2.7.

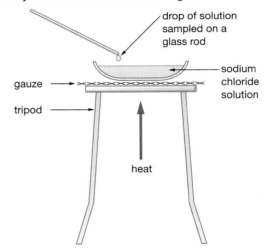

Figure 2.7 Crystallisation can be used to separate a solute from a solution.

## MAKING PURE SALT FROM ROCK SALT

Figure 2.8 Rock salt

We can use filtration and crystallisation to obtain pure salt from rock salt.

Rock salt consists of salt contaminated by various earthy or rocky impurities. These impurities aren't soluble in water.

If you crush the rock salt and mix it with hot water, the salt dissolves, but the impurities don't. The impurities can be filtered off, and remain on the filter paper. The filtrate is then a salt solution. The solid salt can be obtained from the solution by crystallisation.

This is typical of the way you can separate any mixture of two solids, one of which is soluble in water and one of which isn't.

## SIMPLE DISTILLATION

**Simple distillation** can be used to separate the components of a solution. Although we can use crystallisation to separate sodium chloride from a sodium chloride solution, we can also collect the water if we use simple distillation.

The water boils and is condensed back to a liquid by the condenser. The salt remains in the flask.

### KEY POINT

Notice that water is always fed into the condenser at the lower end. That way it fills the condenser jacket better and if the flow of water stops for any reason the condenser jacket remains full of water.

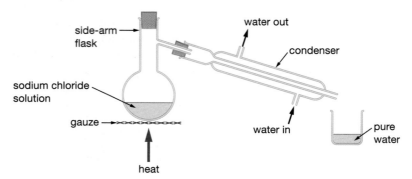

▲ Figure 2.9 Distilling pure water from sodium chloride solution.

You could, of course, collect the salt from the solution as well as collecting pure water. The sodium chloride solution eventually becomes so concentrated that the salt will crystallise out.

## FRACTIONAL DISTILLATION

### EXTENSION WORK

The fractionating column is often packed with glass beads or something similar, although the separation of ethanol and water in the lab works perfectly well just with an empty column. For reasons that are beyond International GCSE, a high surface area in the column helps separation of the two vapours. The ethanol produced by this experiment is about 96% pure. For complicated reasons, again beyond International GCSE, it is impossible to remove the last 4% of water by distillation.

**Fractional distillation** is used to separate a mixture of liquids such as ethanol (alcohol) and water. Ethanol and water are completely miscible with each other. That means you can mix them together in any proportion and they will form a single liquid layer. You can separate them by taking advantage of their different boiling points: water boils at 100 °C, ethanol at 78 °C.

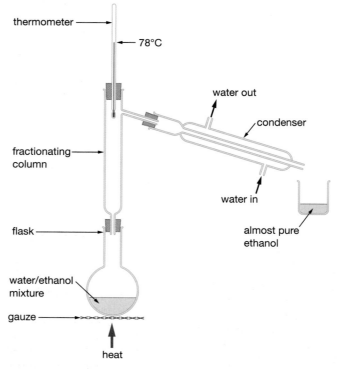

▲ Figure 2.10 Fractional distillation

Both liquids boil, but by careful heating you can control the temperature of the column so that all the water condenses in the column and trickles back into the flask. Only the ethanol remains as a vapour all the way to the top of the fractionating column and out into the condenser.

## PAPER CHROMATOGRAPHY

Paper chromatography can be used to separate a variety of mixtures. However, at International GCSE level we will usually use it to separate mixtures of coloured inks or food colourings. Most inks and food colourings are not just made up of one colour but contain a mixture of dyes.

Paper chromatography can also be used to separate a mixture of colourless substances such as sugars, but then some method must be used to make the spots visible on the paper.

### ACTIVITY 2

▼ **PRACTICAL: INVESTIGATING THE COMPOSITION OF DYE WITH PAPER CHROMATOGRAPHY**

We can investigate the composition of a mixture of coloured dyes using paper chromatography. To do this we carry out the following steps.

1. Draw a line with a pencil across a piece of chromatography paper; this line should be about 1 cm from the bottom of the paper. Do not use a pen as the colours in the ink may move up the chromatography paper with the solvent.

2. Put a spot (use a teat pipette or a capillary tube) of the mixture of dyes on the pencil line and allow it to dry.

3. Suspend the chromatography paper in a beaker that contains a small amount of solvent so that the bottom of the paper goes into the solvent. It is important that the solvent is below the pencil line so that the inks/colourings don't just dissolve in the solvent.

4. Put a lid (such as a watch glass) on the beaker so that the atmosphere becomes saturated with the solvent. This is to stop evaporation of the solvent from the surface of the paper.

5. When the solvent has moved up the paper to about 1 cm from the top, remove the paper from the beaker and draw a pencil line to show where the solvent got to. The highest level of the solvent on the paper at any time is called the *solvent front*.

6. Leave the paper to dry so that all the solvent evaporates.

For the solvent you can use water or a non-aqueous solvent (a solvent other than water). Which solvent you use depends on what substances are present in the mixture. A suitable solvent is usually found by experimenting with different ones.

The dyes that make up the mixture will be different in two important ways:
- the affinity they have for the paper (how well they 'stick' to the paper)
- how soluble they are in the solvent which moves up the paper.

In Figure 2.11 spot C has hardly moved. Either it was not very soluble in the solvent or it has a very high affinity for the paper (or both). On the other hand, spot A has moved almost as far as the solvent. It must be very soluble in the solvent and not have much affinity for the paper. The pattern you get is called a chromatogram.

Safety Note: Avoid skin contact with the solvents and dyes, especially if you have sensitive skin.

## KEY POINT

If the dye does not move from the pencil line during an experiment, then the dye is not at all soluble in the solvent you are using. In this case, you need to find a different solvent. If the dye moves up the paper with the solvent front, the dye is too soluble in that solvent and, again, you have to try a different solvent.

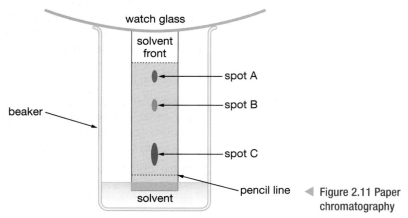

◀ Figure 2.11 Paper chromatography

In this example, the mixture must have contained a minimum of three different dyes. We say a *minimum* of three dyes because there could be more – it is possible that one of the spots is made up of two coloured dyes that by concidence moved the same distance. You could only confirm this by doing the experiment again with a different solvent.

▲ Figure 2.12 A paper chromatography experiment

## USING PAPER CHROMATOGRAPHY IN ANALYSIS

You can use paper chromatography to identify the particular dyes in a mixture. If you think that your mixture (m) could contain dyes d1, d2, d3 and d4, you can carry out an experiment to determine this.

A pencil line is drawn on a larger sheet of paper and pencil marks are drawn along the line to show the original positions of the various dyes placed on the line (see Figure 2.13). One spot is your unknown mixture; the others are single, known dyes. The chromatogram is then allowed to develop as before.

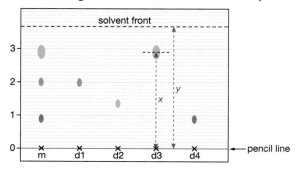

▲ Figure 2.13 Paper chromatography can be used to analyse a mixture. Lines will not be present on your paper, but they have been added here to help you measure the distances.

The mixture (m) has spots corresponding to dyes d1, d3 and d4. They have the same colour as spots in the mixture, and have travelled the same distance on the paper. Although dye d2 is the same colour as one of the spots in the mixture, it has travelled a different distance and so must be a different compound.

Instead of just saying the spots move different distances we can use the $R_f$ value to describe how far the spots move. $R_f$ stands for **retardation factor**. Each time we do a chromatography experiment the solvent (and therefore the spots) will move different distances along the paper. This means we can't just report the distance moved by a particular spot so we have to work out a ratio instead.

$$R_f = \frac{\text{distance moved by a spot (from the pencil line)}}{\text{distance moved by the solvent front (from the pencil line)}}$$

In Figure 2.13 $R_f = \dfrac{x}{y}$.

So in Figure 2.13 the $R_f$ value for dye d3 is:

$$R_f = \frac{2.9\,\text{cm}}{3.6\,\text{cm}} = 0.81$$

**HINT**

Measure to the centre of the spot.

The $R_f$ values of the dyes in mixture m are:

blue spot: $\qquad R_f = \dfrac{0.9}{3.6} = 0.25$

orange spot: $\quad R_f = \dfrac{2.0}{3.6} = 0.56$

green spot: $\quad R_f = \dfrac{2.9}{3.6} = 0.81$

The $R_f$ values of dyes d1 to d4 are:

d1: $\quad R_f = 0.56$

d2: $\quad R_f = 0.36$

d3: $\quad R_f = 0.81$

d4: $\quad R_f = 0.25$

Because the spots in mixture m have the same $R_f$ values as d1, d3 and d4, we can conclude that the mixture contains these dyes.

An $R_f$ value must be between 0 and 1. If you get a number bigger than 1 you have probably divided the numbers the wrong way round. An $R_f$ value has no units.

You have to be careful when using $R_f$ values as they depend on the solvent used and on the type of paper. There was no problem in the experiment described above because the mixture and the individual dyes were all put on the same piece of paper. However, if the mixture was put on one piece of chromatography paper and the individual dyes on a separate piece, you can still compare $R_f$ values as long as you use the same type of paper and the same solvent.

**CHAPTER QUESTIONS**

SKILLS  CRITICAL THINKING

1  Classify each of the following substances as an element, compound or mixture:

| | | |
|---|---|---|
| sea water | hydrogen | honey |
| magnesium oxide | copper(II) sulfate | blood |
| calcium | mud | potassium iodide solution |

SKILLS  ANALYSIS

2  Look at the diagrams below and classify each one as an element, compound or mixture.

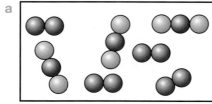

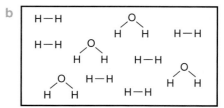

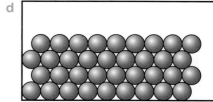

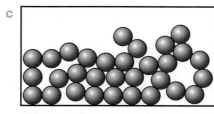

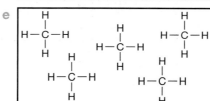

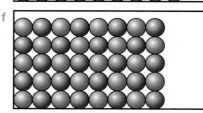

SKILLS  REASONING,
PROBLEM SOLVING

3  A teacher has found two white powders on a desk in the chemistry laboratory. She wants to test to see if they are pure substances, so she measures the melting points. Substance X melts at 122 °C and substance Y melts between 87 and 93 °C. Explain which one is the pure substance.

SKILLS  DECISION MAKING

4  State which separation method you would use to carry out the following separations:

a  Potassium iodide from a potassium iodide solution.

b  Water from a potassium iodide solution.

c  Ethanol from a mixture of ethanol and water.

d  Red dye from a mixture of red and blue dyes.

e  Calcium carbonate (insoluble in water) from a mixture of calcium carbonate and water.

SKILLS  CREATIVITY,
DECISION MAKING

5  Suppose you had a valuable collection of small diamonds, which you kept safe from thieves by mixing them with white sugar crystals. You store the mixture in a jar labelled 'sugar'. Now you want to sell the diamonds. Describe how you would separate all the diamonds from the sugar.

6 In order to identify the writer of an anonymous letter, a sample of ink from the letter was dissolved in a solvent and then placed on some chromatography paper. Spots of ink from the pens of five possible writers, **G**, **M**, **P**, **R** and **T**, were placed next to the sample on the chromatography paper. The final chromatogram looked like this:

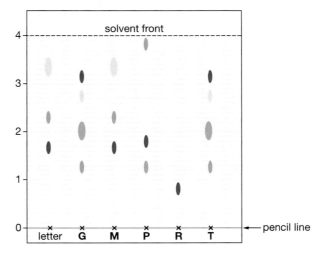

a Which of the five writers is using ink that matches the sample from the letter?

b Which of the writers is using a pen that contains ink made from a single dye?

c What is the $R_f$ value of the blue dye in suspect **P**'s pen?

d Which two of the five writers are using pens containing the same ink?

e Whose pen contained the dye that was most soluble in the solvent?

# 3 ATOMIC STRUCTURE

This chapter explores the nature of atoms and how they differ from element to element. The 118 elements are the building blocks from which everything is made, from a simple substance, such as carbon, to a more complex one, such as DNA.

## LEARNING OBJECTIVES

- Know what is meant by the terms atom and molecule

- Know the structure of an atom in terms of the positions, relative masses and relative charges of sub-atomic particles

- Know what is meant by the terms atomic number, mass number, isotopes and relative atomic mass ($A_r$)

- Be able to calculate the relative atomic mass of an element ($A_r$) from isotopic abundances

Copper is an element. If you tried to cut it up into smaller and smaller pieces, the final result would be the smallest possible piece of copper. At that stage, you would have an individual copper atom. You can, of course, split that atom into smaller pieces (protons, neutrons and electrons), but you would no longer have copper. Therefore, an **atom** is the smallest piece of an element that can still be recognised as that element.

▲ Figure 3.1 New atoms are produced in stars . . .

▲ Figure 3.2 . . . or in nuclear processes such as nuclear bombs, nuclear reactors or radioactive decay.

## ATOMS AND MOLECULES

Atoms can be joined together to make molecules. A **molecule** consists of two or more atoms chemically bonded (by covalent bonds). The atoms that make up a molecule can be from the same elements or different elements. A hydrogen ($H_2$) molecule (Figure 3.3a) consists of 2 hydrogen atoms chemically bonded together. A water ($H_2O$) molecule (Figure 3.3b) consists of 2 hydrogen atoms and an oxygen atom chemically bonded.

▲ Figure 3.3 (a) A $H_2$ molecule and (b) a $H_2O$ molecule. The lines between the atoms represent chemical bonds.

KEY POINT

KEY POINT

You may have come across diagrams of the atom in which the electrons are drawn orbiting the nucleus rather like planets around the sun. This can be misleading.

Electrons are constantly moving in the atom and it is impossible to know exactly where they are at any moment in time. You can only identify that they have a particular energy and that they are likely to be found in a certain region of space at some particular distance from the nucleus. Electrons with different energies are found at different distances from the nucleus.

# THE STRUCTURE OF THE ATOM

Atoms are made of protons, neutrons and electrons. These particles are sometimes called *sub-atomic particles* because they are smaller than an atom.

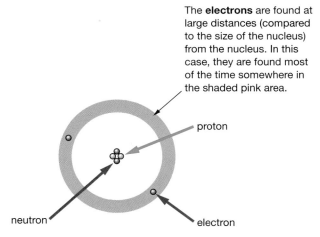

The **electrons** are found at large distances (compared to the size of the nucleus) from the nucleus. In this case, they are found most of the time somewhere in the shaded pink area.

proton

neutron

electron

▲ Figure 3.4 The structure of a helium atom

The nucleus of the atom contains protons and neutrons, and is shown highly magnified in Figure 3.4. In reality, if you scale up a helium atom to the size of a sports hall the nucleus would be no more than the size of a grain of sand.

The relative masses and charges of protons, neutrons and electrons are shown in Table 3.1.

Table 3.1 The properties of protons, neutrons and electrons

| Particle | Relative mass | Relative charge |
| --- | --- | --- |
| proton | 1 | +1 |
| neutron | 1 | 0 |
| electron | 1/1836 | −1 |

HINT

$\frac{1}{1836}$ is approximately 0.0005.

Virtually all the mass of the atom is concentrated in the nucleus because electrons have a much smaller mass than protons and neutrons.

The masses and charges are measured relative to each other because the actual values are incredibly small. For example, it would take about $600\,000\,000\,000\,000\,000\,000\,000$ ($6 \times 10^{23}$) protons to weigh 1 g.

# ATOMIC NUMBER AND MASS NUMBER

The number of protons in an atom's nucleus is called its atomic number or proton number. Each of the 118 different elements has a different number of protons. For example, if an atom has 8 protons it must be an oxygen atom:

atomic number = number of protons

KEY POINT

The atomic number defines an element and is unique to that element. We can identify an element by its atomic number instead of its name. We could talk about a wristwatch made from the element with atomic number 79 instead of talking about 'a gold wristwatch', or say that the element with atomic number 17 is poisonous instead of saying 'chlorine is poisonous'. However, these are more complicated ways of describing things!

The mass number (sometimes known as the nucleon number) counts the total number of protons and neutrons in the nucleus of the atom:

mass number = number of protons + number of neutrons

For any particular atom, this information can be shown as, for example:

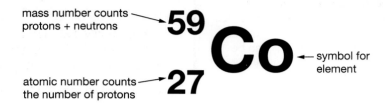

mass number counts
protons + neutrons

atomic number counts
the number of protons

symbol for
element

This particular atom of cobalt contains 27 protons. To make the total number of protons and neutrons up to 59, there must also be 32 neutrons.

You can see from this that:

number of neutrons = mass number – atomic number

## ISOTOPES

The number of neutrons in an atom can vary slightly. For example, there are three kinds of carbon atom called carbon-12, carbon-13 and carbon-14. They all have the same number of protons (because all carbon atoms have 6 protons, its atomic number), but the number of neutrons varies. These different atoms of carbon are called isotopes.

*Isotopes are atoms (of the same element) which have the same atomic number but different mass numbers. They have the same number of protons but different numbers of neutrons.*

The fact that they have varying numbers of neutrons makes no difference to their chemical reactions. The chemical properties (how something reacts) are controlled by the number and arrangement of the electrons, and that is identical for all three isotopes.

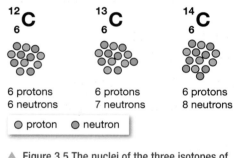

6 protons
6 neutrons

6 protons
7 neutrons

6 protons
8 neutrons

○ proton    ○ neutron

▲ Figure 3.5 The nuclei of the three isotopes of carbon

## RELATIVE ATOMIC MASS

You might have seen the following in a Periodic Table:

35.5

**Cl**

chlorine

17

Chlorine appears to have a mass number of 35.5. If you calculate the number of neutrons for chlorine you obtain:

number of neutrons = 35.5 – 17 = 18.5

It is not possible to have half a neutron and so there must be something wrong with this. The number 35.5 is not actually the mass number for chlorine but rather the **relative atomic mass** ($A_r$). Chlorine consists of two isotopes, $^{35}Cl$ and $^{37}Cl$, and a naturally occurring sample contains a mixture of these.

KEY POINT

This type of average is called a **weighted average** or weighted mean.

Relative atomic mass is the average mass of an atom, taking into account the amount of each isotope present in a naturally occurring sample of an element. It is explained in more detail in Chapter 5.

You can probably see that a naturally occurring sample of chlorine must contain more of the $^{35}Cl$ isotope than the $^{37}Cl$ isotope. This is because the relative atomic mass is closer to 35 than to 37.

We can calculate the relative atomic mass of an element by knowing how much of each isotope is present in a sample (the isotopic abundances) of that element, and then working out the average mass of an atom. This is done in exactly the same way as you would calculate a weighted average in maths. It can be understood more easily by looking at a worked example.

### EXAMPLE 1

A naturally occurring sample of the element boron contains 20% $^{10}B$ and 80% $^{11}B$. Calculate the relative atomic mass.

If we imagine there are 100 atoms we can work out that 20% of them, that is 20, will have mass 10 and 80 will have mass 11.

The total mass of the 20 atoms with mass 10 is $20 \times 10$.

The total mass of the 80 atoms with mass 11 is $80 \times 11$.

The total mass of all the atoms in the sample is $20 \times 10 + 80 \times 11$.

There are 100 atoms so we can work out the average by dividing the total mass by the total number of atoms (100):

$$\text{relative atomic mass} = \frac{20 \times 10 + 80 \times 11}{100} = 10.8$$

Therefore, the relative atomic mass of boron is 10.8.

Even if there are three or four different isotopes, you still do the calculation in the same way: calculate the total mass of 100 atoms, then divide the answer by 100.

## THE ELECTRONS

### COUNTING THE NUMBER OF ELECTRONS IN AN ATOM

Atoms are electrically neutral (they have no overall charge). The charge on a proton (+1) is equal but opposite to the charge on an electron (–1), and therefore in an atom:

number of electrons = number of protons

So, if an oxygen atom (atomic number = 8) has 8 protons, it must also have 8 electrons; if a chlorine atom (atomic number = 17) has 17 protons, it must also have 17 electrons.

HINT

Remember that the number of protons is the same as the atomic number of the element.

You will see that the key feature in this is knowing the atomic number. You can find the atomic number from the Periodic Table.

The number of protons in an atom is equal to the number of electrons. However, the atomic number is defined in terms of the number of protons because the number of electrons can change in chemical reactions, for example when atoms form ions (see Chapter 7).

## THE PERIODIC TABLE

Chapter 4 deals in detail with what you need to know about the Periodic Table for International GCSE purposes.

Atoms are arranged in the Periodic Table in order of increasing atomic number. You will find a full version of the Periodic Table in Appendix A on page 320. Most Periodic Tables have two numbers against each symbol; be careful to choose the right one. The atomic number will always be the smaller number. The other number will either be the mass number of the most common isotope of the element or the relative atomic mass of the element. The Periodic Table will clarify this.

You can use a Periodic Table to find out the number of protons, neutrons and electrons in an atom. Remember:

- the number of protons in an atom is equal to the atomic number
- the number of electrons in an atom is equal to the number of protons
- the number of neutrons in an atom = mass number – atomic number.

EXAMPLE 2

The symbol for uranium is given in a Periodic Table as:

$$
\begin{array}{c}
238 \\
\mathbf{U} \\
\text{Uranium} \\
92
\end{array}
$$

Calculate the number of protons, neutrons and electrons in an atom of uranium.

The atomic number is the smaller number, so the atomic number of uranium is 92. The atomic number tells us the number of protons, therefore a uranium atom contains 92 protons.

The number of protons is equal to the number of electrons, therefore a uranium atom contains 92 electrons.

The number of neutrons = mass number – atomic number.

The number of neutrons = 238 – 92 = 146.

## CHAPTER QUESTIONS

You will need to use the Periodic Table in Appendix A on page 320.

SKILLS    CRITICAL THINKING

1  Atoms contain three types of particle: proton, neutron and electron.

   a  State where the protons and neutrons are in an atom.

   b  State which type of particle in the atom orbits the nucleus.

   c  State which one of the particles has a positive charge.

   d  State which two particles have approximately the same mass.

2  Fluorine atoms have a mass number of 19.

   a  Use the Periodic Table to find the atomic number of fluorine.

   b  Explain what *mass number* means.

   c  State the number of protons, neutrons and electrons in a fluorine atom.

   d  Explain why the number of protons in an atom must always equal the number of electrons.

3  Work out the numbers of protons, neutrons and electrons in each of the following atoms:

   a  $^{56}_{26}$Fe          b  $^{93}_{41}$Nb          c  $^{235}_{92}$U

4  Chlorine has two isotopes, chlorine-35 and chlorine-37.

   a  Explain what *isotopes* are.

   b  State the numbers of protons, neutrons and electrons in the two isotopes.

5  Lithium has two naturally occurring isotopes, $^6$Li (abundance 7%) and $^7$Li (abundance 93%). Calculate the relative atomic mass of lithium, giving your answer to 2 decimal places.

6  Magnesium has three naturally occurring stable isotopes, $^{24}$Mg (abundance 78.99%), $^{25}$Mg (abundance 10.00%) and $^{26}$Mg (abundance 11.01%). Calculate the relative atomic mass of magnesium, giving your answer to 2 decimal places.

7  Lead has four naturally occurring stable isotopes. Calculate the relative atomic mass of lead given the data in the table.

| Mass number | Natural abundance/% |
|---|---|
| 204 | 1.4 |
| 206 | 24.1 |
| 207 | 22.1 |
| 208 | 52.4 |

8  Iridium has two naturally occurring isotopes, $^{191}$Ir and $^{193}$Ir.

   a  State the number of protons, neutrons and electrons in an $^{191}$Ir atom.

   b  Explain the difference between the two isotopes.

   c  The relative atomic mass of iridium is 192.22. Explain whether a naturally occurring sample of iridium contains more $^{191}$Ir or $^{193}$Ir.

9  Use the Periodic Table to explain whether the following statement is true or false.

   *Considering only the most common isotope of each element, there is only one element that has more protons than neutrons.*

# 4 THE PERIODIC TABLE

The Periodic Table shows all the elements in the universe and is one of the most important tools that a chemist has. The arrangement of the elements allows us to understand trends in properties and make predictions. The modern Periodic Table was first presented in 1869 by a famous Russian chemist, Dmitri Mendeleev (left). This chapter explores some of the features of the Periodic Table.

## LEARNING OBJECTIVES

- Understand how elements are arranged in the Periodic Table:
    - in order of atomic number
    - in groups and periods
- Understand how to deduce the electronic configurations of the first 20 elements from their positions in the Periodic Table
- Understand how to use electrical conductivity and the acid–base character of oxides to classify elements as metals or non-metals

- Identify an element as a metal or a non-metal according to its position in the Periodic Table
- Understand how the electronic configuration of a main group element is related to its position in the Periodic Table
- Understand why elements in the same group of the Periodic Table have similar chemical properties
- Understand why the noble gases (Group 0) do not readily react

## THE PERIODIC TABLE

The search for patterns in chemistry during the 19th century resulted in the modern Periodic Table. *The elements are arranged in order of atomic number –* the number of protons in the nuclei of the atoms.

The vertical columns are called **groups**

| | 1 | 2 | | | | | | | | | | | 3 | 4 | 5 | 6 | 7 | 0 |
|---|---|---|---|---|---|---|---|---|---|---|---|---|---|---|---|---|---|---|
| 1 | H | | | | | | | | | | | | | | | | | He |
| 2 | Li | Be | | | | | | | | | | | | B | C | N | O | F | Ne |
| 3 | Na | Mg | | | | | transition metals | | | | | | | Al | Si | P | S | Cl | Ar |
| 4 | K | Ca | Sc | Ti | V | Cr | Mn | Fe | Co | Ni | Cu | Zn | Ga | Ge | As | Se | Br | Kr |
| 5 | Rb | Sr | Y | Zr | Nb | Mo | Tc | Ru | Rh | Pd | Ag | Cd | In | Sn | Sb | Te | I | Xe |
| 6 | Cs | Ba | • | Hf | Ta | W | Re | Os | Ir | Pt | Au | Hg | Tl | Pb | Bi | Po | At | Rn |
| 7 | Fr | Ra | : | Rf | Db | Sg | Bh | Hs | Mt | Ds | Rg | Cn | Nh | Fl | Mc | Lv | Ts | Og |

The horizontal rows are called **periods**

lanthanoids

| • | La | Ce | Pr | Nd | Pm | Sm | Eu | Gd | Tb | Dy | Ho | Er | Tm | Yb | Lu |
|---|---|---|---|---|---|---|---|---|---|---|---|---|---|---|---|
| : | Ac | Th | Pa | U | Np | Pu | Am | Cm | Bk | Cf | Es | Fm | Md | No | Lr |

actinoids

▲ Figure 4.1 The Periodic Table

The vertical columns in the Periodic Table are called groups. The first seven groups are numbered from 1 to 7 and the final group is numbered 0. The elements in orange in Figure 4.1 are called the transition metals or transition elements. At this level, they are not usually included in the numbering of the groups. Some of the groups have names, e.g. Group 1 is the *alkali metals*, Group 7 is the *halogens* and Group 0 is the *noble gases*.

The horizontal rows in the Periodic Table are called periods. It is important to remember that hydrogen and helium make up Period 1.

The lanthanoids and actinoids are usually dropped out of their proper places and written separately at the bottom of the Periodic Table. There is a good reason for this. If you put them where they should be (as in Figure 4.2), everything has to be drawn slightly smaller to fit on the page. That makes it more difficult to read.

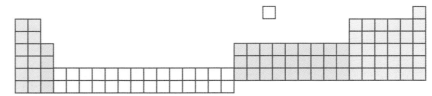

▲ Figure 4.2 The real shape of the Periodic Table

## THE PERIODIC TABLE AND THE NUMBER OF PROTONS, NEUTRONS AND ELECTRONS

Most Periodic Tables have two numbers against each symbol. The atomic number will always be the smaller number. The other number will either be the mass number of the most common isotope of the element or the relative atomic mass of the element. The Periodic Table will tell you which.

You can use a Periodic Table to work out the number of protons, neutrons and electrons there are in atoms. Remember:

- the number of protons in an atom is equal to the atomic number
- the number of electrons in an atom is equal to the number of protons
- the number of neutrons in an atom = mass number – atomic number.

## THE ARRANGEMENT OF THE ELECTRONS IN AN ATOM

The electrons move around the nucleus in a series of levels called energy levels or shells. Each energy level (shell) can only hold a certain number of electrons. Lower energy levels are always filled before higher ones. The lowest energy level is the closest one to the nucleus.

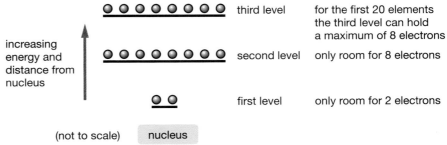

▲ Figure 4.3 The different energy levels for electrons in an atom showing the maximum number of electrons that each energy level (shell) can hold.

## HOW TO WORK OUT THE ARRANGEMENT OF ELECTRONS IN AN ATOM

The arrangement of electrons in an atom is called its **electronic configuration**.

We will use chlorine as an example.

Look up the atomic number in the Periodic Table. (Make sure that you don't use the wrong number if you have a choice. The atomic number will always be the smaller one.)

The Periodic Table tells you that chlorine's atomic number is 17.

This tells you the number of protons. The number of electrons is equal to the number of protons.

There are 17 protons and so 17 electrons in a neutral chlorine atom.

Arrange the electrons in shells (energy levels), always completing an inner shell (lower energy level) before you go to an outer one. Remember that the first shell (lowest energy level) can take up to 2 electrons, the second one can take up to 8, and the third one also takes up to 8.

For chlorine the electrons will be arranged as follows: 2 in the first shell, 8 in the second shell and 7 in the third shell. This is written as 2, 8, 7. When you have finished, always check to make sure that the electrons add up to the right number, in this case 17.

## DRAWING DIAGRAMS OF ELECTRONIC CONFIGURATIONS

When we draw a diagram of an atom we usually draw circles to represent the shells (energy levels); dots or crosses are then drawn on the circles to represent the electrons. You can choose to draw dots or crosses.

Hydrogen has 1 electron and helium has 2 in the first shell (lowest energy level).

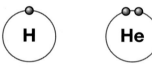

▲ Figure 4.4 Electronic arrangements of hydrogen and helium

The helium electrons are sometimes shown as a pair (as here), and sometimes as two separate electrons on opposite sides of the circle. Either form is acceptable.

The next 8 atoms are drawn like this:

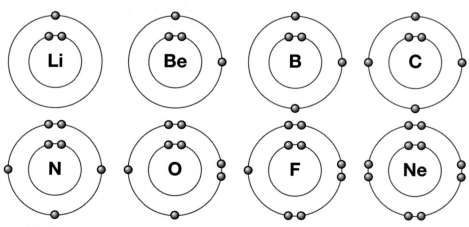

▲ Figure 4.5 The electronic arrangements of the elements in Period 2

It does not matter at this level whether you draw the electrons singly or in pairs.

The atoms in the Periodic Table from sodium to argon fill the third shell in exactly the same way, and potassium and calcium start to fill the fourth shell.

Potassium and calcium will look like this:

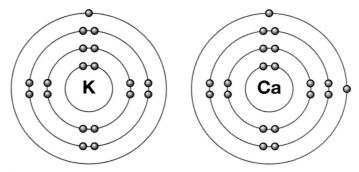

▲ Figure 4.6 Electronic arrangements of potassium and calcium

## ELECTRONIC CONFIGURATIONS AND THE PERIODIC TABLE

The electronic configurations of the first 20 elements in the Periodic Table are shown in Figure 4.7.

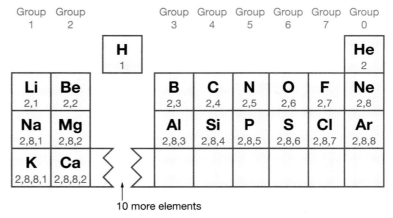

▲ Figure 4.7 The electronic configurations of the first 20 elements in the Periodic Table

Two important facts:

■ *Elements in the same group in the Periodic Table have the same number of electrons in their outer shell.*

■ *The number of electrons in the outer shell is the same as the group number for Groups 1 to 7.*

So if you know that barium is in Group 2, you know it has 2 electrons in its outer shell, the same as all the other elements in Group 2. Iodine (Group 7) has 7 electrons in its outer shell. Lead (Group 4) has 4 electrons in its outer shell. Working out what is in their inner shells is much more difficult – the simple patterns we have described don't work beyond calcium.

*The period number gives the number of occupied shells or the highest occupied shell.*

So, calcium (electronic configuration = 2, 8, 8, 2) is in Period 4 and has four shells occupied; the outermost electron is in the fourth shell. Iodine is in Period 5 and has five occupied shells, and because it is also in Group 7 we can deduce that it has 7 electrons in the fifth shell.

## ELEMENTS IN THE SAME GROUP IN THE PERIODIC TABLE HAVE SIMILAR CHEMICAL PROPERTIES

Groups in the Periodic Table contain elements with similar chemical properties – they react in the same way.

For example:

- all the elements in Group 1 react vigorously with water to form hydrogen and hydroxides with similar formulae: LiOH (lithium hydroxide), NaOH (sodium hydroxide), KOH (potassium hydroxide)

- all the elements in Group 7 react with hydrogen to form compounds with similar formulae: HF (hydrogen fluoride), HCl (hydrogen chloride), HBr (hydrogen bromide)

- all the elements in Group 2 form chlorides with similar formulae: $MgCl_2$, $CaCl_2$.

The reactions of atoms depend on how many electrons there are in their outer shell. These are the electrons which normally get involved when elements bond to other elements. Elements in the same group (apart from helium in Group 0) have the same number of electrons in their outer shell, therefore they react in similar ways.

## THE NOBLE GASES

The Group 0 elements are known as the **noble gases** because they are almost completely unreactive, in fact the two at the top of the group, helium and neon, don't react with anything. The elements in Group 0 have 8 electrons in their outer shell (apart from helium, which has 2).

The lack of reactivity of the elements in Group 0 is associated with their electronic configurations. The noble gases are unreactive because *the outer shell is full, and so there is no tendency to lose, gain or share electrons in a chemical reaction*. You will learn more about how elements form compounds in Chapters 7 and 8.

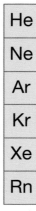

| He |
| Ne |
| Ar |
| Kr |
| Xe |
| Rn |

▲ Figure 4.8 The noble gases

**KEY POINT**

The reason the noble gas group is usually called Group 0 and not Group 8 (which would seem more sensible because most of the elements have 8 electrons in their outer shell) is because when they were first discovered it was believed that noble gases did not combine with anything, they had zero combining power (valency). This is the only group where the group number does not tell you the number of electrons in the outer shell! The key point is that the noble gases (except helium) have 8 electrons in their outer shell. You will see in Chapters 7 and 8 that atoms tend to form compounds by losing/gaining or sharing electrons so that they have 8 electrons in their outer shell. The noble gases already have 8 electrons in their outer shell so they do not do that.

**EXTENSION WORK**

In reality, the outer shells of argon, krypton, xenon and radon are not full; in fact only helium and neon have full outer shells. The third shell can actually hold up to 18 electrons and the fourth up to 32 electrons.

## METALS AND NON-METALS IN THE PERIODIC TABLE

*The metals are on the left-hand side of the Periodic Table and the non-metals are on the right-hand side.* Although the division into metals and non-metals is shown clearly in Figure 4.9, in practice there is a lot of uncertainty on the dividing line. For example, arsenic (As) has properties of both metals and non-metals.

**Groups**

| 1 | 2 | | | | | | | | | | | 3 | 4 | 5 | 6 | 7 | 0 |
|---|---|---|---|---|---|---|---|---|---|---|---|---|---|---|---|---|---|
| | | | | | H | | | | | | | | | | | | He |
| Li | Be | | | | | | | | | | | B | C | N | O | F | Ne |
| Na | Mg | | | | | | | | | | | Al | Si | P | S | Cl | Ar |
| K | Ca | Sc | Ti | V | Cr | Mn | Fe | Co | Ni | Cu | Zn | Ga | Ge | As | Se | Br | Kr |
| Rb | Sr | Y | Zr | Nb | Mo | Tc | Ru | Rh | Pd | Ag | Cd | In | Sn | Sb | Te | I | Xe |
| Cs | Ba | La· | Hf | Ta | W | Re | Os | Ir | Pt | Au | Hg | Tl | Pb | Bi | Po | At | Rn |
| Fr | Ra | Ac: | | | | | | | | | | | | | | | |

☐ metal

☐ non-metal

| ·Ce | Pr | Nd | Pm | Sm | Eu | Gd | Tb | Dy | Ho | Er | Tm | Yb | Lu |
|---|---|---|---|---|---|---|---|---|---|---|---|---|---|
| :Th | Pa | U | Np | Pu | Am | Cm | Bk | Cf | Es | Fm | Md | No | Lr |

▲ Figure 4.9 Metals and non-metals

**KEY POINT**

You may have noticed by now that hydrogen does not really fit into the Periodic Table properly. Although it has 1 electron in its outer shell, it is not in Group 1 because it does not have similar properties to the other elements in Group 1: it is not a metal and does not have similar chemical properties to the alkali metals.

**EXTENSION WORK**

If you are interested you could try an internet search for 'metallic hydrogen'.

▲ Figure 4.10 Copper, like all other metals, conducts electricity.

**KEY POINT**

If a basic oxide is soluble in water it will dissolve to form an alkali. For example, sodium oxide reacts with water to form sodium hydroxide solution, an alkali. If an acidic oxide is soluble in water it will dissolve to form an acidic solution. For example, sulfur(IV) oxide reacts with water to form sulfurous acid.

**KEY POINT**

There are some exceptions to the rules. For example, some metals form amphoteric oxides (e.g. $Al_2O_3$), which react with acids and bases, and some non-metal oxides (e.g. CO) are neutral.

## DIFFERENCES BETWEEN THE PROPERTIES OF METALS AND NON-METALS

There are many differences between the properties of metals and non-metals. Here, we will use two main ones to classify them: electrical conductivity and the acid–base character of their oxides.

***Metals conduct electricity and non-metals do not generally conduct electricity***. The reason that metals conduct electricity will be explained in Chapter 9 (it is due to the presence of delocalised electrons that are free to move). You may be familiar with the use of metals such as copper in electrical wires.

***Non-metals do not conduct electricity*** (because there are no electrons that are free to move or mobile ions) but there are a few exceptions, such as graphite (a form of carbon) and silicon.

***Metals generally form basic oxides***. A basic oxide is one which reacts with acids to form salts. For example, copper forms copper(II) oxide (CuO). This reacts, for example, with sulfuric acid to form the salt copper(II) sulfate:

$$CuO + H_2SO_4 \rightarrow CuSO_4 + H_2O$$
copper(II) oxide + sulfuric acid → copper(II) sulfate + water

The reactions between bases and acids will be discussed in more detail in Chapter 17.

***Non-metals generally form acidic oxides***. You may be familiar, for example, with sulfur(IV) oxide ($SO_2$) being one of the gases responsible for acid rain. Acidic oxides react with bases/alkalis to form salts. For example, carbon dioxide, an acidic oxide, reacts with sodium hydroxide, an alkali:

$$CO_2 + 2NaOH \rightarrow Na_2CO_3 + H_2O$$
carbon dioxide + sodium hydroxide → sodium carbonate + water

Some other properties you might also associate with metals and non-metals are included in the list on the next page.

▲ Figure 4.11 Mercury has most of the properties of a metal (high density, shiny, conducts electricity, forms positive ions); the exception is that it is a liquid at room temperature.

▲ Figure 4.12 Sulfur crystals are shiny, but you wouldn't mistake them for a metal.

Metals:

■ tend to be solids with high melting and boiling points, and with relatively high densities (but as with several of the properties in this list, there are exceptions, for example mercury is a liquid)

■ are shiny (have a metallic lustre) when they are polished or freshly cut

■ are **malleable** (can be hammered into shape)

■ are **ductile** (can be drawn into wires)

■ are good conductors of electricity and heat

■ form ionic compounds (see Chapter 7)

■ form positive ions in their compounds (see Chapter 7).

Non-metals:

■ tend to have low melting and boiling points (there are some exceptions, e.g. carbon and silicon)

■ tend to be brittle when they are solids

■ don't have the same type of shine as metals

■ don't usually conduct electricity; carbon (in the form of graphite) and silicon are again exceptions

■ are poor conductors of heat (diamond is an exception; it is the best conductor of heat of all the elements)

■ form both ionic and covalent compounds (see Chapters 7 and 8)

■ tend to form negative ions in ionic compounds (see Chapter 7).

**CHAPTER QUESTIONS**

You will need to use the Periodic Table in Appendix A on page 320.

**SKILLS** CRITICAL THINKING

1 Answer the questions that follow using only the elements in this list:

caesium, chlorine, molybdenum, neon, nickel, nitrogen, strontium, tin.

a State the name of an element which is:

i in Group 2

ii in the same period as silicon

iii in the same group as phosphorus

iv in Period 6

v a noble gas

b Divide the list of elements at the beginning of the question into two groups, metals and non-metals.

**SKILLS** INTERPRETATION

2 Draw diagrams to show the arrangement of the electrons in:

a sodium

b silicon

c sulfur

3 State the electronic configurations of the following atoms:

a fluorine    1-

b aluminium    3+

c calcium    2+

4 Find each of the following elements in the Periodic Table and state the number of electrons in its outer shell:

a arsenic, As    5

b bromine, Br

c tin, Sn

d xenon, Xe

5 The questions refer to the electronic configurations below. Don't worry if some of these are unfamiliar to you. All of these are the electronic configurations of neutral atoms.

    **A** 2, 4

    **B** 2, 8, 8

    **C** 2, 8, 18, 18, 7

    **D** 2, 8, 18, 18, 8

    **E** 2, 8, 8, 2

    **F** 2, 8, 18, 32, 18, 4

a Explain which of these atoms are in Group 4 of the Periodic Table.

b State which of these electronic configurations represents carbon.

c Explain which atoms are in Period 5 of the Periodic Table.

d Explain which of these electronic configurations represents an element in Group 7 of the Periodic Table.

e State which of these electronic configurations represent noble gases.

f State the name of element **E** and explain how you arrived at your answer.

g State how many protons are present in an atom of element **F**. State the name of the element.

h Element **G** has one more electron than element **B**. Draw a diagram to show how the electrons are arranged in an atom of **G**.

6 Predict two properties of the element palladium, Pd (atomic number 46), or its compounds. The properties can be either physical or chemical.

7 Helium and neon do not form any compounds. Explain why the noble gases are unreactive.

8 The elements in the Periodic Table are arranged in order of atomic number. If they were arranged in order of mass number give the names of two elements that would be in different positions. Explain why this would cause a problem.

# 5 CHEMICAL FORMULAE, EQUATIONS AND CALCULATIONS: PART 1

It is often important in chemistry to know how much of one substance reacts with a certain amount of another. To do this, we will learn about moles in this chapter.

▲ Figure 5.1 Each gold bar contains almost $4 \times 10^{25}$ gold atoms. That's 4 followed by 25 noughts! Learning about the mole allows us to work out how many atoms are in something.

## LEARNING OBJECTIVES

- Write word equations and balanced chemical equations (including state symbols):
  - for reactions studied in this course
  - for unfamiliar reactions where suitable information is provided

- Calculate relative formula masses (including relative molecular masses) ($M_r$) from relative atomic masses ($A_r$)

- Know that the mole (mol) is the unit for the amount of a substance

- Understand how to carry out calculations involving amount of substance, relative atomic mass ($A_r$) and relative formula mass ($M_r$)

- Calculate reacting masses using experimental data and chemical equations

- Calculate percentage yield

- Understand how the formulae of simple compounds can be obtained experimentally, including metal oxides, water and salts containing water of crystallisation

- Know what is meant by the terms empirical formula and molecular formula

- Calculate empirical and molecular formulae from experimental data

- Practical: Know how to determine the formula of a metal oxide by combustion (e.g. magnesium oxide) or by reduction (e.g. copper(II) oxide)

## WRITING EQUATIONS

There are two types of chemical equation that you could be asked to write: word equations and symbol equations. Symbol equations are usually called *chemical equations* and you should only write a word equation if you are specifically asked to. All chemical equations must be balanced.

## WHAT ALL THE NUMBERS MEAN

An example of a balanced chemical equation is:

$$CaCO_3 + 2HCl \rightarrow CaCl_2 + CO_2 + H_2O$$

When you write equations, it is important to be able to count how many of each type of atom you have. In particular, you must understand the difference between big numbers written in front of formulae (sometimes called coefficients), such as the **2** in **2HCl**, and the smaller, subscript (slightly lower on the line) numbers, such as the **3** in $CaCO_3$.

Another chemical equation is $2Cl \rightarrow Cl_2$. Here, **2Cl** represents 2 separate Cl atoms and $Cl_2$ means that the atoms are joined together in a molecule.

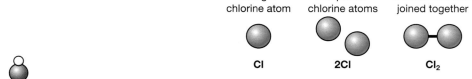

a single chlorine atom    2 separate chlorine atoms    2 chlorine atoms joined together

**Cl**     **2Cl**     **Cl₂**

▲ Figure 5.2 The difference between 2Cl and $Cl_2$

Another balanced chemical equation is $H_2S_2O_7 + H_2O \rightarrow 2H_2SO_4$.

Look at the way the numbers work in $2H_2SO_4$. The big number in front tells you that you have 2 sulfuric acid ($H_2SO_4$) molecules. The subscript 4 tells you that you have 4 oxygen atoms in each molecule. A small, subscript number in a formula applies only to the atom immediately before it in the formula. If you count the atoms in $2H_2SO_4$ you will find 4 hydrogens, 2 sulfurs and 8 oxygens.

If you have brackets in a formula, the small number refers to everything inside the brackets. For example, in the formula $Ca(OH)_2$, the 2 applies to both the oxygen and the hydrogen. The formula shows 1 calcium, 2 oxygens and 2 hydrogens. You will learn more about how to work out the formulae of compounds in Chapter 7.

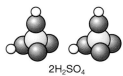

$2H_2SO_4$

▲ Figure 5.3 $2H_2SO_4$. The sulfur atoms are shown in yellow, the oxygens in red and the hydrogens in white.

## BALANCING EQUATIONS

Chemical reactions involve taking elements or compounds and moving their atoms around into new combinations. It follows that you must always end up with the same number of atoms that you started with.

Imagine you had to write an equation for the reaction between methane, $CH_4$, and oxygen, $O_2$. Methane burns in oxygen to form carbon dioxide and water. Think of this in terms of rearranging the atoms in some models.

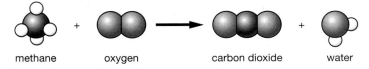

methane     oxygen     carbon dioxide     water

▲ Figure 5.4 Rearranging the atoms in methane and oxygen

If you count the atoms you had at the beginning (on the left-hand side of the arrow) and the atoms you have at the end (on the right-hand side of the arrow), you can see that this can't be right! During the rearrangement, we seem to have gained an oxygen atom and lost two hydrogens. The reaction must be more complicated than this. Since the substances are all correct, the proportions must be wrong.

Try again:

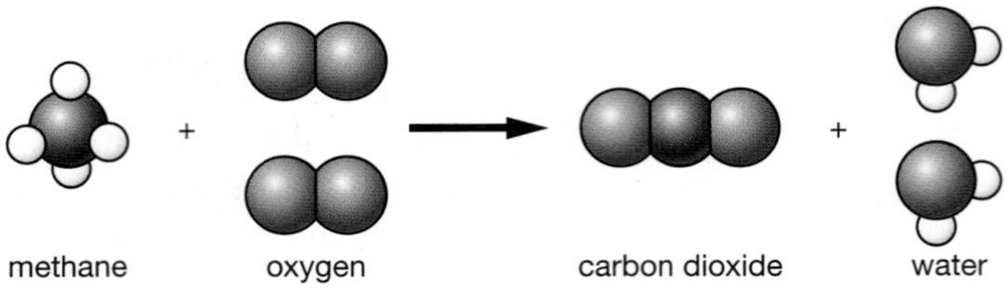

methane     oxygen     carbon dioxide     water

▲ Figure 5.5 A balanced equation for the reaction between methane and oxygen.

There are now the same number of each type of atom before and after. This is called **balancing the equation**.

In symbols, this equation would be:

$$CH_4 + 2O_2 \rightarrow CO_2 + 2H_2O$$

Think of each symbol (C or H or O) as representing one atom of that element. Count them up in the equation, and check that there is the same number of atoms on both sides.

## HOW TO BALANCE EQUATIONS

In order to balance equations you should adopt a systematic approach.

- Work across the equation from left to right, checking one element after another, except if an element appears in several places in the equation. In that case, leave the element until the end and you will often find that it has sorted itself out.

- If you have a group of atoms (like a sulfate group ($SO_4$), for example), which is unchanged from one side of the equation to the other, there is no reason why you can't just count that up as a whole group, rather than counting individual sulfurs and oxygens. It saves time.

- Check everything at the end to make sure you haven't changed something that you have already counted.

### EXAMPLE 1

Balance the equation for the reaction between zinc and hydrochloric acid:

$$Zn + HCl \rightarrow ZnCl_2 + H_2$$

Work from left to right. Count the zinc atoms: 1 on each side; no problem!

Count the hydrogen atoms: 1 on the left, 2 on the right. If you have 2 at the end, you must have started with 2. The only way of achieving this is to have 2HCl. (You must not change the formula to $H_2Cl$ because this substance does not exist.)

$$Zn + 2HCl \rightarrow ZnCl_2 + H_2$$

Now count the chlorines: there are 2 on each side. Good! Finally check everything again to make sure and you've finished.

|  | Zn + 2HCl |  | → | $ZnCl_2 + H_2$ |  |
|---|---|---|---|---|---|
| numbers of atoms | Zn | 1 |  | Zn | 1 |
|  | H | 2 |  | H | 2 |
|  | Cl | 2 |  | Cl | 2 |

### HINT

This is really important! You must never, never change a formula when balancing an equation. All you are allowed to do is to write big numbers in front of the formula.

Balance the equation for the combustion of ethane:

$$C_2H_6 + O_2 \rightarrow CO_2 + H_2O$$

Starting from the left, balance the carbons:

$$C_2H_6 + O_2 \rightarrow 2CO_2 + H_2O$$

Now the hydrogens:

$$C_2H_6 + O_2 \rightarrow 2CO_2 + 3H_2O$$

Finally the oxygens: there are 7 oxygens $((2 \times 2) + 3)$ on the right-hand side, but only 2 on the left. The problem is that the oxygens have to go around in pairs. So how can you obtain an odd number (7) of oxygens on the left-hand side?

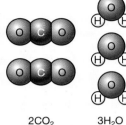

$2CO_2$    $3H_2O$

▲ Figure 5.6 There are 7 O atoms in $2CO_2 + 3H_2O$

The trick with this is to allow yourself to have halves in your equation. 7 oxygen atoms, O, is the same as $3\frac{1}{2}$ oxygen molecules, $O_2$.

$$C_2H_6 + 3\frac{1}{2}O_2 \rightarrow 2CO_2 + 3H_2O$$

Now double everything to get rid of the half:

$$2C_2H_6 + 7O_2 \rightarrow 4CO_2 + 6H_2O$$

## KEY POINT

In fact, it is acceptable to have halves in equations, but you don't usually come across them at Internaional GCSE.

You might reasonably argue that you can't have half an oxygen molecule, but to remove that problem all you have to do is double everything.

## HINT

Don't worry if this chemistry is new to you, or if at this stage you don't know what the state symbols should be. That is not important at the moment.

## HINT

Remember that water is a liquid (l), not an aqueous solution (aq). An aqueous solution is formed when something is *dissolved in water*.

## STATE SYMBOLS

State symbols are often, but not always, written after the formulae of the various substances in an equation to show what physical state everything is in. You need to know four different state symbols:

(s) solid    (l) liquid    (g) gas    (aq) in *aqueous* solution (dissolved in water)

So an equation might look like this:

$$2K(s) + 2H_2O(l) \rightarrow 2KOH(aq) + H_2(g)$$

This shows that *solid* potassium reacts with *liquid* water to make a *solution* of potassium hydroxide in water and hydrogen *gas*.

## HOW MUCH OF EACH SUBSTANCE REACTS IN A CHEMICAL REACTION?

You can make iron(II) sulfide by heating a mixture of iron and sulfur:

$$Fe + S \rightarrow FeS$$

How do you know what proportions to mix them in? You can't just mix equal masses of them because iron and sulfur atoms don't weigh the same. Iron atoms contain more protons and neutrons than sulfur atoms, so an iron atom is one and three-quarter times heavier than a sulfur atom. In this or any other reaction, you can get the right proportions only if you know about the masses of the individual atoms that take part in the reaction.

## RELATIVE ATOMIC MASS ($A_r$)

We have already looked at how to calculate relative atomic masses from the isotopic abundances in Chapter 3. Here we will look a little more closely at what exactly the relative atomic mass is.

Atoms are amazingly small. The mass of a hydrogen atom is about $1.67 \times 10^{-24}$ g (0.00000000000000000000000167 g). It is really difficult to use numbers such as this and so we use a scale of *relative* masses instead. The masses of atoms (and molecules) are compared with the mass of an atom of the carbon-12 isotope. We call this the carbon-12 scale. On this scale, one atom of the carbon-12 isotope weighs *exactly* 12 units and the mass of the most common hydrogen isotope is 1, which is a much simpler number to use!

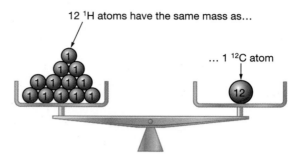

12 $^1$H atoms have the same mass as...

... 1 $^{12}$C atom

▲ Figure 5.7 The most common hydrogen atom weighs one-twelfth as much as a $^{12}$C atom.

The basic unit on the carbon-12 scale is $\frac{1}{12}$ of the mass of a $^{12}$C atom, which is approximately the mass of the most common hydrogen atom. A fluorine-19 atom has a relative mass of 19 because its atoms have a mass 19 times that of $\frac{1}{12}$ of a $^{12}$C atom. An atom of the most common isotope of magnesium weighs 24 times as much as $\frac{1}{12}$ of a $^{12}$C atom, and is therefore said to have a relative mass of 24.

The masses we are talking about here are the masses of individual *isotopes*, but samples of an actual element contain different isotopes and so we need a measure of the average mass of an *atom* taking into account the different isotopes. This is the relative atomic mass.

The relative atomic mass of an element (as opposed to one of its isotopes) is given the symbol $A_r$ and it is defined like this:

*The relative atomic mass of an element is the weighted average mass of the isotopes of the element. It is measured on a scale on which a carbon-12 ($^{12}$C) atom has a mass of exactly 12.*

Because we are talking here about *relative* masses they have no units.

On this scale, the relative atomic mass of chlorine is 35.45, that of lithium is 6.94 and that of sodium is 22.99 because we are taking into account the different isotopes, and are quoting an average mass for an atom. Although all elements consist of a mixture of isotopes, at International GCSE we only use relative atomic masses, including decimal places for Cl (35.5) and Cu (63.5), therefore we will take the relative atomic mass of lithium as 7 and that of sodium as 23.

# RELATIVE FORMULA MASS ($M_r$)

You can measure the masses of compounds on the same carbon-12 scale. For example, a water molecule, $H_2O$, has a mass of 18 on the carbon-12 scale. This means that a water molecule has 18 times the mass of $\frac{1}{12}$ of a $^{12}$C atom.

If you are talking about compounds, you use the term *relative formula mass*. Relative formula mass is sometimes called relative molecular mass.

Relative formula mass is given the symbol $M_r$.

## CALCULATING SOME RELATIVE FORMULA MASSES

### To find the relative formula mass ($M_r$) of magnesium carbonate, $MgCO_3$

Relative atomic masses: C = 12, O = 16, Mg = 24.

Add up the relative atomic masses to give the relative formula mass of the whole compound. In this case, you need to add up the masses of 1 × Mg, 1 × C and 3 × O.

$$M_r = \underset{Mg}{24} + \underset{C}{12} + \underset{(3 \times O)}{(3 \times 16)}$$

$$M_r = 84$$

### To find the relative formula mass of calcium hydroxide, $Ca(OH)_2$

Relative atomic masses: H = 1, O = 16, Ca = 40.

$$M_r = \underset{Ca}{40} + (\underset{O}{16} + \underset{H}{1}) \times 2$$

$$M_r = 74$$

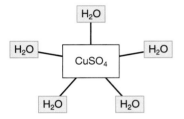

▲ Figure 5.8 A schematic diagram showing the formula of copper(II) sulfate crystals: there are five water molecules bonded to each copper(II) sulfate unit.

### To find the $M_r$ of copper(II) sulfate crystals, $CuSO_4 \cdot 5H_2O$

This formula looks very different and a lot more complicated than the ones we have looked at above. When some substances crystallise from solution, water becomes chemically bound up with the salt. This is called water of crystallisation. The $5H_2O$ is water of crystallisation. The water is necessary to form crystals of copper(II) sulfate (and some other substances). There are always 5 water molecules associated with 1 $CuSO_4$ unit and they are part of the formula. Salts containing water of crystallisation are said to be hydrated. Other examples include $Na_2CO_3 \cdot 10H_2O$ and $MgCl_2 \cdot 6H_2O$.

Relative atomic masses: H = 1, O = 16, S = 32, Cu = 63.5.

It is easiest to work out the relative formula mass of water first:

$$H_2O: \quad M_r = 2 \times 1 + 16 = 18$$

Now add the correct number of waters on to the rest of the formula:

$$CuSO_4 \cdot 5H_2O: \quad M_r = \underset{Cu}{63.5} + \underset{S}{32} + \underset{(4 \times O)}{(4 \times 16)} + \underset{(5 \times H_2O)}{(5 \times 18)} = 249.5$$

## USING RELATIVE FORMULA MASS TO FIND PERCENTAGE COMPOSITION

Having found the relative formula mass of a compound, we can work out the percentage by mass of any part of it. Examples make this clear.

### To find the percentage by mass of copper in copper(II) oxide, CuO

Relative atomic masses: O = 16, Cu = 63.5.

$$M_r \text{ of } CuO = 63.5 + 16 = 79.5$$

Of this, 63.5 is copper.

$$\text{Percentage of copper} = \frac{63.5}{79.5} \times 100$$

$$= 79.9\%$$

### To find the percentage by mass of oxygen in sodium carbonate, $Na_2CO_3$

Relative atomic masses: Na = 23, C = 12, O = 16.

$$M_r = (2 \times 23) + 12 + (3 \times 16)$$

$$= 106$$

48 of this 106 is due to the oxygen (remember that there are 3 oxygens in the formula, so the total mass of oxygen is $3 \times 16 = 48$).

$$\text{Percentage of oxygen} = \frac{3 \times 16}{106} \times 100$$
$$= 45.3\%$$

## THE MOLE

In chemistry, the **mole** is a unit of the amount of substance. We can talk about an amount of substance in grams or an amount of substance in moles. The difference between expressing the amount of substance in moles or in grams is that 1 mole of any substance has its own particular mass. For example, 1 mole of water has a mass of 18 g, 1 mole of sulfur has a mass of 32 g and 1 mole of magnesium oxide has a mass of 40 g. The abbreviation for mole is **mol**.

We can therefore talk, for example, about an amount of water as 36 g or 2 mol, since 1 mol of water has a mass of 18 g. 36 g of sulfur, however, would only be 1.125 mol and 36 g of magnesium would be 1.5 mol because the mass of 1 mol of sulfur is 32 g and that of 1 mol of magnesium is 24 g.

Some people talk about the *amount in moles* and others talk about the *number of moles*. This is like talking about the *mass of a substance* or the *number of grams*. For example, we could say that the amount in grams of water is 36 g or the number of grams of water is 36 g. Similarly, we can say that the amount in moles of water is 2 mol or the number of moles of water is 2 mol.

▲ Figure 5.9 1-mole quantities (clockwise from upper left) of carbon (12 g), sulfur (32 g), iron (56 g), copper (63.5 g) and magnesium (24 g).

## CALCULATING THE MASSES OF A MOLE OF SUBSTANCE

You find the mass of 1 mole of a substance by calculating the relative formula mass ($M_r$) and attaching the units, grams.

**1 mole of oxygen gas, $O_2$**
Relative atomic mass: O = 16.
$$M_r \text{ of } O_2 = 2 \times 16$$
$$= 32$$
1 mole of oxygen, $O_2$, has a mass of 32 g.

**1 mole of calcium chloride, $CaCl_2$**
Relative atomic masses: Cl = 35.5, Ca = 40.
$$M_r \text{ of } CaCl_2 = 40 + (2 \times 35.5)$$
$$= 111$$
1 mole of calcium chloride has a mass of 111 g.

**1 mole of iron(II) sulfate crystals, $FeSO_4 \cdot 7H_2O$**
Relative atomic masses: H = 1, O = 16, S = 32, Fe = 56.
$$M_r \text{ of crystals} = 56 + 32 + (4 \times 16) + \{7 \times [(2 \times 1) + 16]\}$$
1 mole of iron(II) sulfate crystals has a mass of 278 g.

## THE IMPORTANCE OF QUOTING THE FORMULA

Whenever you talk about a mole of something, you *must* quote its formula, otherwise there is a risk of confusion.

For example, if you talk about 1 mole of oxygen, this could mean:

1 mole of oxygen atoms, O, with a mass of 16 g

1 mole of oxygen molecules, $O_2$, with a mass of 32 g.

**HINT**

We often use the term 'molar mass' instead of 'mass of 1 mole'.

**EXTENSION WORK**

To be 100% correct 'molar mass' and 'mass of 1 mole' are not exactly the same thing and have different units: molar mass has units of g/mol and the mass of 1 mole has units of g. You are unlikely to come across the distinction between these at International GCSE.

## SIMPLE CALCULATIONS WITH MOLES

You need to be able to interconvert between a mass in grams and an amount in moles for a given substance. There is a simple formula that you can learn:

$$\text{number of moles} = \frac{\text{mass (g)}}{\text{mass of 1 mole (g)}}$$

You can rearrange this formula to find whatever you want. If rearranging this expression causes you problems, you can learn a simple triangular arrangement to help you.

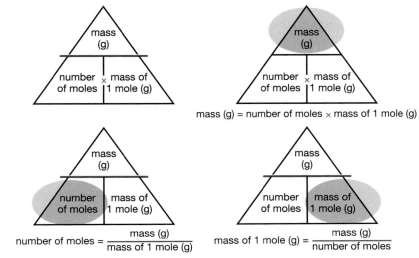

$$\text{mass (g)} = \text{number of moles} \times \text{mass of 1 mole (g)}$$

$$\text{number of moles} = \frac{\text{mass (g)}}{\text{mass of 1 mole (g)}} \qquad \text{mass of 1 mole (g)} = \frac{\text{mass (g)}}{\text{number of moles}}$$

▲ Figure 5.10 Converting between mass in grams and number of moles.

Look at this carefully and make sure that you understand how you can use it to deduce the three equations that you might need.

**Finding the mass of 0.2 mol of calcium carbonate, $CaCO_3$**

Relative atomic masses: C = 12, O = 16, Ca = 40.

First find the relative formula mass of calcium carbonate.

$$M_r \text{ of } CaCO_3 = 40 + 12 + (3 \times 16)$$
$$= 100$$

1 mol of $CaCO_3$ has a mass of 100 g.

$$\text{Mass (g)} = \text{number of moles} \times \text{mass of 1 mole (g)}$$
$$= 0.2 \times 100$$
$$= 20 \text{ g}$$

0.2 mol of $CaCO_3$ has a mass of 20 g.

**Finding the number of moles in 54 g of water, $H_2O$**

Relative atomic masses: H = 1, O = 16.

1 mol of $H_2O$ has a mass of 18 g.

$$\text{Number of moles} = \frac{\text{mass (g)}}{\text{mass of 1 mole (g)}}$$
$$= \frac{54}{18}$$
$$= 3 \text{ mol}$$

54 g of water is 3 mol.

**HINT**

Don't get too reliant on learning formulae for doing simple calculations. If 1 mole weighs 100 g, it should be fairly obvious to you that 2 moles will weigh twice as much, 10 moles will weigh 10 times as much and 0.2 moles will weigh 0.2 times as much.

**HINT**

If 1 mol weighs 18 g, then you need to find out how many times 18 goes into 54 to find out how many moles you have. On the other hand, if you feel safer using a formula, use it.

**EXTENSION WORK**

## Moles and the Avogadro constant

Imagine you had 1 mole of $^{12}C$. It would have a mass of 12 g and contain a very large number of carbon atoms, in fact about $6 \times 10^{23}$ carbon atoms, which is 6 followed by 23 noughts. This number of atoms in 12 g of $^{12}C$ is called the Avogadro constant.

1 mole of anything else contains this same number of particles. For example:

1 mole of magnesium contains $6 \times 10^{23}$ magnesium atoms, Mg, and has a mass of 24 g.

1 mole of water contains $6 \times 10^{23}$ water molecules, $H_2O$, and has a mass of 18 g.

This is the reason that we use moles: if we know the number of moles we also know how many particles are present. If we know how many particles are present, we can work out how much of one substance reacts with a certain amount of another.

▲ Figure 5.11 Approximately $1 \times 10^{32}$ water molecules go over Niagara Falls every second during the summer. That's 100 million million million million million water molecules per second.

# FORMULAE

The formula of sulfur dioxide is $SO_2$. We can see that there are 2 O atoms for each S atom. If we had 1 mol $SO_2$ we would still have twice as many O atoms as S atoms; there would be 1 mol S atoms and 2 mol O atoms. If we know that in a certain sample of sulfur dioxide we have 0.1 mol of S atoms we also know that we must have 0.2 mol O atoms. 0.2 mol contains twice as many atoms as 0.1 mol.

If we did an experiment and found that a compound contained 0.2 mol Ca and 0.2 mol O, then we could work out that the formula must be CaO because there are the same number of Ca atoms as O atoms.

If a compound contains 0.4 mol Mn and 0.8 mol O then the formula must be $MnO_2$ because there are twice as many O atoms as Mn atoms.

The formulae that we have found here are called *empirical formulae*.

**The empirical formula *shows the simplest whole number ratio of the atoms present in a compound.***

There is another type of formula that we will come across, called the *molecular formula*.

**The molecular formula *shows the actual number of atoms of each element present in a molecule (covalent compound) or formula unit (ionic compound) of a compound.***

The molecular formula can be the same as the empirical formula or a multiple of the empirical formula. The empirical formula of calcium oxide is CaO and the molecular formula is also CaO. The empirical formula of hydrogen peroxide is HO and the molecular formula is $H_2O_2$.

In order to work out the molecular formula from the empirical formula we need more information – the $M_r$ of the compound. We will look at this again below.

## WORKING OUT EMPIRICAL FORMULAE

In order to find out the empirical formula of a compound such as copper oxide, we need to know how many atoms of copper combine with how many atoms of oxygen. We can work out the number of atoms from the number of moles. If we know the ratio between the number of copper atoms and oxygen atoms in the compound we know the formula.

### EXAMPLE 3

A sample of a compound contains 1.27 g of Cu and 0.16 g of O. Calculate the empirical formula. ($A_r$ of Cu = 63.5, $A_r$ of O = 16)

It is easiest to do this in columns using a table:

| | Cu | O |
|---|---|---|
| masses/g | 1.27 | 0.16 |
| find the number of moles of atoms by dividing the mass by the mass of 1 mole of atoms | 1.27/63.5 | 0.16/16 |
| number of moles of atoms | 0.02 | 0.01 |
| divide by the smaller number to find the ratio | 0.02/0.01 | 0.01/0.01 |
| ratio of moles | 2 | 1 |
| empirical formula | $Cu_2O$ | |

**KEY POINT**

0.02 mol of atoms is twice as many atoms as 0.01 mol of atoms.

**HINT**

More significant figures should have been used for the number of moles in the examples on this page but these have been omitted for simplicity.

From calculating the number of moles of Cu and O we can see that there are twice as many moles of Cu and therefore there must be twice as many Cu atoms as O atoms in the compound.

It is important to remember in this calculation that we are working out how many *atoms* of copper combine with how many *atoms* of oxygen so we divide the mass of oxygen by 16 (the mass of 1 mole of oxygen atoms) rather than by 32 (the mass of 1 mole of oxygen molecules).

### EXAMPLE 4

A sample of a compound contains 0.78 g of K, 1.10 g of Mn and 1.28 g of O. ($A_r$ of K = 39, $A_r$ of Mn = 55, $A_r$ of O = 16)

Again, we use a table:

| | K | Mn | O |
|---|---|---|---|
| masses/g | 0.78 | 1.10 | 1.28 |
| find the number of moles of atoms by dividing the mass by the mass of 1 mole of atoms | 0.78/39 | 1.10/55 | 1.28/16 |
| number of moles of atoms | 0.02 | 0.02 | 0.08 |
| divide by the smallest number to find the ratio | 0.02/0.02 | 0.02/0.02 | 0.08/0.02 |
| ratio of moles | 1 | 1 | 4 |
| empirical formula | $KMnO_4$ | | |

The empirical formula of this compound is $KMnO_4$.

Safety Note: Wear eye protection and take care not to get burned when raising the crucible lid with tongs. The crucible will stay very hot for some time.

## ACTIVITY 3

### ▼ PRACTICAL: INVESTIGATING THE FORMULA OF A METAL OXIDE BY COMBUSTION

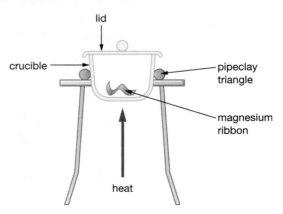

▲ Figure 5.12 Apparatus for determining the formula of magnesium oxide

We can find the formula of magnesium oxide by burning magnesium in oxygen and looking at how the mass changes. The following procedure is usually used:

- Weigh a crucible with a lid.
- Place a piece of magnesium ribbon about 10 cm long in the crucible and weigh the crucible and contents.
- Set up the apparatus as shown in Figure 5.12.
- Heat the crucible strongly (a roaring flame).
- Lift the lid every few seconds.
- When the reaction is finished, allow the crucible and contents to cool.
- Weigh the crucible and contents.

When the magnesium burns it does so with a bright white flame. Magnesium oxide, a white powder, is produced in the reaction.

A lid is placed on the crucible to prevent the white powder (magnesium oxide) escaping but the lid must be lifted every few seconds to allow oxygen into the crucible to react with the magnesium.

A set of results for this practical could be:

| | |
|---|---|
| mass of empty crucible/g | 32.46 |
| mass of crucible + magnesium/g | 32.70 |
| mass of crucible + contents at end of experiment/g | 32.86 |

We can work out the mass of magnesium by subtracting the mass of the crucible from the mass of the crucible + magnesium:

$$\text{mass of magnesium} = 32.70 - 32.46 = 0.24 \text{ g}$$

The reason that the mass increases is because the magnesium combines with the oxygen in the air. The mass of magnesium oxide is greater than the mass of just magnesium due to the extra oxygen.

**Remember**: Number of moles is mass in grams divided by the mass of 1 mole in grams.
The overall reaction that occurred was:
$$2Mg + O_2 \rightarrow 2MgO$$
magnesium + oxygen → magnesium oxide
The type of reaction that occurred was combustion.
When we do this experiment in practice, it is not very often that the ratio comes out as 1:1. There are things that may go wrong with this experiment which could affect the results, such as:
- some of the magnesium oxide powder might escape when the lid is lifted
- not all the magnesium might have reacted
- the magnesium can also react with nitrogen in the air.

The original masses were measured to 2 significant figures so the number of moles should also be given to 2 significant figures (0.010). If you write 0.01 for the number of moles of atoms in the exam, this will be fine and still give you the correct answer.

Safety Note: The teacher demonstrating needs to wear a face shield and use safety screens. The pupils require eye protection and should be no closer than 2 metres. If a drying agent is needed anhydrous calcium chloride should be used NOT concentrated sulfuric acid.

We must always make sure that the tube is filled with hydrogen before lighting the stream of hydrogen because a mixture of hydrogen and oxygen is explosive! It is therefore important to let the stream of hydrogen gas flow through the tube for a little while (to flush out all the oxygen) before lighting it or lighting the Bunsen burner.

We can work out the mass of oxygen that combines with the magnesium:

mass of oxygen = 32.86 − 32.70 = 0.16 g

In order to find the formula of magnesium oxide we need to work out how many moles of magnesium atoms combine with how many moles of oxygen atoms.

|  | Mg | O |
|---|---|---|
| masses/g | 0.24 | 0.16 |
| find the number of moles of atoms by dividing the mass by the mass of 1 mole | 0.24/24 | 0.16/16 |
| number of moles of atoms | 0.010 | 0.010 |
| divide by the smaller number to find the ratio | 0.010/0.010 | 0.010/0.010 |
| ratio of moles | 1 | 1 |
| empirical formula | MgO | |

The empirical formula of magnesium oxide is MgO.

# THE FORMULA FOR COPPER OXIDE

## ACTIVITY 4

### ▼ PRACTICAL: INVESTIGATING THE FORMULA OF A METAL OXIDE BY REDUCTION

We can also find the formula of an oxide by removing the oxygen from it and looking at how the mass changes.

The following procedure can be used to find the formula of copper oxide:
- Weigh a ceramic dish.
- Put about 3 g of copper oxide in the ceramic dish and weigh the dish again.
- Place the ceramic dish in a tube as shown in Figure 5.13.
- Pass hydrogen gas over the copper oxide.
- Ignite the excess hydrogen, which comes out of the small hole in the boiling tube.
- Heat the copper oxide strongly until the reaction is finished (pink-brown copper metal will be seen).

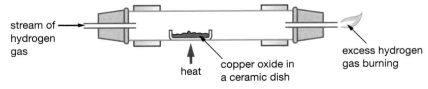

stream of hydrogen gas

heat

copper oxide in a ceramic dish

excess hydrogen gas burning

▲ Figure 5.13 The experimental set-up for finding the formula of copper oxide.

A set of results for this practical could be:

| mass of empty dish/g | 23.78 |
| --- | --- |
| mass of dish + copper oxide/g | 26.96 |
| mass of dish + contents at end of experiment/g | 26.32 |

The reason that the mass decreases is because the hydrogen combines with the oxygen from the copper oxide to form water. The oxygen is removed from the copper oxide and we are left with only copper in the dish at the end of the experiment. Because oxygen is removed from the copper oxide we say that the copper oxide has been reduced. Reduction is explained in Chapter 14.

From this data we can calculate the mass of copper oxide at the beginning by subtracting the mass of the dish from the mass of the dish + copper oxide:

mass of copper oxide = 26.96 − 23.78 = 3.18 g

We can work out the mass of copper remaining at the end by subtracting the mass of the dish from the mass of the dish + copper at the end.

mass of copper = 26.32 − 23.78 = 2.54 g

The mass has decreased because the oxygen has been removed from the copper oxide and we can calculate the mass of oxygen by subtracting the mass of copper at the end from the mass of copper oxide:

mass of oxygen = 3.18 − 2.54 = 0.64 g

We now know that 2.54 g of copper combines with 0.64 g of oxygen in copper oxide and can deduce the empirical formula:

| | Cu | O |
| --- | --- | --- |
| masses/g | 2.54 | 0.64 |
| find the number of moles of atoms by dividing the mass by the mass of 1 mole | 2.54/63.5 | 0.64/16 |
| number of moles of atoms | 0.0400 | 0.040 |
| divide by the smaller number to find the ratio | 0.0400/0.040 | 0.040/0.040 |
| ratio of moles | 1 | 1 |
| empirical formula | CuO | |

**KEY POINT**

This is a displacement (or competition) reaction. Hydrogen is more reactive than copper and displaces it from copper oxide. Displacement reactions are discussed in Chapter 14.

**KEY POINT**

The equation for the reaction that occurs in this experiment is:

$CuO$ + $H_2$ → $Cu$ + $H_2O$
copper(II) + hydrogen → copper + water
oxide

As well as a displacement reaction, this type of reaction can also be called a redox reaction. Redox reactions are discussed in Chapter 14.

Safety Note: The teacher demonstrating needs to wear a face shield and use safety screens. The pupils require eye protection and should be no closer than 2 metres. If a drying agent is needed anhydrous calcium chloride should be used NOT concentrated sulfuric acid.

## DETERMINING THE FORMULA OF WATER

We can modify the apparatus in Figure 5.13 to allow us to determine the formula of water.

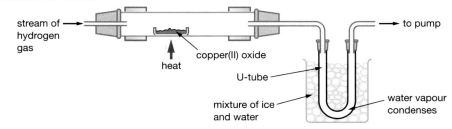

▲ Figure 5.14 Apparatus that can be used to determine the formula of water.

The experiment is conducted in the same way except that this time the water vapour that is produced from the reaction between copper(II) oxide and hydrogen is condensed. The contents of the dish at the beginning and the end of the experiment are again weighed but this time the mass of water that collects in the U-tube must also be measured.

| | |
|---|---|
| mass of empty dish/g | 23.78 |
| mass of dish + copper oxide/g | 26.96 |
| mass of dish + contents at end of experiment/g | 26.32 |
| mass of water/g | 0.72 |

Here we are using the same results as above. The mass of oxygen lost from the copper(II) oxide is 0.64g.

All the oxygen lost from the copper(II) oxide combines with hydrogen to form water. This means that the water contains 0.64g of oxygen. 0.72g of water was collected so the mass of hydrogen in the water must be $0.72 - 0.64 = 0.08$g. We can now determine the empirical formula of water:

| | H | O |
|---|---|---|
| masses/g | 0.08 | 0.64 |
| number of moles of atoms | 0.08/1 | 0.64/16 |
| number of moles of atoms | 0.08 | 0.040 |
| divide by the smaller number to find the ratio | 0.08/0.040 | 0.040/0.040 |
| ratio of moles | 2 | 1 |
| empirical formula | $H_2O$ | |

## WORKING OUT FORMULAE USING PERCENTAGE COMPOSITION FIGURES

In the worked examples and practical examples above, we have determined the empirical formulae of compounds using masses. However, we are often given percentages by mass instead of just masses.

**EXAMPLE 5**

Find the empirical formula of a compound containing 82.7% C and 17.3% H by mass ($A_r$ of H = 1, $A_r$ of C = 12).

The percentage figures apply to any amount of substance you choose, so let's choose 100 g. In this case the percentages convert simply into masses: 82.7% of 100 g is 82.7 g (Table 5.1).

Table 5.1 Finding the ratio from percentage by mass

| | C | H |
|---|---|---|
| percentages/% | 82.7 | 17.3 |
| masses in 100 g/g | 82.7 | 17.3 |
| number of moles of atoms | 82.7/12 | 17.3/1 |
| number of moles | 6.89 | 17.3 |
| divide by smallest to get ratio | 6.89/6.89 | 17.3/6.89 |
| ratio of moles | 1 | 2.5 |

From this we can see that there are 2.5 mol of H atoms for every mole of C atoms. The empirical formula, however, is the *whole number ratio* of the elements present in a compound. To obtain a whole number here we multiply both numbers by 2 to get $C_2H_5$, which is the empirical formula.

## CONVERTING EMPIRICAL FORMULAE INTO MOLECULAR FORMULAE

When you have learned a bit more organic chemistry in Unit 4 you will realise that, in the example we have just looked at, $C_2H_5$ can't possibly be the real formula of the compound. The molecular formula (the actual number of each atom present in a molecule) has to be a multiple of $C_2H_5$, like $C_4H_{10}$.

We can find the molecular formula if we know the relative formula mass of the compound (or the mass of 1 mole, which is just the $M_r$ in grams).

In the previous example, suppose you were told that the relative formula mass of the compound was 58.

$C_2H_5$ has a relative formula mass of 29 ($A_r$ of H = 1, $A_r$ of C = 12).

All you need to find out is how many times 29 goes into 58.

$58/29 = 2$

So you need 2 lots of $C_2H_5$, in other words, $C_4H_{10}$.

The molecular formula is $C_4H_{10}$.

> **HINT**
>
> When you calculate 17.3/6.89 you actually obtain 2.51. When doing these calculations it is fine to round 2.51 to 2.5 or 1.01 to 1 but what you must not do is round 2.51 to 3!
>
> If you got a ratio of C:H of 1:1.33 you would not round this down to 1:1 but rather multiply by 3 to obtain $C_3H_4$.

> **EXAMPLE 6**
>
> A compound has the empirical formula $CH_2$. If the relative formula mass is 56, work out the molecular formula.
>
> The relative formula mass of $CH_2$ is 12 + (2 × 1) = 14.
>
> $56/14 = 4$
>
> Therefore there must be 4 lots of $CH_2$ in the actual molecule and the molecular formula is $C_4H_8$.
>
> Remember, the ratio of elements in the molecular formula must be the same as in the empirical formula: $C_4H_8$ cancels down to $CH_2$.

## EMPIRICAL FORMULA CALCULATIONS INVOLVING WATER OF CRYSTALLISATION

### FINDING THE $n$ IN $BaCl_2 \cdot nH_2O$

#### REMINDER

Remember that when some substances crystallise from solution, water becomes chemically bound up with the salt. This is called **water of crystallisation**. The salt is said to be **hydrated**.

#### REMINDER

Anhydrous means without water.

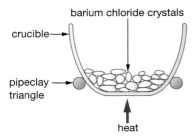

crucible→

barium chloride crystals

pipeclay → triangle

heat

▲ Figure 5.15 Heating barium chloride crystals in a crucible

 Safety Note: Barium chloride is toxic but a Bunsen flame is not hot enough to release it into the atmosphere.

#### HINT

If you are doing this experiment, how can you be sure that all the water has been driven off? The best way to do this is to heat it to constant mass. You heat the crucible and weigh it. Keep doing this until the mass remains constant. If the mass has gone down after heating there may still be water present, but when the mass remains constant after heating you know that all the water has been driven off.

### INTERPRETING SYMBOLS IN EQUATIONS IN TERMS OF MOLES

#### KEY POINT

The big numbers in front of the formulae tell us the number of moles of each substance that react. Although there does not appear to be a big number in front of the $CH_4$ and $CO_2$, there should actually be a '1' in front of each but we just don't write it.

Usually, when you heat a salt that contains water of crystallisation, the water is driven off, leaving the **anhydrous** salt behind. Hydrated barium chloride is a commonly used example because the barium chloride itself doesn't decompose even on quite strong heating.

If you heated barium chloride crystals in a crucible you might obtain these results:

| | |
|---|---|
| Mass of crucible | $= 30.00\,g$ |
| Mass of crucible + barium chloride crystals, $BaCl_2 \cdot nH_2O$ | $= 32.44\,g$ |
| Mass of crucible + anhydrous barium chloride, $BaCl_2$ | $= 32.08\,g$ |

To find $n$, you need to find the ratio of the number of moles of $BaCl_2$ to the number of moles of water. It's just another empirical formula calculation ($A_r$ of H = 1, $A_r$ of O = 16, $A_r$ of Cl = 35.5, $A_r$ of Ba = 137).

Mass of $BaCl_2$ = 32.08 − 30.00 = 2.08 g
Mass of water = 32.44 − 32.08 = 0.36 g

Table 5.2 Finding the $n$ in $BaCl_2 \cdot nH_2O$

| | $BaCl_2$ | $H_2O$ |
|---|---|---|
| masses/g | 2.08 | 0.36 |
| divide by $M_r$ to find the number of moles | 2.08/208 | 0.36/18 |
| number of moles | 0.0100 | 0.020 |
| ratio of moles | 1 | 2 |
| empirical formula | $BaCl_2 \cdot 2H_2O$ | |

## CALCULATIONS USING MOLES, CHEMICAL EQUATIONS AND MASSES OF SUBSTANCES

If we write a chemical equation:

$$CH_4 + 2O_2 \rightarrow CO_2 + 2H_2O$$

we can say that 1 molecule of $CH_4$ combines with 2 molecules of $O_2$ to form 1 molecule of $CO_2$ and 2 molecules of water. When doing calculations, however, it is often more useful to think about this in terms of moles and take symbols as meaning moles of each substance. So here we can say:

1 mol $CH_4$ reacts with 2 mol $O_2$ to form 1 mol $CO_2$ and 2 mol $H_2O$

We have not changed the meaning of the equation as the ratios are still the same.

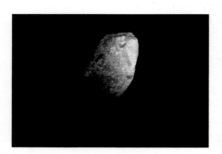

▲ Figure 5.16 A glowing piece of limestone. Limestone is impure calcium carbonate – it decomposes when heated to form calcium oxide (quicklime) and carbon dioxide. This is a thermal decomposition reaction.

We can therefore tell from this equation that 16 g of $CH_4$ (1 mol) reacts exactly with 64 g of $O_2$ (2 mol) to form 44 g of $CO_2$ (1 mol) and 36 g of $H_2O$ (2 mol).

## CALCULATIONS INVOLVING ONLY MASSES

Typical calculations will give you a mass of starting material and ask you to calculate how much product you are likely to obtain. You will also find examples done in reverse, where you are told the mass of the product and are asked to find out how much of the starting material you would need. In almost all the cases you will see at International GCSE you will be given the equation for the reaction.

## A PROBLEM INVOLVING HEATING CALCIUM CARBONATE

When calcium carbonate, $CaCO_3$, is heated, calcium oxide is formed. Imagine you wanted to calculate the mass of calcium oxide produced by heating 25 g of calcium carbonate ($A_r$: C = 12, O = 16, Ca = 40).

$$CaCO_3 \rightarrow CaO + CO_2$$

### THE CALCULATION

The method we will use has three steps:

*Step 1*: Calculate the number of moles using the mass that you have been given.

*Step 2*: Use the chemical equation to work out the number of moles of the substance you are interested in.

*Step 3*: Convert the number of moles to a mass.

In this reaction we have 25 g of $CaCO_3$ and the number of moles can be calculated as:

$$\text{number of moles} = \frac{\text{mass}}{\text{mass of 1 mole}} = \frac{25}{100} = 0.25 \, \text{mol}$$

The chemical equation for this reaction is

$$CaCO_3 \rightarrow CaO + CO_2$$

We can see from the equation that 1 mole of $CaCO_3$ will decompose to produce 1 mole of CaO. In other words, the equation is telling us that if we start with a certain number of moles of $CaCO_3$ we will obtain the same number of moles of CaO at the end.

Therefore, we can deduce that 0.25 mol $CaCO_3$ will decompose to produce 0.25 mol CaO. Now that we know the number of moles of CaO ($M_r = 56$) we can convert this to a mass:

$$\text{mass} = \text{number of moles} \times \text{mass of 1 mole}$$

$$\text{mass} = 0.25 \times 56 = 14 \, \text{g}$$

Therefore, the reaction will produce 14 g of calcium oxide.

## ALTERNATIVE METHOD

### HINT

You do not have to learn both methods, just learn whichever method you are happier with.

Your maths may be so good that you don't need to take all these steps to find the answer. If you can do it more quickly, that's fine. However, you must still show all your calculations.

When you have finished a chemistry calculation, the impression should nearly always be that there are a lot of words with a few numbers between them, not vice versa.

The calculation can also be done in a different way using ratios:

$$CaCO_3(s) \rightarrow CaO(s) + CO_2(g)$$

Interpret the equation in terms of moles:

1 mol $CaCO_3$ produces 1 mol CaO (and 1 mol $CO_2$)

Substitute masses where relevant:

100 g (1 mol) $CaCO_3$ produces 56 g (1 mol) CaO

Do a proportion calculation:

If 100 g of calcium carbonate gives 56 g of calcium oxide

1 g of calcium carbonate gives $\dfrac{56}{100}$ g of calcium oxide = 0.56 g

25 g of calcium carbonate gives 25 × 0.56 g of calcium oxide = 14 g of calcium oxide

## A PROBLEM ABOUT EXTRACTING IRON

The equation below is for a reaction that occurs in the extraction of iron:

$$Fe_2O_3 + 3C \rightarrow 2Fe + 3CO$$

Calculate the mass of iron which can be formed from 1000 g of iron oxide.

We are given the mass of $Fe_2O_3$ in the question and so we can calculate the number of moles of $Fe_2O_3$:

$$\text{number of moles} = \frac{\text{mass}}{\text{mass of 1 mole}}$$

$$\text{number of moles} = \frac{1000}{160} = 6.25 \, \text{mol}$$

### KEY POINT

The $A_r$ of Fe is 56.
The $M_r$ of $Fe_2O_3$ = 2 × 56 + 3 × 18 = 160.

From the chemical equation we can see that 1 mol $Fe_2O_3$ produces 2 mol Fe. We know this because of the big numbers in front of the formulae.

So we know that, if we start with a certain number of moles of $Fe_2O_3$, we will obtain twice as many moles of Fe.

So 6.25 mol $Fe_2O_3$ produces 2 × 6.25 = 12.5 mol Fe:

$$\text{mass} = \text{number of moles} \times \text{mass of 1 mole}$$

The mass of 12.5 mol Fe is 12.5 × 56 = 700 g.

### HINT

To find the mass of Fe we multiply the number of moles of Fe by the mass of 1 mole of Fe (56 g). A common mistake is to multiply the number of moles by the mass of 2Fe (112 g), but we have already used the '2' when we worked out the number of moles of Fe.

## ALTERNATIVE METHOD IN TERMS OF RATIOS

First interpret the equation in terms of moles:

1 mol $Fe_2O_3$ reacts with C to form 2 mol Fe (and $CO_2$).

We are only looking at how much iron is produced, so let's introduce just these masses.

160 g (1 mol) of $Fe_2O_3$ produces 2 × 56 g (2 mol) of Fe.

That is, 160 g of $Fe_2O_3$ produces 112 g of Fe.

From this we can calculate that 1 g of $Fe_2O_3$ will produce 112/160 g of Fe, and therefore 1000 g of Fe will produce 1000 × 112/160 = 700 g.

## A PROBLEM INVOLVING THE EXTRACTION OF LEAD

Lead is extracted from galena, PbS. The ore is roasted in air to produce lead(II) oxide, PbO:

$$2PbS(s) + 3O_2(g) \rightarrow 2PbO(s) + 2SO_2(g)$$

The lead(II) oxide is then converted to lead by heating it with carbon in a blast furnace:

$$PbO(s) + C(s) \rightarrow Pb(l) + CO(g)$$

The molten lead is tapped from the bottom of the furnace.

Calculate the mass of lead that would be produced from 1 tonne of galena.

($A_r$: O = 16, S = 32, Pb = 207)

▲ Figure 5.17 Molten lead tapped from the bottom of a furnace.

We have been given the mass of PbS so we can calculate the number of moles. However, we have to be careful because the mass is in tonnes but to calculate a number of moles we need to know the mass in grams, so we must convert tonnes to grams. The conversion for this is shown in Figure 5.18. Just write down this relationship: *1 tonne is equivalent to 1 000 000 g* so, to change from tonnes to grams, the number has to *increase* (from 1 to 1 000 000), so we *multiply* by 1 000 000.

Therefore we have 1 000 000 g PbS.

$$\text{Number of moles} = \frac{\text{mass}}{\text{mass of 1 mole}}$$

The number of moles of PbS $= \dfrac{1\,000\,000}{239} = 4184\,\text{mol}$.

From the first equation:

$$2PbS(s) + 3O_2(g) \rightarrow 2PbO(s) + 2SO_2(g)$$

we can see that 2 mol PbS produce 2 mol PbO, therefore we can work out that 4184 mol PbS will produce 4184 mol PbO.

From the second equation:

$$PbO(s) + C(s) \rightarrow Pb(l) + CO(g)$$

we can see that 1 mol PbO produces 1 mol Pb. But we are starting this reaction with 4184 mol PbO and so we can work out that we will produce 4184 mol Pb.

The mass of 4184 mol Pb is 4184 × 207 = 866 000 g.

We can convert this to tonnes by dividing by 1 000 000:

866 000/1 000 000 = 0.866 tonne

**HINT**

You should be told how many grams there are in a tonne if you need this information in the exam.

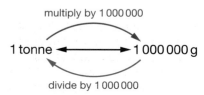

multiply by 1 000 000

1 tonne ◄──────► 1 000 000 g

divide by 1 000 000

▲ Figure 5.18 How to convert between grams and tonnes

**HINT**

1 000 000 might be given to you as $10^6$ or $1 \times 10^6$. These are both the same.

## ALTERNATIVE METHOD

### KEY POINT

We've doubled the second equation so that we can trace what happens to all the 2PbO from the first one.

This could also be simplified to 1 mol PbS produces 1 mol Pb. This would save you doing some arithmetic as you would not have to multiply everything by 2. However, in the end it does not make any difference to the answer, so you do not have to simplify it if you don't want to.

### HINT

There are advantages and disadvantages of both methods. For the first method, we can use the same three steps in all the calculations we do below. However, a bit more maths will be involved: converting tonnes to grams and back again. The alternative method has fewer steps but we cannot use it in the same way when doing some of the calculations involving solutions in the next chapter.

$$2PbS(s) + 3O_2(g) \rightarrow 2PbO(s) + 2SO_2(g)$$

$$PbO(s) + C(s) \rightarrow Pb(l) + CO(g)$$

Interpret the equation in terms of moles and trace the lead through the equations:

2 mol PbS produces 2 mol PbO

If we double the second equation:

$$2PbO(s) + 2C(s) \rightarrow 2Pb(l) + 2CO(g)$$

we can see that 2 mol PbO produces 2 mol Pb.

In other words, every 2 mol of PbS produces 2 mol of Pb.

Substitute masses where relevant. In this case, the relevant masses are only the PbS and the Pb:

2 × 239 g PbS produces 2 × 207 g Pb

478 g PbS produces 414 g Pb

Now there seems to be a problem. The question is asking about tonnes and not grams. You could calculate how many grams there are in a tonne. However, it's much easier to think a bit, and realise that the ratio is always going to be the same, whatever the units, which means that:

478 tonnes PbS produces 414 tonnes Pb

Do the proportion calculation:

If 478 tonnes PbS produces 414 tonnes Pb

then 1 tonne PbS gives $\dfrac{417}{478}$ tonne Pb = 0.866 tonne

0.866 tonne of lead is produced from 1 tonne of galena.

## CALCULATING PERCENTAGE YIELDS

### WHAT IS A PERCENTAGE YIELD?

If you calculate how much product a reaction might produce, in real life you rarely obtain as much as you expected. If you expect to get 100 g, but only get 80 g, your percentage yield is 80%. The rest of it has been lost in some way. This could be due to spillages, or losses when you transfer a liquid from one container to another. Or it may be that there are all sorts of side reactions going on, so that some of your starting materials are changed into unwanted products. That happens a lot during reactions in organic chemistry.

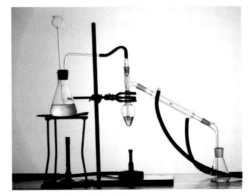

▶ Figure 5.19 Experiments done by students to make and purify organic compounds rarely give a high percentage yield.

## CALCULATING THE PERCENTAGE YIELD

**HINT**

The percentage yield should always come out as less than 100%. If it comes out as more than 100% either you have made a mistake in the calculation or the question will continue by asking about what might have gone wrong in the experiment to give a value greater than 100%. If you are making a salt (see Chapter 17), for instance, a possible reason you could get a percentage yield greater than 100% is that it was not completely dry.

Suppose you do a calculation and work out that you should get 12.5 g of copper sulfate crystals, but you only get 11.2 g when you actually do the experiment. 12.5 g is the *theoretical yield* and can be obtained by carrying out a moles calculation. 11.2 g is the *actual yield* – this is what is obtained in the actual experiment and will be given to you in the question.

$$\text{Percentage yield} = \frac{\text{actual yield}}{\text{theoretical yield}} \times 100$$

The percentage yield is $\frac{11.2}{12.5} \times 100 = 89.6\%$.

**EXAMPLE 7**

A student reacted 2.40 g of copper(II) oxide (CuO) with hot sulfuric acid. She made 5.21 g of copper(II) sulfate crystals ($CuSO_4 \cdot 5H_2O$).

The equations for the reactions are:

$$CuO(s) + H_2SO_4(aq) \rightarrow CuSO_4(aq) + H_2O(l)$$
$$CuSO_4(aq) + 5H_2O(l) \rightarrow CuSO_4 \cdot 5H_2O(l)$$

First we need to calculate the theoretical yield. This is done by carrying out a moles calculation. We have enough information to calculate the number of moles of copper(II) oxide:

$$\text{number of moles of CuO} = \frac{2.40}{79.5} = 0.0302\,\text{mol}$$

From the first equation we can deduce that 0.0302 mol CuO will produce 0.0302 mol $CuSO_4$.

From the second equation we can deduce that 0.0302 mol $CuSO_4$ will produce 0.0302 mol $CuSO_4 \cdot 5H_2O$.

Now we need to calculate the mass of $CuSO_4 \cdot 5H_2O$ by multiplying the number of moles by the mass of 1 mole:

$$\text{mass} = 0.0302 \times 249.5 = 7.53\,\text{g}$$

This is the theoretical yield of copper(II) sulfate crystals.

The actual yield is what was obtained in the experiment. This was 5.21 g.

The percentage yield = (actual yield/theoretical yield) × 100

The percentage yield is 5.21/7.53 × 100 = 69.2%.

## CALCULATIONS IN WHICH YOU HAVE TO CALCULATE WHICH SUBSTANCE IS IN EXCESS

When you make something in the lab, you rarely mix things together in exactly the right proportions. Usually, something is in excess, and you remove the excess (by filtering, for example) when the reaction is complete. If something is in excess there is more than enough of it to react with the other substance and so we don't need to worry about it.

In all the examples we have done above, we have only ever been given enough information to calculate the number of moles of one thing at the beginning of the problem. We have used this number of moles throughout the calculations. But what would happen if we were given the following question?

Magnesium reacts with hydrochloric acid according to the following equation:

$$Mg(s) + 2HCl(aq) \rightarrow MgCl_2(aq) + H_2(g)$$

0.2 mol Mg is reacted with 0.2 mol HCl. Calculate the mass of hydrogen gas produced.

We might be tempted to say that 0.2 mol Mg produces 0.2 mol $H_2(g)$ and the mass of $H_2 = 0.2 \times 2 = 0.4\,g$. Unfortunately this is not the correct answer!

We can understand why it is not correct by looking at the chemical equation. 1 mol Mg reacts with 2 mol HCl and therefore 0.2 mol Mg would react with $2 \times 0.2 = 0.4\,mol$ HCl.

But we only used 0.2 mol HCl and so we did not have enough to react with all the Mg. Because all the Mg did not react we cannot use the number of moles of Mg to work out how much hydrogen is produced.

If we try starting with the number of moles of HCl, we can deduce from the chemical equation that 2 mol HCl reacts with 1 mol Mg, therefore 0.2 mol HCl reacts with 0.1 mol Mg. We added 0.2 mol Mg and so there is more than enough to react with all the HCl – the Mg is in excess.

Because we know that all the HCl reacts we can use the number of moles of HCl to work out how much $H_2$ is produced.

2 mol HCl produce 1 mol $H_2$: the number of moles of $H_2$ is half the number of moles of HCl.

Therefore 0.2 mol HCl produce 0.1 mol $H_2$.

The mass of $H_2 = 0.1 \times 2 = 0.2\,g$.

This is the correct answer!

To deduce which substance is in excess we try both numbers of moles in the chemical equation and compare the numbers.

**EXAMPLE 8**

Copper reacts with concentrated nitric acid according to the equation:

$$Cu(s) + 4HNO_3(aq) \rightarrow Cu(NO_3)_2(aq) + 2H_2O(l) + 2NO_2(g)$$

3.2 g of copper is reacted with 0.40 mol concentrated nitric acid. Work out which reagent is in excess.

We need to convert the mass of Cu to a number of moles:

$$\text{number of moles of Cu} = \frac{3.2}{63.5} = 0.050\,mol$$

Looking at the chemical equation we can see that 1 mol Cu reacts with 4 mol $HNO_3$, so we can calculate that:

$$0.050\,mol\ Cu\ reacts\ with\ 4 \times 0.050 = 0.20\,mol\ HNO_3$$

We have more than 0.20 mol $HNO_3$ therefore $HNO_3$ is in excess.

If the question went further and asked us to calculate the amount in moles of $NO_2$ produced using these quantities we would have to use the number of moles of Cu because not all the $HNO_3$ is used up. We would obtain $2 \times 0.050 = 0.10\,mol\ NO_2$.

## CHAPTER QUESTIONS

**SKILLS** PROBLEM SOLVING

1 Balance the following equations:

a $Fe + HCl \rightarrow FeCl_2 + H_2$

b $Zn + H_2SO_4 \rightarrow ZnSO_4 + H_2$

c $Ca + H_2O \rightarrow Ca(OH)_2 + H_2$

d $Al + Cr_2O_3 \rightarrow Al_2O_3 + Cr$

e $Fe_2O_3 + CO \rightarrow Fe + CO_2$

f $NaHCO_3 + H_2SO_4 \rightarrow Na_2SO_4 + CO_2 + H_2O$

g $C_8H_{18} + O_2 \rightarrow CO_2 + H_2O$

h $Fe_3O_4 + H_2 \rightarrow Fe + H_2O$

i $Pb + AgNO_3 \rightarrow Pb(NO_3)_2 + Ag$

j $AgNO_3 + MgCl_2 \rightarrow Mg(NO_3)_2 + AgCl$

k $C_3H_8 + O_2 \rightarrow CO_2 + H_2O$

2 Calculate the relative formula masses of the following compounds:

a $CO_2$

b $CH_3CO_2H$

c $Na_2SO_4$

d $(NH_4)_2SO_4$

e $Na_2CO_3 \cdot 10H_2O$

f $Cr_2(SO_4)_3$

($A_r$: H = 1, C = 12, N = 14, O = 16, Na = 23, S = 32, Ca = 40, Cr = 52, Fe = 56)

3 Calculate the percentage of nitrogen in each of the following substances (all used as nitrogen fertilisers):

a potassium nitrate, $KNO_3$

b ammonium nitrate, $NH_4NO_3$

c ammonium sulfate, $(NH_4)_2SO_4$

($A_r$: H = 1, N = 14, O = 16, S = 32, K = 39)

4 Calculate the mass of the following:

a 1 mol of HCN

b 1 mol of lead(II) nitrate, $Pb(NO_3)_2$

c 4.30 mol of methane, $CH_4$

d 0.70 mol of $Na_2O$

e 0.015 mol of $NaNO_3$

f 0.24 mol of sodium carbonate crystals, $Na_2CO_3 \cdot 10H_2O$

($A_r$: H = 1, C = 12, N = 14, O = 16, Na = 23, Pb = 207)

5 Calculate the number of moles represented by each of the following:

a 20 g of magnesium oxide, MgO

b 3.20 g of iron(III) oxide, $Fe_2O_3$

c 2 kg of copper(II) oxide, CuO

d 50 g of copper(II) sulfate crystals, $CuSO_4 \cdot 5H_2O$

e 1 tonne of iron, Fe (1 tonne is 1 000 000 g)

f 0.032 g of sulfur dioxide, $SO_2$

($A_r$: H = 1, O = 16, S = 32, Mg = 24, Fe = 56, Cu = 63.5)

6 The following calculations use $A_r$: H = 1, O = 16, Na = 23, Cl = 35.5, Ca = 40, Cu = 63.5.

a Calculate the mass of 4 mol of sodium chloride, NaCl.

b Calculate how many moles is 37 g of calcium hydroxide, $Ca(OH)_2$.

c Calculate how many moles is 1 kg (1000 g) of calcium, Ca.

d Calculate the mass of 0.125 mol of copper(II) oxide, CuO.

e 0.1 mol of a substance has a mass of 4 g. Calculate the mass of 1 mole.

f 0.004 mol of a substance has a mass of 1 g. Calculate the relative formula mass of the compound.

7 Determine the empirical formulae of the compounds which contain:

a 9.39 g P, 0.61 g H

b 5.85 g K, 2.10 g N, 4.80 g O

c 3.22 g Na, 4.48 g S, 3.36 g O

d 22.0% C, 4.6% H, 73.4% Br (by mass)

($A_r$: H = 1, C = 12, N = 14, O = 16, Na = 23, S = 32, K = 39, Br = 80)

8 1.24 g of phosphorus was burned completely in oxygen to give 2.84 g of phosphorus oxide. Find:

a the empirical formula of the oxide

b the molecular formula of the oxide given that 1 mole of the oxide has a mass of 284 g.

($A_r$: O = 16, P = 31)

9 An organic compound contained C 66.7%, H 11.1%, O 22.2% by mass. Its relative formula mass was 72. Find:

a the empirical formula of the compound

b the molecular formula of the compound.

($A_r$: H = 1, C = 12, O = 16)

10 In an experiment to find the number of molecules of water of crystallisation in sodium sulfate crystals, $Na_2SO_4 \cdot nH_2O$, 3.22 g of sodium sulfate crystals were heated gently. When all the water of crystallisation had been driven off, 1.42 g of anhydrous sodium sulfate was left. Find the value of $n$ in the formula. ($A_r$: H = 1, O = 16, Na = 23, S = 32)

**11** Gypsum is hydrated calcium sulfate, $CaSO_4 \cdot nH_2O$. A sample of gypsum was heated in a crucible until all the water of crystallisation had been driven off. The following results were obtained:

Mass of crucible = 37.34 g

Mass of crucible + gypsum, $CaSO_4 \cdot nH_2O$ = 45.94 g

Mass of crucible + anhydrous calcium sulfate, $CaSO_4$ = 44.14 g

Calculate the value of $n$ in the formula $CaSO_4 \cdot nH_2O$.

($A_r$: H = 1, O = 16, S = 32, Ca = 40)

**12 a** Calculate the amount in moles of $SO_3$ formed when 0.36 mol $SO_2$ reacts with excess $O_2$.

$$2SO_2 + O_2 \rightarrow 2SO_3$$

**b** Calculate the amount in moles of HCl that reacts with 0.4 mol $CaCO_3$.

$$CaCO_3(s) + 2HCl(aq) \rightarrow CaCl_2(aq) + H_2O(l) + CO_2(g)$$

**c** Calculate the amount in moles of $H_2S$ formed when 0.4 mol of HCl reacts with excess $Sb_2S_3$.

$$Sb_2S_3 + 6HCl \rightarrow 2SbCl_3 + 3H_2S$$

**d** Calculate the amount in moles of iron formed when 0.9 mol carbon monoxide reacts with excess iron(III) oxide.

$$Fe_2O_3 + 3CO \rightarrow 2Fe + 3CO_2$$

**e** Calculate the amount in moles of hydrogen that would be required to make 0.8 mol $NH_3$.

$$N_2 + 3H_2 \rightarrow 2NH_3$$

**13** The reaction between iron and bromine is

$$2Fe + 3Br_2 \rightarrow 2FeBr_3$$

10 g of iron is reacted with bromine.

**a** Calculate the amount in moles of iron that reacted.

**b** Calculate the amount in moles of bromine that the iron reacted with.

**c** Calculate the amount in moles of $FeBr_3$ that will be formed.

**d** Calculate the mass of $FeBr_3$ formed.

($A_r$: Fe = 56, Br = 80)

**14** Titanium is manufactured by heating titanium(IV) chloride with sodium.

$$TiCl_4(g) + 4Na(l) \rightarrow Ti(s) + 4NaCl(s)$$

1.0 g of $TiCl_4$ is reacted with excess sodium

($A_r$: Na = 23, Ti = 48, Cl = 35.5)

**a** Calculate the amount in moles of $TiCl_4$ reacted.

**b** Calculate the amount in moles of Ti formed.

**c** Calculate the mass of Ti formed.

**d** Calculate the mass of NaCl formed.

**e** Calculate the mass of Ti formed when 1 tonne of $TiCl_4$ reacts with excess sodium (1 tonne is 1 000 000 g).

**SKILLS** PROBLEM SOLVING

15   2.67 g of aluminium chloride was dissolved in water and an excess of silver nitrate solution was added to give a precipitate of silver chloride (AgCl).

$$AlCl_3(aq) + 3AgNO_3(aq) \rightarrow Al(NO_3)_3(aq) + 3AgCl(s)$$

Calculate the mass of silver chloride that would be formed.

($A_r$: Al = 27, Cl = 35.5, Ag = 108)

16   Chromium is manufactured by heating a mixture of chromium(III) oxide with aluminium powder.

$$Cr_2O_3(s) + 2Al(s) \rightarrow 2Cr(s) + Al_2O_3(s)$$

a   Calculate the mass of aluminium needed to react with 50 g of $Cr_2O_3$.

b   Calculate the mass of chromium produced from 50 g of $Cr_2O_3$.

c   Calculate the mass of chromium produced from 5 kg of $Cr_2O_3$.

d   Calculate the mass of chromium produced from 5 tonnes of $Cr_2O_3$.

(1 tonne is 1 000 000 g)

($A_r$: Cr = 52, O = 16, Al = 27)

17   Copper(II) sulfate crystals, $CuSO_4 \cdot 5H_2O$, can be made by heating copper(II) oxide with dilute sulfuric acid and then crystallising the solution formed.

a   Calculate the maximum mass of crystals that could be made from 4.00 g of copper(II) oxide using an excess of sulfuric acid.

$$CuO(s) + H_2SO_4(aq) \rightarrow CuSO_4(aq) + H_2O(l)$$
$$CuSO_4(aq) + 5H_2O(l) \rightarrow CuSO_4 \cdot 5H_2O(s)$$

b   If the actual mass of copper(II) sulfate collected at the end of the experiment was 11.25 g, calculate the percentage yield.

($A_r$: H = 1, O = 16, S = 32, Cu = 63.5)

18   Ethanol reacts with ethanoic acid to form ethyl ethanoate and water:

$$C_2H_5OH(l) + CH_3COOH(l) \rightarrow CH_3COOC_2H_5(l) + H_2O(l)$$
ethanol     ethanoic acid     ethyl ethanoate     water

a   Calculate the mass of ethyl ethanoate produced when 20.0 g of ethanol reacts with excess ethanoic acid.

b   The actual yield in this experiment is 30.0 g of ethyl ethanoate. Calculate the percentage yield.

($A_r$: C = 12, H = 1, O = 16)

**SKILLS** REASONING

19   Determine which reactant is in excess in each case.

a   0.5 mol of $Na_2CO_3$ is reacted with 0.5 mol hydrochloric acid:

$$Na_2CO_3(s) + 2HCl(aq) \rightarrow 2NaCl(aq) + H_2O(l) + CO_2(g)$$

b   0.01 mol $C_3H_8$ is reacted with 0.02 mol $O_2$:

$$C_3H_8 + 5O_2 \rightarrow 3CO_2 + 4H_2O$$

c   28 g CO is reacted with 0.6 mol $Fe_2O_3$:

$$Fe_2O_3 + 3CO \rightarrow 2Fe + 3CO_2$$

d   16 g $O_2$ is reacted with 16 g $SO_2$:

$$2SO_2 + O_2 \rightarrow 2SO_3$$

($A_r$ : C = 12, O = 16, S = 32)

20   1.0 g of $CaCO_3$ is reacted with 0.015 mol of HCl:

$$CaCO_3(s) + 2HCl(aq) \rightarrow CaCl_2(aq) + H_2O(l) + CO_2(g)$$

a   Determine which reagent is present in excess.

**SKILLS** PROBLEM SOLVING

b   Calculate the mass of $CO_2$ produced.

($A_r$: Ca = 40, C = 12, O = 16)

**CHEMISTRY ONLY**

# 6 CHEMICAL FORMULAE, EQUATIONS AND CALCULATIONS: PART 2

In this chapter we will look at how to extend our understanding of moles to carry out calculations involving gases and solutions. You need to have studied the material covered in Chapter 5 before starting on this chapter.

▶ Figure 6.1 When magnesium reacts with hydrochloric acid, hydrogen gas is formed. In this chapter you will learn to calculate what volume of gas is produced.

**LEARNING OBJECTIVES**

■ Understand how to carry out calculations involving amount of substance, volume and concentration (in $mol/dm^3$) of solution

■ Understand how to carry out calculations involving gas volumes and the molar volume of a gas ($24\,dm^3$ and $24\,000\,cm^3$ at room temperature and pressure (rtp))

## CALCULATIONS INVOLVING GAS VOLUMES

▲ Figure 6.2 Three identical flasks containing different gases at the same temperature and pressure all contain equal numbers of molecules.

**KEY POINT**

For helium and the other noble gases we would use the word 'atoms' instead of 'molecules'.

Avogadro's law:

***Equal volumes of gases at the same temperature and pressure contain equal numbers of molecules.***

This means that if you have $100\,cm^3$ of hydrogen at some temperature and pressure, it contains exactly the same number of molecules as there are in $100\,cm^3$ of $Cl_2$, or any other gas under those conditions, irrespective of the size of the molecules.

You can see how this works in a chemical reaction. Methane ($CH_4$) reacts with oxygen to form carbon dioxide and water. The equation for the reaction is:

$$CH_4(g) + 2O_2(g) \rightarrow CO_2(g) + 2H_2O(l)$$

We need two $O_2$ molecules to react with each $CH_4$ molecule, so if we react $100\,cm^3$ of $CH_4$ we will need twice the volume of $O_2$ to obtain the correct number of particles. In this equation, $100\,cm^3$ of methane will react exactly with $200\,cm^3$ of oxygen gas and $100\,cm^3$ of carbon dioxide gas will be formed.

## UNITS OF VOLUME

Volumes (of gases or liquids) are measured in
- cubic centimetres ($cm^3$)
- or cubic decimetres ($dm^3$)
- or litres (l)

$$1 \text{ litre} = 1 \text{ dm}^3 = 1000 \text{ cm}^3$$

The most important thing you need to be able to do is to convert between $cm^3$ and $dm^3$. Figure 6.3 shows how you can do this.

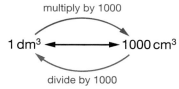

multiply by 1000

$$1 \text{ dm}^3 \longleftrightarrow 1000 \text{ cm}^3$$

divide by 1000

▲ Figure 6.3 Because 1 dm³ is equivalent to 1000 cm³, to convert dm³ to cm³ we have to go from 1 to 1000, therefore we need to multiply by 1000. To convert cm³ to dm³ we need to go from 1000 down to 1, therefore we divide by 1000.

## THE VOLUME OCCUPIED BY 1 MOLE OF A GAS

1 mole of any gas contains the same number of molecules and so occupies the same volume as 1 mole of any other gas at the same temperature and pressure.

At room temperature and pressure, the volume occupied by 1 mole of any gas is approximately 24 dm³ (24 000 cm³). The volume occupied by 1 mole of a gas is often called the **molar volume**.

You can use the triangles shown in Figures 6.4 and 6.5 to convert between volumes and numbers of moles.

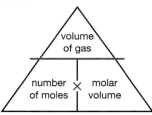

▲ Figure 6.4 This triangle can be used to convert between volume of a gas and number of moles.

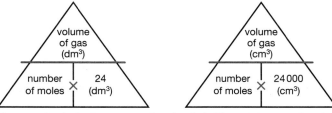

▲ Figure 6.5 These are basically the same as the triangle in Figure 6.4 except that the molar volume is shown as a number. Be careful with the units of volume.

## CALCULATIONS WITH MOLAR VOLUME

Calculate the volume in dm³ of 0.20 mol $CO_2$ at rtp:

$$\text{volume} = \text{number of moles} \times \text{molar volume}$$

Because we want the volume in dm³ we use 24 dm³ as the molar volume:

$$\text{volume} = 0.20 \times 24 = 4.8 \text{ dm}^3$$

## CALCULATING THE VOLUME OF A GIVEN MASS OF GAS

Calculate the volume (in $cm^3$) of 0.01 g of hydrogen at rtp ($A_r$: H = 1).

1 mol $H_2$ has a mass of 2 g:

$$\text{number of moles} = \frac{\text{mass}}{\text{mass of 1 mol}}$$

0.01 g of hydrogen is $\frac{0.01}{2}$ mol = 0.005 mol.

Because we want the volume in $cm^3$ we use the molar volume as 24 000 $cm^3$:

$$\text{volume} = \text{number of moles} \times \text{molar volume}$$

0.005 mol of hydrogen occupies 0.005 × 24 000 = 120 $cm^3$

## CALCULATING THE NUMBER OF MOLES FROM A VOLUME

You have to be careful that the units are correct when you are calculating a number of moles from a volume. Look carefully, is the volume of gas given in $cm^3$ or $dm^3$? If the volume of the gas is given in $cm^3$ then use the molar volume of the gas as 24 000 $cm^3$, if it is given in $dm^3$ then use 24 $dm^3$.

Calculate the amount of moles in 120 $cm^3$ of carbon dioxide.

We can see that the volume is given in $cm^3$ therefore we use 24 000 $cm^3$ as the molar volume:

$$\text{number of moles} = \frac{\text{volume of gas}}{\text{molar volume}}$$

$$\text{number of moles} = \frac{120}{24\,000}$$

$$\text{number of moles} = 0.005 \text{ mol}$$

## USING THE MOLAR VOLUME IN CALCULATIONS WITH CHEMICAL EQUATIONS

In order to do these calculations, we follow basically the same method as we used for masses in Chapter 5.

■ Calculate the number of moles of anything you can.

■ Use the chemical equation to deduce the number of moles of what you want.

■ Convert the number of moles to the required quantity, e.g. a mass or a volume.

### KEY POINT

Excess means that more than enough acid has been added to react with all the calcium carbonate.

Calculate the volume of carbon dioxide produced at room temperature and pressure when an excess of dilute hydrochloric acid is added to 1.00 g of calcium carbonate. ($A_r$: C = 12, O = 16, Ca = 40; molar volume = 24 $dm^3$ at rtp.)

The equation for the reaction is

$$CaCO_3(s) + 2HCl(aq) \rightarrow CaCl_2(aq) + CO_2(g) + H_2O(l)$$

We have been given the mass of calcium carbonate ($CaCO_3$) and so we can calculate the number of moles of calcium carbonate:

$$\text{number of moles} = \frac{\text{mass}}{\text{mass of 1 mole}}$$

The mass of 1 mole of $CaCO_3$ is the $M_r$ in grams. The $M_r$ is 40 + 12 + (3 × 16) = 100.

The mass of 1 mole of $CaCO_3$ is 100 g.

The number of moles in 1.00 g of $CaCO_3$ = $\frac{1.00}{100}$ = 0.0100 mol.

▲ Figure 6.6 Shells of sea creatures such as limpets are made of calcium carbonate, $CaCO_3$. If they were 100% pure, 1 g of shells would react with hydrochloric acid to give 0.24 dm³ of carbon dioxide.

**HINT**

Another way of thinking about this is that to go from 3 ($H_2$) to 2 (Al) we divide the 3 by 3 to get 1 and then multiply this by 2 to get 2. In other words, to go from the number of moles of $H_2$ to the number of moles of Al, we divide the number of moles of $H_2$ by 3 and then multiply by 2.

**REMINDER**

The equation to calculate the mass is the number of moles × mass of 1 mol. It is important to multiply by 27 here and not 54. Although it is 2Al in the equation, we have already used the 2 in calculating the number of moles of Al. Also, the mass of 0.00278 mol of Al must always be the same, it cannot depend on what it is reacting with. 0.00278 mol of Al always contains the same number of atoms.

**CONCENTRATIONS OF SOLUTIONS**

**KEY POINT**

You may also find the symbol M used. For example, dilute hydrochloric acid might have a concentration quoted as 2 M. M means 'mol/dm³' and is described as the **molarity** of the solution. You can also read 2 M as '2 molar'.

From the chemical equation we can see that 1 mol of $CaCO_3$ produces 1 mol of $CO_2$ so we know that 0.0100 mol of $CaCO_3$ will produce the same number of moles of $CO_2$, that is, 0.0100 mol.

We now need to work out the volume occupied by 0.0100 mol of $CO_2$:

volume of gas = number of moles × molar volume

$$= 0.0100 \times 24 = 0.24 \, dm^3$$

We have used the molar volume in dm³ and so our answer will be in dm³. It is really important to include the units here because the examiner will not know whether you mean cm³ or dm³ unless you write it down.

## A PROBLEM INVOLVING MAKING HYDROGEN

Aluminium reacts with dilute hydrochloric acid according to the equation

$$2Al(s) + 6HCl(aq) \rightarrow 2AlCl_3(aq) + 3H_2(g)$$

What mass of aluminium would you need to add to an excess of dilute hydrochloric acid so that you produced 100 cm³ of hydrogen at room temperature and pressure? ($A_r$ of Al = 27; molar volume = 24 000 cm³ at rtp.)

We can only calculate the number of moles of the hydrogen gas. We do not know the mass of aluminium, that is what we are trying to find.

The volume of hydrogen gas is given in cm³ so we use the molar volume of a gas as 24 000 cm³:

$$\text{number of moles of gas} = \frac{\text{volume of gas}}{\text{molar volume}} = \frac{100}{24\,000} = 0.00417 \, mol$$

We could also write this number in standard form as $4.17 \times 10^{-3}$ mol.

Looking at the chemical equation we can see that 3 mol of $H_2$ is produced from 2 mol of Al (we are only looking at the big numbers in front of the formulae). So we can see that the number of moles of Al will be $\frac{2}{3}$ times the number of moles of $H_2$.

The number of moles of aluminium required to produce 0.00417 mol of $H_2$ is $0.00417 \times \frac{2}{3} = 0.00278$ mol.

We can convert this to a mass by using the equation:

mass = number of moles × mass of 1 mol

$$= 0.00278 \times 27 = 0.075 \, g \, Al$$

## WORKING WITH SOLUTION CONCENTRATIONS

Concentrations can be measured in either

- g/dm³

or

- mol/dm³.

These can also be written as $g\,dm^{-3}$ and $mol\,dm^{-3}$. You read them as 'grams per cubic decimetre' and 'moles per cubic decimetre'. Remember: 1 cubic decimetre is the same as 1 litre.

You have to be able to convert between g/dm³ and mol/dm³. This is no different from converting moles into grams and *vice versa*. When you are doing this

▲ Figure 6.7 Sea water contains about 0.6 moles of NaCl per cubic decimetre.

conversion in concentration calculations, it does not make any difference that the amount of substance you are talking about is dissolved in $1\,dm^3$ of solution.

A sample of sea water has a concentration of sodium chloride of $35.1\,g/dm^3$. Find its concentration in $mol/dm^3$. ($A_r$: Na = 23, Cl = 35.5)

1 mol NaCl has a mass of 58.5 g.

There are 35.1 g in every $dm^3$ of solution

$$\text{number of moles} = \frac{\text{mass(g)}}{\text{mass of 1 mole(g)}}$$

$35.1\,g$ is $\dfrac{35.1}{58.5}\,mol = 0.600\,mol$

Therefore there are 0.600 mol in every $dm^3$ of solution and the concentration of the NaCl is $0.600\,mol/dm^3$.

**EXAMPLE 1**

What is the concentration of a $0.050\,mol/dm^3$ solution of sodium carbonate, $Na_2CO_3$, in $g/dm^3$? ($A_r$: C = 12, O = 16, Na = 23)

$1\,dm^3$ of solution contains $0.050\,mol\ Na_2CO_3$.

1 mol $Na_2CO_3$ weighs 106 g.

0.050 mol weighs $0.050 \times 106 = 5.3\,g$.

$1\,dm^3$ of solution contains $5.3\,g\ Na_2CO_3$, therefore the concentration is $5.3\,g/dm^3$.

**EXAMPLE 2**

What is the concentration in $mol/dm^3$ of a solution containing 2.1 g of sodium hydrogencarbonate, $NaHCO_3$, in $250\,cm^3$ of solution? ($A_r$: H = 1, C = 12, O = 16, Na = 23)

1 mol $NaHCO_3$ has a mass of 84 g.

$2.1\,g$ is $\dfrac{2.1}{84} = 0.025\,mol$

This is in a volume of $250\,cm^3$ but we need to find out how much there is in $1\,dm^3$ ($1000\,cm^3$). There are four lots of $250\,cm^3$ in $1000\,cm^3$. Each portion of $250\,cm^3$ contains 0.025 mol, therefore there must be $4 \times 0.025 = 0.10\,mol$ in $1000\,cm^3$.

The concentration is $0.10\,mol/dm^3$.

Another way of doing this is to use the triangle shown in Figure 6.8.

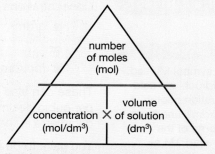

▲ Figure 6.8 We can use the triangle for calculations involving solutions.

## WORKING OUT A NUMBER OF MOLES FROM A VOLUME AND A CONCENTRATION

When doing calculations involving solutions you will most often be asked to calculate a number of moles given the volume of the solution and its concentration. The equation for doing this is:

number of moles = volume of solution ($dm^3$) × concentration (mol/$dm^3$)

Because concentration is usually given in units of mol/$dm^3$ but volumes of solutions are often given in $cm^3$, we usually have to convert the volume in $cm^3$ to $dm^3$ by dividing by 1000.

### EXAMPLE 3

Calculate the number of moles of NaOH in 50 $cm^3$ of 0.10 mol/$dm^3$ solution.

Converting the volume to $dm^3$ we get 50/1000 = 0.050 $dm^3$.

Number of moles = volume of solution ($dm^3$) × concentration (mol/$dm^3$).

Number of moles = 0.050 × 0.10 = 0.0050 mol.

# CALCULATIONS WITH EQUATIONS INVOLVING SOLUTIONS

## A CALCULATION INVOLVING HARD WATER

### HINT

Don't be scared that some of this is new to you! As long as you realise that $CH_3COOH$ is ethanoic acid, that's all you need to worry about for this calculation.

Limescale can be removed from, for example, electric kettles by reacting it with a dilute acid such as ethanoic acid, which is present in vinegar:

$$CaCO_3(s) + 2CH_3COOH(aq) \rightarrow (CH_3COO)_2Ca(aq) + CO_2(g) + H_2O(l)$$

What mass of calcium carbonate can be removed by 50 $cm^3$ of a solution of ethanoic acid that has a concentration of 2 mol/$dm^3$?
($A_r$: C = 12, O = 16, Ca = 40)

Again, we will use our three stages:

■ calculate the number of moles of what we can

■ use the chemical equation to work out the number of moles of what we want

■ convert the number of moles to what is required in the question.

We know the volume and concentration of ethanoic acid and so we can calculate the number of moles of ethanoic acid:

number of moles = volume of solution ($dm^3$) × concentration (mol/$dm^3$)

We need to be careful to convert the volume of ethanoic acid to $dm^3$ by dividing by 1000:

$$\text{number of moles of ethanoic acid} = \frac{50}{1000} \times 2 = 0.1\,\text{mol}$$

From the chemical equation we can see that there is a 2 in front of the $CH_3COOH$ (ethanoic acid) but no number (which means a 1) in front of the $CaCO_3$, so we can deduce that 2 mol of $CH_3COOH$ react with 1 mol of $CaCO_3$, in other words the number of moles of $CaCO_3$ is half the number of moles of $CH_3COOH$. Therefore, we can deduce that 0.1 mol of ethanoic acid will react with 0.1/2 = 0.05 mol of $CaCO_3$.

We can now calculate the mass of $CaCO_3$ ($M_r$ = 100):

mass = number of moles × mass of 1 mol = 0.05 × 100 = 5 g

Therefore the ethanoic acid reacts with 5 g of calcium carbonate.

## CALCULATIONS FROM TITRATIONS

### ACID–ALKALI TITRATIONS

**HINT**

The acid and alkali can also be added the other way round, with the alkali in the burette.

Titration is a technique that is used to find out how much of one solution reacts with a certain volume of another solution of known concentration. A solution of an alkali is measured into a conical flask using a pipette. An acid is run in from a burette, swirling the flask constantly. Towards the end, the acid is run in a drop at a time until the indicator just changes colour. If you know the concentration of either the acid or the alkali you can use the results of the titration to find the concentration of the solution you reacted it with.

▲ Figure 6.9 The endpoint of a titration using methyl orange as indicator.

Acid–alkali titrations are discussed in Chapters 16 and 17.

## THE STANDARD CALCULATION

A standard titration problem will look like this:

25.00 cm³ of 0.100 mol/dm³ sodium hydroxide solution required 23.50 cm³ of dilute hydrochloric acid for neutralisation. Calculate the concentration of the hydrochloric acid.

$$NaOH(aq) + HCl(aq) \rightarrow NaCl(aq) + H_2O(l)$$

We will use the same three-step method as we used above to calculate the concentration of the hydrochloric acid.

We know the volume (25.00 cm³) and concentration (0.100 mol/dm³) of the sodium hydroxide solution and so we can calculate the number of moles:

$$\text{number of moles of NaOH} = \frac{25.00}{1000} \times 0.100 = 0.00250 \text{ mol}$$

**HINT**

Remember to divide the volume by 1000 to convert to dm³.

From the chemical equation we know that 1 mol of NaOH reacts with 1 mol of HCl.

Therefore 0.00250 mol NaOH reacts with 0.00250 mol HCl.

Now that we have the number of moles of HCl we need to convert it to a concentration.

We know that we need 0.00250 mol of HCl to neutralise the NaOH. The volume of HCl we added was 23.50 cm³. So we know that this volume must contain 0.00250 mol of HCl.

We have the volume and number of moles of HCl and can work out the concentration:

$$\text{concentration (mol/dm}^3) = \frac{\text{number of moles (mol)}}{\text{volume (dm}^3)}$$

Because we want to calculate concentration in $mol/dm^3$ we have to convert the volume of HCl to $dm^3$ by dividing by 1000:

$$\text{volume of HCl} = \frac{23.50}{1000} = 0.02350\,dm^3$$

$$\text{concentration} = \frac{0.00250}{0.02350} = 0.106\,mol/dm^3$$

The concentration of the hydrochloric acid is $0.106\,mol/dm^3$.

**A VERY SLIGHTLY HARDER CALCULATION**

$25.0\,cm^3$ of sodium hydroxide solution of unknown concentration was titrated with dilute sulfuric acid of concentration $0.050\,mol/dm^3$. $20.0\,cm^3$ of the acid was required to neutralise the alkali. Find the concentration of the sodium hydroxide solution in $mol/dm^3$.

$$2NaOH(aq) + H_2SO_4(aq) \rightarrow Na_2SO_4(aq) + 2H_2O(l)$$

This time, we know everything we need about the sulfuric acid and can calculate the number of moles.

The experiment used $20.0\,cm^3$ of $0.050\,mol/dm^3$ $H_2SO_4$:

$$\text{number of moles of sulfuric acid used} = \frac{20.0}{1000} \times 0.050$$

$$= 0.0010\,mol$$

The equation proportions aren't 1:1 this time. That's what makes the calculation slightly different from the last one. The equation shows that 1 mol of sulfuric acid reacts with 2 mol of sodium hydroxide. So the number of moles of sodium hydroxide is twice the number of moles of sulfuric acid:

$$\text{number of moles of sodium hydroxide} = 2 \times 0.0010 = 0.0020\,mol$$

That 0.0020 mol must have been in the $25.0\,cm^3$ ($25/1000 = 0.025\,dm^3$) of sodium hydroxide solution:

$$\text{concentration (mol/dm}^3) = \frac{\text{number of moles (mol)}}{\text{volume (dm}^3)}$$

$$\text{concentration} = 0.0020/0.025 = 0.080\,mol/dm^3$$

Therefore the concentration of the sodium hydroxide is $0.080\,mol/dm^3$.

## REVERSING THE CALCULATIONS

Instead of calculating the concentration of a solution using titration results, you may be asked to calculate what volume of a solution is needed to neutralise something else. Here is an example.

**EXAMPLE 4**

Calculate the volume of $0.100\,mol/dm^3$ sodium hydrogencarbonate ($NaHCO_3$) solution needed to neutralise $20.0\,cm^3$ of $0.125\,mol/dm^3$ hydrochloric acid (HCl).

$$NaHCO_3(aq) + HCl(aq) \rightarrow NaCl(aq) + CO_2(g) + H_2O(l)$$

We have been given the volume and concentration of the hydrochloric acid and so we can calculate the number of moles of this. We do not have enough information to calcuate the number of moles of anything else, so must start here:

$$\text{number of moles of HCl} = \frac{20.0}{1000} \times 0.125 = 0.00250\,mol$$

The equation shows that you will need the same number of moles of sodium hydrogencarbonate. Therefore we know that we need 0.00250 mol of $NaHCO_3$.

To calculate the volume this is in, we just need to rearrange our concentration equation to obtain:

$$\text{volume (dm}^3) = \frac{\text{number of moles (mol)}}{\text{concentration (mol/dm}^3)}$$

$$\text{volume} = \frac{0.00250}{0.100} = 0.0250 \text{ dm}^3$$

The volume comes out in $dm^3$ because the concentration is in $mol/dm^3$. We can convert to $cm^3$ by multiplying by 1000:

$$\text{volume of NaHCO}_3 \text{ solution} = 25.0 \text{ cm}^3$$

You will need 25.0 $cm^3$ of the sodium hydrogencarbonate solution to neutralise the hydrochloric acid.

## CHAPTER QUESTIONS

**SKILLS**   PROBLEM SOLVING  

In questions 1 to 4, take the molar volume to be 24 $dm^3$ (24 000 $cm^3$) at rtp.

1 Calculate the amount in moles of each of the following:

    a   2.4 $dm^3$ of $O_2$ at rtp        b   480 $dm^3$ of He at rtp

    c   100 $cm^3$ of $CO_2$ at rtp        d   1500 $cm^3$ of $N_2$ at rtp

2 Calculate the volume of each of the following gases at rtp:

    a   2.0 mol $H_2$       b   0.10 mol $SO_2$       c   $1.0 \times 10^{-3}$ mol CO

3   a   Calculate the mass of 200 $cm^3$ of chlorine gas ($Cl_2$) at rtp. ($A_r$: Cl = 35.5)

    b   Calculate the volume occupied by 0.16 g of oxygen ($O_2$) at rtp. ($A_r$: O = 16)

    c   If 1 $dm^3$ of a gas at rtp has a mass of 1.42 g, calculate the mass of 1 mole of the gas.

4 0.240 g of magnesium is reacted with an excess of dilute sulfuric acid. ($A_r$: Mg = 24)

$$Mg(s) + H_2SO_4(aq) \rightarrow MgSO_4(aq) + H_2(g)$$

Calculate the amount in moles of Mg which reacted.

Calculate the number of moles of hydrogen produced in the reaction.

Calculate the volume of hydrogen (measured at rtp) produced in the reaction.

5 Potassium nitrate decomposes when heated to produce oxygen gas:

$$2KNO_3(s) \rightarrow 2KNO_2(s) + O_2(g)$$

In an experiment 1.00 $dm^3$ (measured at rtp) of oxygen gas was collected.

($A_r$: N = 14, O = 16, K = 39; molar volume = 24 $dm^3$ at rtp)

   Calculate the amount in moles of oxygen gas collected.

   Calculate the amount in moles of $KNO_3$ that reacted.

   Calculate the mass of $KNO_3$ that reacted.

**SKILLS**   PROBLEM SOLVING

6 Chlorine can be prepared by heating manganese(IV) oxide with an excess of concentrated hydrochloric acid. What is the maximum volume of chlorine (measured at room temperature and pressure) that could be obtained from 2.00 g of manganese(IV) oxide? ($A_r$: O = 16, Mn = 55; molar volume = 24 000 cm³ at rtp)

$$MnO_2(s) + 4HCl(aq) \rightarrow MnCl_2(aq) + Cl_2(g) + H_2O(l)$$

7 Some dilute sulfuric acid, $H_2SO_4$, had a concentration of 4.90 g/dm³. What is its concentration in mol/dm³? ($A_r$: H = 1, O = 16, S = 32)

8 What is the concentration in g/dm³ of potassium hydroxide, KOH, solution with a concentration of 0.200 mol/dm³? ($A_r$: H = 1, O = 16, K = 39)

9 Calculate the amount in moles in each of the following:

  a 25.0 cm³ of 0.100 mol/dm³ NaCl(aq)

  b 200 cm³ of 0.200 mol/dm³ $H_2SO_4$(aq)

  c 75.0 cm³ of 0.150 mol/dm³ HCl(aq)

  d 22.4 cm³ of 0.280 mol/dm³ $HNO_3$(aq)

10 Calculate the concentration in mol/dm³ of each of the following solutions:

  a 2.00 dm³ of sodium hydroxide solution containing 0.100 mol sodium hydroxide

  b 25.0 cm³ of sulfuric acid containing 0.0200 mol sulfuric acid

  c 27.8 cm³ of hydrochloric acid containing 0.00150 mol hydrochloric acid

11 Calculate the volume in cm³ of each of the following solutions:

  a the volume of 0.100 mol/dm³ $H_2SO_4$(aq) that contains 0.500 mol

  b the volume of 0.0200 mol/dm³ NaOH(aq) that contains 0.00500 mol

  c the volume of 0.500 mol/dm³ $MgCl_2$(aq) that contains 0.0200 mol

12 When barium chloride solution is added to copper(II) sulfate solution a precipitate of barium sulfate ($BaSO_4$) is formed.

$$BaCl_2(aq) + CuSO_4(aq) \rightarrow BaSO_4(s) + CuCl_2(aq)$$

Excess barium chloride solution is added to 20.0 cm³ of copper(II) sulfate solution of concentration 0.100 mol/dm³. ($A_r$: O = 16, S = 32, Ba = 137)

Calculate the number of moles of copper(II) sulfate.

Calculate the number of moles of barium sulfate formed.

Calculate the mass of barium sulfate formed.

13 Calcium carbonate reacts with hydrochloric acid:

$$CaCO_3(s) + 2HCl(aq) \rightarrow CaCl_2(aq) + H_2O(l) + CO_2(g)$$

Calcium carbonate is added to 25.0 cm³ of 2.00 mol/dm³ hydrochloric acid. ($A_r$: C = 12, O = 16, Ca = 40)

Calculate the amount in moles of hydrochloric acid.

Calculate the amount in moles of $CaCO_3$ that reacts with the acid.

Calculate the mass of $CaCO_3$ that reacts with the acid.

Calculate the volume of $CO_2$ (measured at rtp) produced.

(Molar volume at rtp is 24 000 cm³.)

14 Hydrogen peroxide solution decomposes to form oxygen gas:

$$2H_2O_2(aq) \rightarrow 2H_2O(l) + O_2(g)$$

Calculate the volume of oxygen gas (measured at rtp) produced when $30.0\,cm^3$ of $0.0200\,mol/dm^3$ hydrogen peroxide decomposes to form oxygen gas. (Molar volume at rtp is $24\,000\,cm^3$.)

15 a Calculate the volume of $0.200\,mol/dm^3$ sulfuric acid needed to neutralise $25.0\,cm^3$ of $0.400\,mol/dm^3$ sodium hydroxide solution.

$$2NaOH(aq) + H_2SO_4(aq) \rightarrow Na_2SO_4(aq) + 2H_2O(l)$$

  b Calculate the minimum volume of $2.00\,mol/dm^3$ hydrochloric acid needed to react with $10.0\,g$ of calcium carbonate. ($A_r$: C = 12, O = 16, Ca = 40)

$$CaCO_3(s) + 2HCl(aq) \rightarrow CaCl_2(aq) + H_2O(l) + CO_2(g)$$

16 In each of these questions concerning titrations, calculate the unknown concentration in $mol/dm^3$.

  a $25.0\,cm^3$ of $0.100\,mol/dm^3$ sodium hydroxide was neutralised by $20.0\,cm^3$ of dilute nitric acid of unknown concentration.

$$NaOH(aq) + HNO_3(aq) \rightarrow NaNO_3(aq) + H_2O(l)$$

  b $25.0\,cm^3$ of sodium carbonate solution of unknown concentration was neutralised by $30.0\,cm^3$ of $0.100\,mol/dm^3$ nitric acid.

$$Na_2CO_3(aq) + 2HNO_3(aq) \rightarrow 2NaNO_3(aq) + CO_2(g) + H_2O(l)$$

  c $25.0\,cm^3$ of $0.250\,mol/dm^3$ potassium carbonate solution was neutralised by $12.5\,cm^3$ of ethanoic acid of unknown concentration.

$$2CH_3COOH(aq) + K_2CO_3(aq) \rightarrow 2CH_3COOK(aq) + CO_2(g) + H_2O(l)$$

17 Limewater is calcium hydroxide solution. In an experiment to find the concentration of calcium hydroxide in limewater, $25.0\,cm^3$ of limewater needed $18.8\,cm^3$ of $0.0400\,mol/dm^3$ hydrochloric acid to neutralise it.

$$Ca(OH)_2(aq) + 2HCl(aq) \rightarrow CaCl_2(aq) + 2H_2O(l)$$

Calculate the concentration of the calcium hydroxide in:

a $mol/dm^3$              b $g/dm^3$

($A_r$: H = 1, O = 16, Ca = 40)

18 Sodium carbonate solution reacts with nitric acid according to the following equation:

$$Na_2CO_3(aq) + 2HNO_3(aq) \rightarrow 2NaNO_3(aq) + CO_2(g) + H_2O(l)$$

In each of the following questions work out which reagent is in excess and calculate the volume of $CO_2$ produced at rtp. (Molar volume at rtp is $24000\,cm^3$)

  a $0.1\,mol\ Na_2CO_3$ is reacted with $0.1\,mol\ HNO_3$.

  b $20.0\,cm^3$ of $0.100\,mol/dm^3\ Na_2CO_3$ is reacted with $0.0200\,mol\ HNO_3$.

  c $25.0\,cm^3$ of $0.300\,mol/dm^3\ Na_2CO_3$ is reacted with $20.0\,cm^3$ of $0.400\,mol/dm^3\ HNO_3$.

**END OF CHEMISTRY ONLY**

# 7 IONIC BONDING

We have already looked at the definition of a compound. A compound is formed when two or more elements chemically combine. In this chapter we are going to look at one way in which elements can chemically combine: by the transfer of electrons to form ionic compounds.

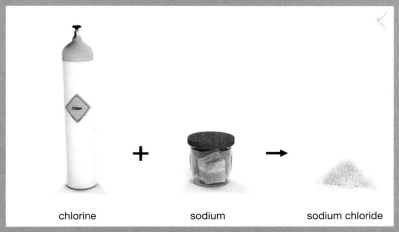

chlorine          +          sodium          →          sodium chloride

▲ Figure 7.1 The properties of a compound are very different from those of the elements. Sodium (an element) is a dangerously reactive metal. It is stored under oil to prevent it reacting with air or water. Chlorine (an element) is a very poisonous, reactive gas. But salt, sodium chloride (an ionic compound), is safe to eat in small quantities.

## LEARNING OBJECTIVES

■ Understand how ions are formed by electron loss or gain

■ Know the charges of these ions:
  ■ metals in Groups 1, 2 and 3
  ■ non-metals in Groups 5, 6 and 7
  ■ $Ag^+$, $Cu^{2+}$, $Fe^{2+}$, $Fe^{3+}$, $Pb^{2+}$, $Zn^{2+}$
  ■ hydrogen ($H^+$), hydroxide ($OH^-$), ammonium ($NH_4^+$), carbonate ($CO_3^{2-}$), nitrate ($NO_3^-$), sulfate ($SO_4^{2-}$)

■ Write formulae for compounds formed between the ions listed above

■ Draw dot-and-cross diagrams to show the formation of ionic compounds by electron transfer, limited to combinations of elements from Groups 1, 2, 3, 5, 6 and 7 *(only outer electrons need be shown)*

■ Understand ionic bonding in terms of electrostatic attractions

■ Understand why compounds with giant ionic lattices have high melting and boiling points

■ Know that ionic compounds do not conduct electricity when solid, but do conduct electricity when molten and in aqueous solution

## IONIC BONDING

**HINT**

There are one or two exceptions to this: there are ionic compounds that do not contain a metal, for example those containing the ammonium ion (such as $NH_4Cl$, $(NH_4)_2SO_4$. We will look at these later.

Sodium chloride is probably the best-known example of an ionic compound – one that is held together by **ionic bonding**. There are lots of other ionic compounds, such as magnesium oxide, calcium fluoride and zinc bromide. What you will realise about all these compounds is that they all contain a metal combined with a non-metal.

*You can recognise ionic compounds because they (usually) contain a metal.*

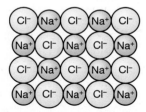

▲ Figure 7.3 Part of the structure of a sodium chloride crystal. The structure is held together by the attraction between positive and negative ions.

When a non-metal such as chlorine combines with a metal such as sodium, the chlorine atom has a stronger attraction for electrons than the sodium atom and *an electron is transferred from the outer shell of the sodium atom to the outer shell of the chlorine atom*. Because an electron has a negative charge, when something gains an electron it becomes negatively charged, and when something loses an electron it becomes positively charged.

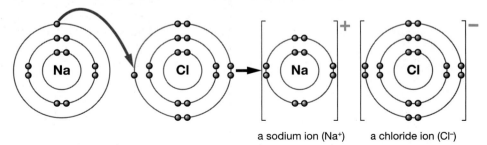

a sodium ion ($Na^+$)          a chloride ion ($Cl^-$)

▲ Figure 7.2 Ionic bonding in sodium chloride. Ionic bonding involves the transfer of electron(s).

The charged particles that are formed are called **ions**.

Ions are charged particles formed when atoms (or groups of atoms) lose or gain electrons. Ions can have either a positive or a negative charge.

- A positive ion is called a **cation**, for example $Na^+$.
- A negative ion is called an **anion**, for example $Cl^-$.

When an ionic compound is formed, electron(s) are transferred from a metal atom to a non-metal atom to form positive and negative ions. Ionic compounds have ionic bonding. *Ionic bonding is the strong electrostatic attraction between positive and negative ions.*

Ionic bonding is often shown using **dot-and-cross diagrams**. Figure 7.4 is an example of a dot-and-cross diagram. Although the electrons are drawn as dots or as crosses, there is absolutely no difference between them in reality. The dots and the crosses simply show that the electrons have come from two different atoms. You could equally well use two different coloured dots (as in Figure 7.2), or two different coloured crosses.

## IONIC BONDING IN MAGNESIUM OXIDE

A dot-and-cross diagram for MgO is shown in Figure 7.4.

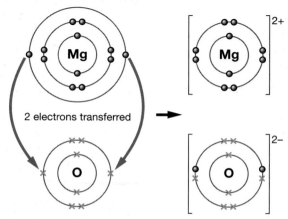

▲ Figure 7.4 A dot-and-cross diagram for magnesium oxide. When drawing these diagrams, don't forget the charges on the ions.

Two electrons are transferred from a magnesium atom to an oxygen atom to form $Mg^{2+}$ and $O^{2-}$ ions.

$Ag^+$  $Cu^{2+}$  $Fe^{2+}$  $Fe^{3+}$

$Pb^{2+}$  $Zn^{2+}$

$H^+$  $OH^-$  $NH_4^+$

$CO_3^{2-}$  $NO_3^-$  $SO_4^{2-}$

## THE SIGNIFICANCE OF NOBLE GAS ELECTRONIC CONFIGURATIONS IN IONIC BONDING

If you look at the electronic arrangements of the ions formed in Figures 7.2, 7.4, 7.5 and 7.6, each of them has a noble gas electronic configuration: [2, 8] (the same as neon) or [2, 8, 8] (the same as argon). For the first 20 elements, atoms lose or gain electrons so that they achieve a noble gas electronic configuration. The elements in Groups 1, 2 and 3 of the Periodic Table will lose their outer shell electrons to form 1+, 2+ and 3+ ions, and the elements in Groups 5, 6 and 7 will gain electrons to form 3–, 2– and 1– ions.

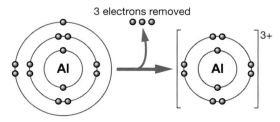

▲ Figure 7.5 An aluminium atom loses its 3 outer shell electrons to form an $Al^{3+}$ ion.

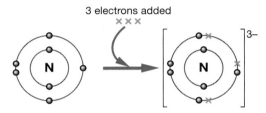

▲ Figure 7.6 A nitrogen atom gains 3 electrons to form the nitride ion ($N^{3-}$).

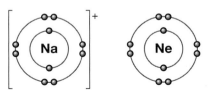

▲ Figure 7.7 $Na^+$ and Ne are isoelectronic – they have the same number of electrons.

Elements in Groups 1, 2, 6 and 7 always form ions that are isoelectronic with the nearest <u>noble gas atom</u>. For example, rubidium (Rb), which is in Group 1, has 1 electron in its outer shell and will lose this outer shell electron to form a 1+ ion, which has the same number of electrons as a krypton (Kr) atom. Iodine is in Group 7, so it has 7 electrons in its outer shell; it will gain 1 electron when it forms an ion to form a 1– ion. This 1– ion has the same number of electrons as a xenon (Xe) atom.

However, there are a lot of common ions that don't have noble gas structures. $Fe^{2+}$, $Fe^{3+}$, $Cu^{2+}$, $Zn^{2+}$, $Ag^+$ and $Pb^{2+}$ are all ions that you will come across during the International GSCE course, although you won't have to write their electronic structures. Not one of them has a noble gas structure.

We sometimes use the word **isoelectronic**, which means 'have the same number of electrons'. So, when sodium forms an ion (electronic configuration [2, 8]) it becomes isoelectronic with neon [2, 8].

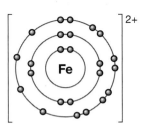

▲ Figure 7.8 An $Fe^{2+}$ ion – definitely not a noble gas structure!

## OTHER EXAMPLES OF IONIC BONDING

Ionic bonds are usually formed only if small numbers of electrons need to be transferred, typically 1 or 2, but occasionally 3.

## LITHIUM FLUORIDE

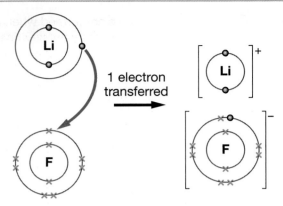

▲ Figure 7.9 A dot-and-cross diagram for lithium fluoride

The lithium atom has 1 electron in its outer shell that is easily lost, and the fluorine has space to receive one. One electron is transferred from the lithium atom to the fluorine atom. Lithium fluoride is held together by the strong electrostatic attractions between positive lithium ($Li^+$) ions and the negative fluoride ($F^-$) ions.

## CALCIUM CHLORIDE

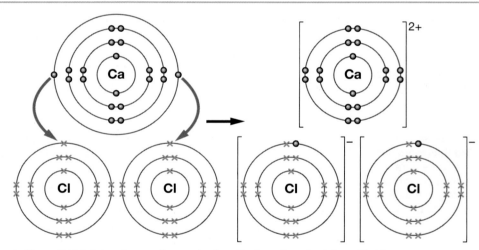

▲ Figure 7.10 A dot-and-cross diagram to show the formation of calcium chloride.

The calcium atom [2, 8, 8, 2] has 2 electrons in its outer shell but each chlorine atom [2, 8, 7] only has room in its outer shell to take one of them. You need two chlorines for every calcium. The 2 electrons are transferred from the outer shell of a calcium atom to two chlorine atoms, one to each. The formula for calcium chloride is therefore $CaCl_2$. There will be very strong electrostatic attractions holding the ions together because of the 2+ charge on the calcium ions.

## FORMULAE FOR IONIC COMPOUNDS

There are so many different ionic compounds that you might encounter at International GSCE that it would be impossible to learn all their formulae. You need a simple way to work them out. You could work out a few from first principles, using their electronic structures, but that would take ages. Others would be too difficult. You need a simple, shortcut method.

## THE NEED FOR EQUAL NUMBERS OF PLUSES AND MINUSES

Ions are formed when atoms, or groups of atoms, lose or gain electrons. They carry an electrical charge, either positive or negative. Compounds are electrically neutral. Therefore in an ionic compound there must be the right number of each sort of ion, so that the total positive charge exactly cancels out the total negative charge. Obviously, then, if you are going to work out a formula, you need to know the charges on the ions.

## CASES WHERE YOU CAN WORK OUT THE CHARGE ON AN ION

Any element in Group 2 has 2 outer electrons, which it will lose to form a 2+ ion. Any element in Group 6 has 6 outer electrons, and it has room to gain 2 more; this leads to a 2– ion. Similar arguments apply in the other groups shown in Table 7.1.

Table 7.1 The charges on an ion in Groups 1–7

| Group in Periodic Table | Charge on ion | Example |
|---|---|---|
| 1 | 1+ | $Na^+$ |
| 2 | 2+ | $Mg^{2+}$ |
| 3 | 3+ | $Al^{3+}$ |
| 5 | 3– | $N^{3-}$ |
| 6 | 2– | $O^{2-}$ |
| 7 | 1– | $Br^-$ |

### HINT

You will always have a copy of the Periodic Table, even in an exam. That means that you can always find out which group an element is in. Elements in Group 4 only form a few ionic compounds and the situation is a bit more complicated. You will need to learn that lead forms a 2+ ion ($Pb^{2+}$).

### REMINDER

Remember that all metals form positive ions.

## CASES WHERE THE NAME TELLS YOU THE CHARGE

All metals form positive ions. Names such as lead(II) oxide, iron(III) chloride or copper(II) sulfate tell you directly about the charge on the metal ion. The number after the metal tells you the number of charges, so:

- lead(II) oxide contains a $Pb^{2+}$ ion
- iron(III) chloride contains an $Fe^{3+}$ ion
- copper(II) sulfate contains a $Cu^{2+}$ ion.

You cannot work out the charges for some ions, you have to learn them. Ions that need to be learned are shown in Table 7.2.

Table 7.2 Ions that you should learn.

| Charge | Substance | Ion | Charge | Substance | Ion |
|---|---|---|---|---|---|
| positive | zinc | $Zn^{2+}$ | negative | nitrate | $NO_3^-$ |
| | silver | $Ag^+$ | | hydroxide | $OH^-$ |
| | hydrogen | $H^+$ | | carbonate | $CO_3^{2-}$ |
| | ammonium | $NH_4^+$ | | sulfate | $SO_4^{2-}$ |

### KEY POINT

Ammonium chloride ($NH_4Cl$) is an example of an ionic compound that does not contain a metal. There is ionic bonding between the $NH_4^+$ and $Cl^-$ ions. There is, however, also covalent bonding (see Chapter 8) in this compound: the $NH_4^+$ ion is held together by covalent bonding.

You will encounter other ions during the course, but these are the important ones for now. The ions in this list are the difficult ones – be sure to learn both the formula and the charge for each ion.

## CONFUSING ENDINGS!

Don't confuse ions such as sulf**ate** and sulf**ide**. A name like copper(II) sulf**ide** means that it contains copper and sulfur **only**. Any 'ide' ending means that there isn't anything complicated there. Sodium chloride, for example, is just sodium and chlorine combined together. So copper(II) sulfide contains $Cu^{2+}$ and $S^{2-}$ ions.

copper(II) sulfide

▲ Figure 7.11 Copper(II) sulfide is CuS.

Once you have an 'ate' ending, it means that there is oxygen (and possibly other things) there as well. So, for example, copper(II) sulfate contains copper, sulfur and oxygen.

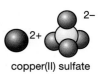

copper(II) sulfate

▲ Figure 7.12 Copper(II) sulfate is $CuSO_4$.

**HINT**

Not looking carefully at word endings is one of the most common mistakes students make when they start to write formulae. Be careful!

## DEDUCING THE FORMULA FOR AN IONIC COMPOUND

**EXAMPLE 1**

### To find the formula for sodium oxide

Sodium is in Group 1, so the ion is $Na^+$.

Oxygen is in Group 6, so the ion is $O^{2-}$.

To have equal numbers of positive and negative charges, you would need two sodium ions to provide the two positive charges to cancel the two negative charges on one oxide ion. In other words, you need:

$Na^+$ $Na^+$ $O^{2-}$

The formula is therefore **$Na_2O$**.

**EXAMPLE 2**

### To find the formula for barium nitrate

Barium is in Group 2, so the ion is $Ba^{2+}$.

Nitrate ions are $NO_3^-$. You will have to remember this.

To have equal numbers of positive and negative charges, you would need two nitrate ions for each barium ion.

The formula is **$Ba(NO_3)_2$**.

Notice the brackets around the nitrate group. *Brackets must be written if you have more than one of these complex ions* (ions containing more than one atom). In any other situation, they are completely unnecessary.

**KEY POINT**

If you didn't write the brackets, the formula would look like this: $BaNO_{32}$. That would read as 1 barium, 1 nitrogen and 32 oxygens!

EXAMPLE 3

**To find the formula for iron(III) sulfate**

Iron(III) tells you that the metal ion is $Fe^{3+}$.

Sulfate ions are $SO_4^{2-}$.

To have equal numbers of positive and negative charges, you would need two iron(III) ions for every three sulfate ions, giving 6+ and 6− in total.

The formula is **$Fe_2(SO_4)_3$**.

A shortcut to working out complicated formulae such as these is to just swap over the numbers in the charges. This is shown in Figure 7.13.

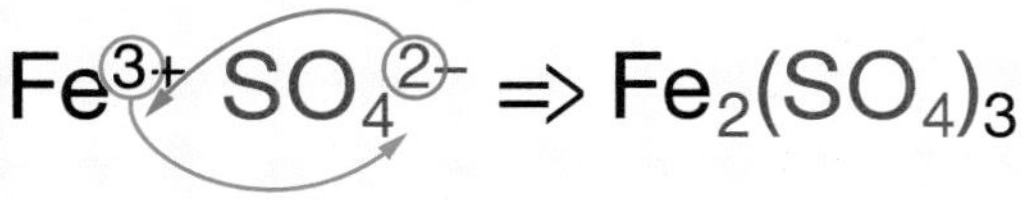

$$Fe^{3+} \quad SO_4{}^{2-} \Rightarrow Fe_2(SO_4)_3$$

▲ Figure 7.13 If you cross over the numbers in the charges you will get the formula.

## CALCIUM CHLORIDE PROVIDES ANOTHER EXAMPLE

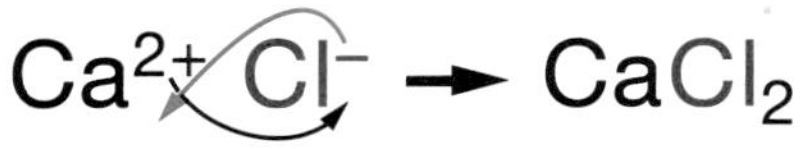

$$Ca^{2+} \quad Cl^- \rightarrow CaCl_2$$

▲ Figure 7.14 To work out the formula of calcium chloride we cross over the numbers in the charges. There is no extra number in front of the charge in $Cl^-$ because we do not tend to write in a 1.

KEY POINT

Why aren't ion charges shown in formulae? Actually, they can be shown. For example, the formula for sodium chloride is NaCl. It is sometimes written $Na^+Cl^-$ if you are trying to make a particular point, but for most purposes the charges are left out. In an ionic compound, the charges are there, whether you write them or not.

You have to be careful using this method because you can get the wrong answer when the charges on the ions are the same. For example, the formula of calcium oxide is CaO and not $Ca_2O_2$. When the charges on the positive and negative ions are the same you can deduce that there will be 1 of each ion in the formula, so there is no need to swap anything over.

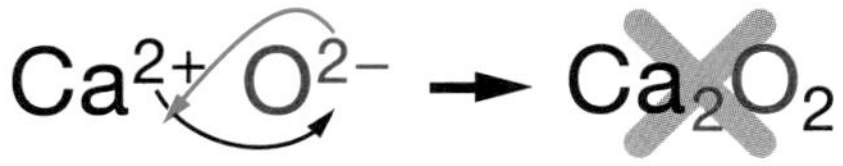

$$Ca^{2+} \quad O^{2-} \rightarrow Ca_2O_2$$

▲ Figure 7.15 The formula of calcium oxide is CaO and not $Ca_2O_2$.

## GIANT IONIC STRUCTURES

All ionic compounds form crystals that consist of lattices of positive and negative ions packed together in a regular way. A **lattice** is a regular array of particles. The lattice is held together by the strong electrostatic attractions between the positively and negatively charged ions.

▲ Figure 7.16 A lattice fence. A lattice is a regular, repeating structure.

## THE STRUCTURE OF SODIUM CHLORIDE

▲ Figure 7.17 A model of a small part of a sodium chloride crystal

Figure 7.17 shows how the ions in a crystal of sodium chloride are arranged.

In diagrams, the ions are usually drawn in an 'exploded' view (Figure 7.18). Each sodium ion is surrounded by 6 chloride ions. In turn, each chloride ion is surrounded by 6 sodium ions. You have to remember that this pattern repeats itself throughout the structure over vast numbers of ions.

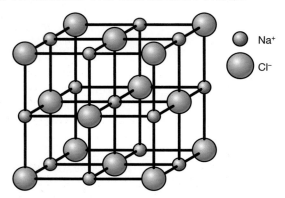

Na$^+$

Cl$^-$

▲ Figure 7.18 An 'exploded' view of sodium chloride. The lines in this diagram are not bonds, they are just there to help show the arrangement of the ions. Those ions joined by lines are touching each other.

The structure of sodium chloride is described as a **giant ionic lattice**. We are using the word '**giant**' here not in the sense of big but rather to describe a structure where there are no individual molecules. All the sodium ions in the structure attract all the chloride ions, we cannot pick out sodium chloride molecules; there are no individual molecules. The bonding in a giant ionic lattice extends throughout the structure in all directions. There is no limit to the number of particles present, all we know is that there must be the same number of sodium and chloride ions.

**HINT**

This is really important: you must not talk about *molecules* of an ionic compound. This will be marked wrong in the exam and you could lose all the marks for a question!

## THE STRUCTURE OF MAGNESIUM OXIDE

Magnesium oxide, MgO, contains magnesium ions, $Mg^{2+}$, and oxide ions, $O^{2-}$. It has exactly the same structure as sodium chloride. The only difference is that the magnesium oxide lattice is held together by stronger forces of attraction. This is because in magnesium oxide, 2+ ions are attracting 2– ions. In sodium chloride, the electrostatic attractions are weaker because they are only between 1+ and 1– ions.

**EXTENSION WORK**

The $Mg^{2+}$ ion is also smaller than the $Na^+$ ion, and the $O^{2-}$ ion is smaller than the $Cl^-$ ion. This causes stronger attractions but the effect of the charge on the ions is more important.

## THE PHYSICAL PROPERTIES OF IONIC SUBSTANCES

*Ionic compounds have high melting points and boiling points* because of the strong electrostatic forces of attraction holding the lattice together. A lot of energy has to be supplied to break the strong electrostatic forces of attraction between oppositely charged ions in the giant lattice structure.

*Ionic compounds tend to be crystalline*. This reflects the regular arrangement of ions in the lattice. Sometimes the crystals are too small to be seen except under powerful microscopes. Magnesium oxide, for example, is always seen as a white powder because the individual crystals are too small to be seen with the naked eye.

▲ Figure 7.19 Sodium chloride is crystalline

*Ionic crystals tend to be brittle.* This is because any small distortion of a crystal will bring ions with the same charge alongside each other. Like charges repel and so the crystal splits itself apart.

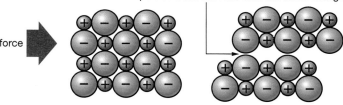

▲ Figure 7.20 Ionic crystals tend to be brittle

*Ionic substances tend to be soluble in water.*

*Ionic compounds tend to be insoluble in organic solvents.*

**EXTENSION WORK**

The reasons that ionic compounds tend to be soluble in water are quite complicated. Water is a covalent molecule (Chapter 8) but the electrons in the bonds are more attracted towards the oxygen end of the bond. This makes the oxygen slightly negative and the hydrogen slightly positive – the molecule is called **polar**. This means that reasonably strong forces can be formed between water molecules and ions, which provide the energy to break the lattice apart.

Not all ionic substances are soluble in water: magnesium oxide isn't soluble in water because the attractions between the water molecules and the ions aren't strong enough to overcome the very strong electrostatic forces of attraction between magnesium and oxide ions.

Hexane is non-polar and does not form strong enough attractions to the ions to break apart the ionic lattice.

# THE ELECTRICAL CONDUCTIVITY OF IONIC SUBSTANCES

**HINT**

Molten just means that the salt has been melted – it is a liquid.

Ionic compounds don't conduct electricity when they are solid because the *ions* are fixed in position and are not free to move around. They do, however, conduct electricity when they are **molten** (have melted) or if they are dissolved in water (in aqueous solution). This happens because the *ions* then become free to move around. It is really important that you use the correct words when explaining this. Do not use the word 'electrons'. You must talk about the **ions** being free to move.

# CHAPTER QUESTIONS

SKILLS ▶ INTERPRETATION

SKILLS ▶ CRITICAL THINKING

1 a Explain what is meant by **i** an ion and **ii** ionic bonding.

b In each of the following cases, write down the electronic configurations of the original atoms and then explain (in words or diagrams) what happens when:
   i sodium bonds with chlorine to make sodium chloride
   ii lithium bonds with oxygen to make lithium oxide
   iii magnesium bonds with fluorine to make magnesium fluoride.

SKILLS ▶ INTERPRETATION

2 Draw dot-and-cross diagrams to show the ions formed (outer electrons only) when:

a potassium combines with fluorine

b calcium combines with bromine

c magnesium combines with iodine.

**SKILLS** CRITICAL THINKING

3 a State the formula of the ion formed by:

|  |  |  |  |
|---|---|---|---|
| i | magnesium | vii | chlorine |
| ii | strontium | viii | iodine |
| iii | potassium | ix | aluminium |
| iv | oxygen | x | calcium |
| v | sulfur | xi | nitrogen |
| vi | caesium | | |

b State the name of each negative ion in a.

4 Work out the formulae of the following compounds:

lead(II) oxide        sodium bromide

magnesium sulfate      zinc chloride

potassium carbonate     ammonium sulfide

calcium nitrate         iron(III) hydroxide

iron(II) sulfate         copper(II) carbonate

aluminium sulfate      calcium hydroxide

cobalt(II) chloride     calcium oxide

silver nitrate          iron(III) fluoride

ammonium nitrate     rubidium iodide

sodium sulfate        chromium(III) oxide

**SKILLS** REASONING

5 Explain why sodium chloride:

a has a high melting point

b does not conduct electricity when solid

c conducts electricity when molten.

6 Predict, giving reasons, whether you would expect potassium chloride to have a higher or lower melting point than calcium oxide.

# 8 COVALENT BONDING

We met ionic compounds in Chapter 7 and in this chapter we will understand what covalent bonding is. There are a lot more covalent compounds than ionic compounds so it is important that you understand how the bonding works.

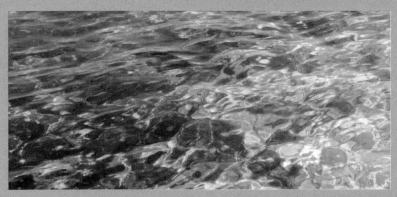

▲ Figure 8.1 Water is a covalent compound but the salt dissolved in sea water is an ionic compound.

## LEARNING OBJECTIVES

- Know that a covalent bond is formed between atoms by the sharing of a pair of electrons

- Understand covalent bonds in terms of electrostatic attractions

- Understand how to use dot-and-cross diagrams to represent covalent bonds in:
  - diatomic molecules, including hydrogen, oxygen, nitrogen, halogens and hydrogen halides
  - inorganic molecules, including water, ammonia and carbon dioxide
  - organic molecules containing up to two carbon atoms, including methane, ethane, ethene and those containing halogen atoms

- Explain why substances with a simple molecular structure are gases or liquids, or solids with low melting and boiling points

- Understand that the term intermolecular forces of attraction can be used to represent all forces between molecules

- Explain why the melting and boiling points of substances with simple molecular structures increase, in general, with increasing relative molecular mass

- Explain why substances with giant covalent structures are solids with high melting and boiling points

- Explain how the structures of diamond, graphite and $C_{60}$ fullerene influence their physical properties, including electrical conductivity and hardness

- Know that covalent compounds do not usually conduct electricity

## COVALENT BONDING

**WHAT IS A COVALENT BOND?**

In any bond, particles are held together by electrostatic attractions between something positively charged and something negatively charged. In a **covalent bond**, *a pair of electrons is shared between two atoms*. What holds the atoms together is *the strong electrostatic attraction between the nuclei (positively charged) of the atoms that make up the bond, and the shared pair of electrons (negatively charged)*.

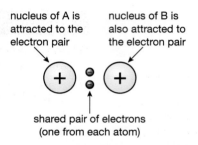

nucleus of A is attracted to the electron pair

nucleus of B is also attracted to the electron pair

shared pair of electrons (one from each atom)

▲ Figure 8.2 The shared pair of electrons is attracted to the nuclei of both atoms.

## COVALENT BONDING IN A HYDROGEN MOLECULE

Covalent bonds are often shown using dot-and-cross diagrams.

Both hydrogen nuclei in Figure 8.3 are strongly attracted to the shared pair of electrons.

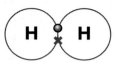

▲ Figure 8.3 A dot-and-cross diagram for $H_2$

Hydrogen atoms form diatomic molecules with the formula $H_2$. The atoms in an $H_2$ molecule are joined together by a covalent bond. The covalent bond between two hydrogen atoms is very strong.

Molecules contain a certain fixed number of atoms, which are joined together by covalent bonds. Hydrogen molecules are said to be **diatomic** because they contain two atoms. Other sorts of molecule may have as many as thousands of atoms joined together, for example proteins and DNA.

## THE SIGNIFICANCE OF NOBLE GAS STRUCTURES IN COVALENT BONDING

In $H_2$, each hydrogen atom has only one electron to share, so it can only form one covalent bond. The shared pair of electrons is in the outer shell of both, therefore each atom has the same number of electrons as a noble gas atom (helium in this case).

Does that mean that the hydrogen has turned into helium? No. The number of protons in the nucleus hasn't changed, it is the number of protons that defines what an atom is.

In virtually all of the molecules you will meet at this level, electrons will be shared so that H atoms have a total of 2 electrons in their outer shell and all other atoms will have 8 electrons in their outer shell. Some people talk about the 'octet rule', referring to this 8. Remember, we are counting shared electrons as belonging to the outer shells of both atoms.

When there is one atom in the middle and other atoms are joined to it (as in $CH_4$ or $PCl_3$) the *outer* atoms will virtually always have 8 electrons in their outer shell (or 2 if they are H). In fact, it is very difficult to think of an example where the outer atoms do not have 8 electrons. There are some molecules where the *central* atom does not have 8 electrons in the outer shell, and we will look at a couple of examples of those later on.

## WHY DOES HYDROGEN FORM MOLECULES?

Whenever a bond is formed (of whatever kind), energy is released, and that makes the things involved more stable than they were before. The more bonds an atom can form, the more energy is released and the more stable the system becomes. The $H_2$ molecule is much more stable than two separate hydrogen atoms.

## COVALENT BONDING IN A HYDROGEN CHLORIDE MOLECULE

A chlorine atom has 7 electrons in its outer shell. By sharing 1 electron with a hydrogen atom, both atoms will have the same number of electrons as the nearest noble gas atom. If you look at the arrangement of electrons around the chlorine atom in the covalently bonded molecule of HCl (Figure 8.4), you will see that its electronic configuration is now [2, 8, 8]. That is the same as an argon atom. Similarly, the hydrogen now has 2 electrons in its outer shell – the same as helium.

Notice in Figure 8.4 that only the electrons in the outer shell of the chlorine are used in bonding. In the examples you will meet at International GCSE, the inner electrons never get used. In fact, the inner electrons are often left out of bonding diagrams. But be careful! In an exam, only leave out the inner electrons if the question tells you to. Another way of representing the covalent bonding in HCl is shown in Figure 8.5.

We also use lines to represent the covalent bonds between atoms, but be careful, the diagram shown in Figure 8.6 is not a dot-and-cross diagram.

In chemistry, we talk about things being more stable or less stable. When we do this we are usually talking about how much energy something has. Generally, the lower the energy something has, the more stable it is. Chemical reactions usually occur so that something becomes more stable. Think about holding a book: if you let go of the book, it will fall to the floor, where it has less (potential) energy, therefore it becomes more stable. When bonds form energy is given out and so the substance formed has less energy, therefore it is more stable.

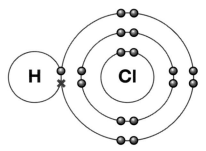

▲ Figure 8.4 A dot-and-cross diagram for HCl.

▲ Figure 8.5 The covalent bonding in HCl showing outer shell electrons only.

▲ Figure 8.6 The line between the atoms represents a covalent bond.

## COVALENT BONDING IN A CHLORINE MOLECULE (Cl₂)

A chlorine atom has 7 electrons in its outer shell. Each Cl shares 1 electron so that both Cl atoms will have 8 electrons in their outer shell.

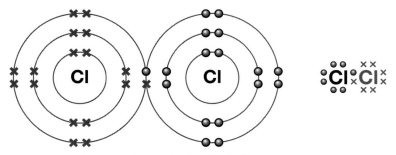

▲ Figure 8.7 Two ways of showing the covalent bonding in Cl₂

## COVALENT BONDING IN METHANE, AMMONIA AND WATER

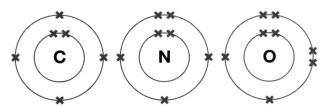

▲ Figure 8.8 The electronic configurations of C, N and O atoms

## COVALENT BONDING IN METHANE

A carbon atom has 4 electrons in its outer shell. By sharing 1 electron with each of 4 hydrogen atoms the C will have 8 electrons in its outer shell and each H will have 2 electrons in its outer shell. Therefore C forms 4 covalent bonds, 1 with each H atom. Methane has the formula $CH_4$.

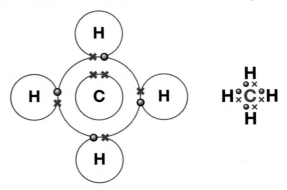

▲ Figure 8.9 Dot-and-cross diagrams for methane

## COVALENT BONDING IN AMMONIA

A nitrogen atom has 5 electrons in its outer shell. It will therefore share 3 other electrons to have 8 electrons in its outer shell. In ammonia, a nitrogen atom forms 3 covalent bonds, 1 with each H atom.

The formula of ammonia is $NH_3$.

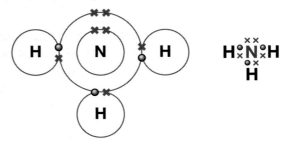

▲ Figure 8.10 Dot-and-cross diagrams for ammonia

## COVALENT BONDING IN WATER

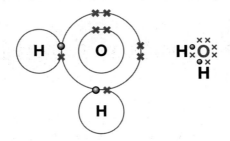

▲ Figure 8.11 Dot-and-cross diagrams for water

An oxygen atom has 6 electrons in its outer shell. It will therefore share 2 other electrons to have 8 electrons in its outer shell. In water, an oxygen atom forms 2 covalent bonds – 1 with each H atom. The formula of water is $H_2O$.

### LOOKING AHEAD

**Shapes of molecules**

▲ Figure 8.12 (a) A methane molecule is tetrahedral and (b) a water molecule is bent.

By understanding the bonding in a covalent molecule it is possible to work out the shape of the molecule. Pairs of electrons in the outer shell of the central atom repel each other and will therefore tend to get as far apart as possible. For example, in a methane molecule there are four pairs of electrons around the central C atom and for these to be as far away from each other as possible, they must be arranged in a tetrahedral shape. A tetrahedron (adjective: tetrahedral) is a triangular pyramid.

There are also four pairs of electrons around the central atom in water: two of these are pairs of electrons involved in covalent bonds and two are pairs of electrons in the outer shell of the oxygen which are not involved in bonding (these are often called *lone pairs of electrons* or just *lone pairs*). These four pairs of electrons are also arranged in a tetrahedral arrangement so the actual shape of a water molecule (how the atoms are arranged) is described as 'bent' or 'V-shaped'. The fact that a water molecule is bent and the electrons are attracted to a different extent by the oxygen and hydrogen atoms means that a water molecule is polar (has a slightly negative and a slightly positive end) and that a stream of water can be bent by an electrically charged object.

▲ Figure 8.13 Water being bent by an electrically charged comb.

## COVALENT BONDING IN A SLIGHTLY MORE COMPLICATED MOLECULE: ETHANE

Ethane has the formula $C_2H_6$. The bonding is similar to methane (Figure 8.9), except that there is a carbon–carbon covalent bond as well as the carbon–hydrogen bonds.

This is called an organic compound. You will learn more about molecules such as this in Unit 4.

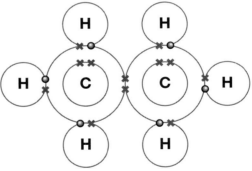

▲ Figure 8.14 A dot-and-cross diagram for ethane

## MULTIPLE COVALENT BONDING

**COVALENT BONDING IN AN OXYGEN MOLECULE: DOUBLE BONDING**

An oxygen atom has 6 electrons in its outer shell and so if two oxygen atoms combine, they will both share 2 electrons each; this means that each atom will have 8 electrons in its outer shell. There are therefore two shared pairs of electrons between the oxygen atoms; this is called a double covalent bond or, usually, just a **double bond**.

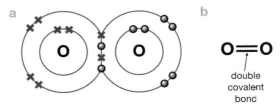

▲ Figure 8.15 (a) A dot-and-cross diagram for $O_2$ and (b) the double covalent bond.

## THE TRIPLE BOND IN A NITROGEN MOLECULE

A nitrogen atom has 5 electrons in its outer shell and so if two nitrogen atoms combine they will both share 3 electrons each; this means that each atom will have 8 electrons in its outer shell. There are, therefore, three shared pairs of electrons between the nitrogen atoms; this is called a triple covalent bond or, usually, just a **triple bond**.

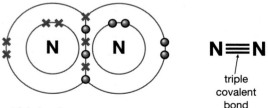

▲ Figure 8.16 $N_2$ has a triple bond

Nitrogen gas consists of nitrogen molecules bonded like this. The triple bond from the sharing of three pairs of electrons between the two nitrogen atoms is very strong and needs a lot of energy to break. That is why nitrogen is relatively unreactive.

## COVALENT DOUBLE BONDING IN CARBON DIOXIDE, $CO_2$

An oxygen atom has 6 electrons in its outer shell and a carbon atom has 4. Each oxygen atom will share 2 electrons with the carbon atom. Two double bonds are formed between the carbon and the two oxygens (Figure 8.17). All atoms have 8 electrons in their outer shells.

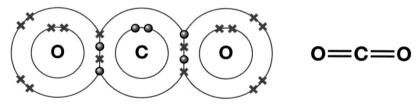

▲ Figure 8.17 $CO_2$ has two double bonds between atoms.

## THE DOUBLE BOND IN ETHENE, $C_2H_4$

Ethene is rather like ethane in Figure 8.14, except that it only has two hydrogen atoms attached to each carbon atom and a double bond between the carbon atoms.

### HINT

With organic compounds such as ethane and ethene you have to look at their names very carefully, even one different letter in the name can matter. Here, for example, ethane and ethene are completely different compounds. You will find out more about this in Unit 4.

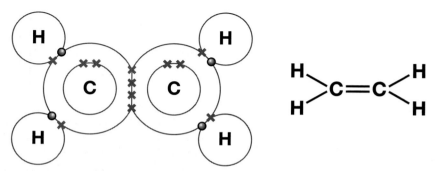

▲ Figure 8.18 Ethene has a double bond between the C atoms.

## ORGANIC MOLECULES CONTAINING HALOGEN ATOMS

Bromomethane has the formula $CH_3Br$: the 3 H atoms and the Br atom are joined to the central C atom. Br has 35 electrons and we have not learned how to work out the electronic configuration of an atom with 35 electrons, but if

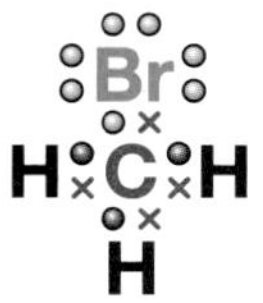

▲ Figure 8.19 A dot-and-cross diagram for $CH_3Br$ showing outer shells only. Although we have used different colours here for clarity, you do not need to use different colours in the exam.

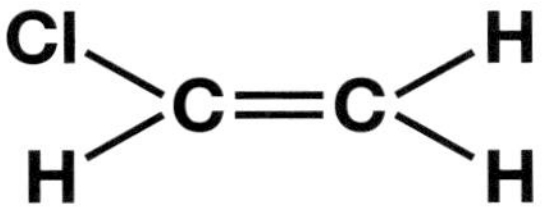

▲ Figure 8.21 The structure of chloroethene showing the covalent bonds. This is not, however, a dot-and-cross diagram and you must not draw this if you are asked to draw a dot-and-cross diagram.

you look at the Periodic Table you will see that Br is in Group 7. Because it is in Group 7 we know that it has 7 electrons in its outer shell, and will share 1 electron so that it has 8 in its outer shell. Therefore, we know that Br will form just 1 covalent bond. When you draw a dot-and-cross diagram for $CH_3Br$ you will only be asked to show the outer electrons (Figure 8.19).

Probably the most complicated molecule for which you could be asked to draw a dot-and-cross diagram would be something like chloroethene ($CH_2CHCl$). When drawing chloroethene, remember that C will form 4 covalent bonds, H will form 1 and Cl will form 1. The dot-and-cross diagram (outer shells only) is shown in Figure 8.20.

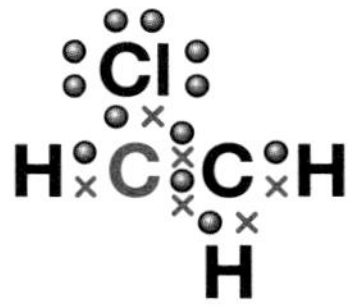

▲ Figure 8.20 There is a double bond between the 2 C atoms. It does not matter where you put the H and Cl atoms relative to the C, and you do not have to use different colours.

The structure of chloroethene can also be shown as illustrated in Figure 8.21.

## SOME MORE DIFFICULT MOLECULES WHERE THE CENTRAL ATOM DOES NOT HAVE 8 ELECTRONS IN ITS OUTER SHELL

Although the *outer* atoms in a molecule will always have 8 (or 2 in the case of hydrogen) electrons in their outer shell, there are a few examples you may encounter where the *central* atom has more or fewer than 8 electrons in the outer shell.

### $BF_3$

Boron is the central atom and the Fs are the outer atoms. Each F will share 1 other electron to have 8 electrons in its outer shell. This means that B only has a total of 6 electrons in its outer shell. Another way of thinking about this is that a B atom only has 3 electrons in its outer shell, and so this is the maximum number that it can share.

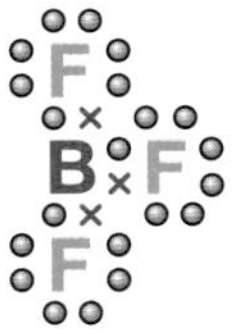

▲ Figure 8.22 A dot-and-cross diagram for $BF_3$

### $SO_2$

#### EXTENSION WORK

Atoms in Periods 3 and below (4, 5, 6, 7) can have more than 8 electrons in their outer shells. The maximum number of electrons that the central atom in a molecule can share is equal to the number of electrons in its outer shell. So, sulfur can form up to 6 bonds (in $SF_6$) and chlorine 7 (in $HClO_4$).

This is called sulfur dioxide or sulfur(IV) oxide. The central atom is S and the outer atoms are O. Each O atom has 6 electrons in its outer shell and so will need to share 2 other electrons to have 8 electrons in the outer shell. So the S atom shares 2 electrons with each of the O atoms to form two double bonds. A sulfur atom originally had 6 electrons in its outer shell and so now, if it shares 4 electrons, it has 10 electrons in its outer shell.

▲ Figure 8.23 There are two double bonds in $SO_2$.

## SIMPLE MOLECULAR STRUCTURES

Molecules contain fixed numbers of atoms joined by strong covalent bonds. If we look closely at liquid water, there are individual water molecules, where the H and O atoms are joined together with strong covalent bonds. But there must also be some forces between water molecules which keep them in the liquid state. These forces are **intermolecular forces**.

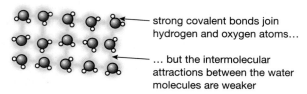

strong covalent bonds join hydrogen and oxygen atoms…

… but the intermolecular attractions between the water molecules are weaker

▲ Figure 8.24 Water is a simple molecular compound.

These intermolecular forces between molecules are much weaker than covalent bonds. When we boil water it is only these weak intermolecular forces of attraction that are broken; covalent bonds are not broken. Looking at Figure 8.25 you can see that there are $H_2O$ molecules in liquid water and in gaseous water. The covalent bonds between the H and O atoms in the molecules have not changed in any way. All that has changed in gaseous water is that there are no intermolecular forces, they have been broken.

When a substance consists of molecules with intermolecular forces of attraction between them, we say that it has a *simple molecular structure*. Examples of things that have simple molecular structures are $H_2O$, $CO_2$, $CH_4$, $NH_3$ and $C_2H_4$. These are all the things that we have drawn dot-and-cross diagrams for above. Virtually all the compounds you will encounter that have covalent bonding will have simple molecular structures.

*Substances with simple molecular structures tend to be gases or liquids or solids with low melting points and boiling points*. The reason for this is that not much energy is required to break the weak intermolecular forces of attractions between molecules. Remember, no covalent bonds are broken, covalent bonds are strong.

**DID YOU KNOW?**
Intermolecular literally means between molecules.

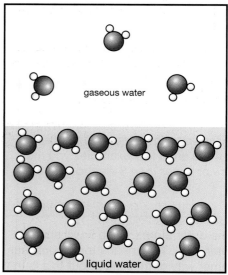

gaseous water

liquid water

▲ Figure 8.25 Only intermolecular forces are broken when water evaporates/boils.

**EXTENSION WORK**

If you continue with chemistry you will learn that there are different types of intermolecular forces. You will come across terms like van der Waals' forces, London dispersion forces and hydrogen bonds. There is a special type of intermolecular force between water molecules called **hydrogen bonds**. Hydrogen bonding gives water some of its very special properties, for example the solid form (ice) is less dense than the liquid form.

## MELTING AND BOILING POINTS INCREASE AS RELATIVE MOLECULAR MASS INCREASES

The halogens (Group 7 in the Periodic Table) all have a simple molecular structure consisting of diatomic molecules with intermolecular forces between them.

You can see from Table 8.1 that the melting points and boiling points increase as the relative molecular mass increases.

Table 8.1 The melting points and boiling points of the halogens.

| Halogen | Formula | Relative molecular mass/$M_r$ | Melting point/°C | Boiling point/°C |
|---|---|---|---|---|
| fluorine | $F_2$ | 38 | −220 | −188 |
| chlorine | $Cl_2$ | 71 | −101 | −34 |
| bromine | $Br_2$ | 160 | −7 | 59 |
| iodine | $I_2$ | 254 | 114 | 184 |

It is not always the case that melting and boiling points increase as the $M_r$ increases and really the rule only applies to sets of very similar substances, such as the halogens or the alkanes (see Chapter 23). Some examples where the rule does not work are water ($M_r = 18$, boiling point = $100\,°C$), ethane ($M_r = 30$, boiling point = $-89\,°C$), $NH_3$ ($M_r = 17$, boiling point = $-33\,°C$) and $PH_3$ ($M_r = 34$, boiling point = $-88\,°C$).

**KEY POINT**

We are using the term covalent molecular compounds to mean covalent compounds with a simple molecular structure.

Remember, as we melt or boil these substances we are only breaking the intermolecular forces of attraction between molecules. The boiling points increase down this group, which means we have to put in more energy to break the intermolecular forces as the relative molecular mass increases. This means that the *intermolecular forces of attraction must become stronger as relative molecular mass increases*.

This is something that we see quite often in chemistry, for example boiling points increase along the series $CH_4$, $C_2H_6$, $C_3H_8$ as relative molecular mass increases. This is discussed further in Chapter 23.

## SOME OTHER PHYSICAL PROPERTIES OF COVALENT COMPOUNDS

**Covalent molecular compounds do not conduct electricity**. This is because the molecules don't have any overall electrical charge (there are no ions) and all the electrons are held tightly in the atoms or in covalent bonds and so are not able to move from molecule to molecule.

*Covalent molecular substances tend to be insoluble in water*. There are some exceptions to this, for example ethanol ($C_2H_5OH$) and substances such as $NH_3$ and HCl that react with water as they dissolve.

*Covalent molecular substances are often soluble in organic solvents.*

## GIANT COVALENT STRUCTURES

**DIAMOND**

**KEY POINT**

A tetrahedron is a triangular-based pyramid. In a tetrahedral arrangement, one atom is at the centre of the tetrahedron, and the atoms it is attached to are at the four corners. Look carefully at the top five atoms in Figure 8.26 to see what this looks like. You will find other similar arrangements in this diagram.
In Figure 8.26 some carbon atoms seem to be forming only two bonds (or even one bond), but that's not really the case. We are only showing a small part of the whole structure. The structure continues in three dimensions, and each of the atoms drawn here is attached to four others. Each of the lines in this diagram represents a covalent bond.

Diamond is a form of pure carbon.

Each carbon atom has four electrons in its outer shell and it therefore forms four covalent bonds. In diamond, each carbon bonds strongly to four other carbon atoms in a *tetrahedral* arrangement. Figure 8.26 shows enough of the structure to see what is happening.

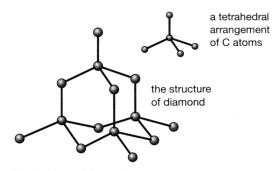

a tetrahedral arrangement of C atoms

the structure of diamond

▲ Figure 8.26 Part of the structure of diamond

This is a giant covalent structure; it continues on and on in three dimensions. It is *not* a molecule because the number of atoms joined up in a real diamond is completely variable and depends on the size of the crystal. Molecules always contain *fixed numbers* of atoms joined by covalent bonds.

*Diamond has a very high melting and boiling point*. This is because of the very strong carbon–carbon covalent bonds, which extend throughout the whole crystal in three dimensions. A lot of energy has to be supplied to break these strong covalent bonds, therefore diamond has very high melting and boiling points. It is important to realise how this is different from the simple molecular structures that we saw above. In order to melt or boil a substance with a simple molecular structure, such as $CH_4$, we only had to supply enough

▲ Figure 8.27 Diamond (a form of carbon) is crystalline, and is the hardest naturally occurring substance.

energy to break the relatively weak intermolecular forces of attraction. In diamond there are no intermolecular forces (it has a giant structure, there are no molecules). Covalent bonds, which are very strong, must be broken in order to melt or boil it.

In general, *all substances with giant covalent structures are solids with high melting and boiling points* because a lot of energy has to be supplied to break all the strong covalent bonds throughout the giant structure. Other substances with giant covalent structures include graphite (discussed below) and silicon dioxide ($SiO_2$).

*Diamond is very hard*. Again, a lot of energy has to be supplied to break the strong covalent bonds in the giant structure. Drill bits can be tipped with diamonds for drills used on stone and rock.

*Diamond doesn't conduct electricity*. All the electrons in the outer shells of the carbon atoms are tightly held in covalent bonds between the atoms. None are free to move around.

> **EXTENSION WORK**
>
> *Diamond doesn't dissolve in water or in any other solvent*. This is again because of the strong covalent bonds between the carbon atoms. If the diamond dissolved, these bonds would have to be broken.
> *Diamond conducts heat very well (better than any other element)*. As one end of the crystal is heated the atoms vibrate more. The strong bonds throughout the giant structure mean that these vibrations are quickly transmitted from one end of the crystal to the other.

## GRAPHITE

Graphite is also a form of carbon, but the atoms are arranged differently, although it still has a giant structure. Graphite has a *layer structure*, rather like a pack of playing cards. In a pack of cards, each card is strong but the individual cards are easily separated. The same is true in graphite.

▲ Figure 8.28 Graphite has a structure rather like a pack of playing cards.

▲ Figure 8.29 Graphite is soft with a layer structure.

atoms in a layer of graphite

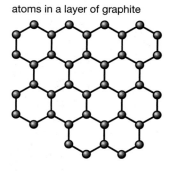

edge-on view of the layers

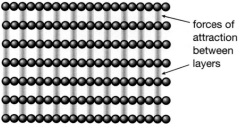

forces of attraction between layers

the gaps between the layers are much bigger than the distances between the atoms in the layers

▲ Figure 8.30 The structure of graphite. Some forces of attraction between layers have been shown. These are not bonds and could have been drawn anywhere between the layers.

Actually, the reason that the layers slide over each other fairly easily is more complicated than this and graphite is not a lubricant in a vacuum. Graphite being a lubricant relies on water molecules sticking to the surface; this does not happen in a vacuum.

*Graphite is a soft material.* Although the forces holding the atoms together in each layer are very strong, the attractions between the layers are much weaker and not much energy is needed to overcome them. Layers slide over each other and can easily be flaked off.

Graphite (mixed with clay to make it harder) is used in pencils. When you write with a pencil, you are leaving a trail of graphite layers behind on the paper. Pure graphite is so slippery that it is used as a dry lubricant, for example powdered graphite is used to lubricate locks.

*Graphite has high melting and boiling points.* To melt/boil graphite, you don't just have to separate the layers, you have to break up the whole structure, including the covalent bonds. That needs very large amounts of energy because the covalent bonds are so strong.

*Graphite conducts electricity.* If you look back at Figure 8.30, you will see that each carbon atom is joined to only three others. Each carbon atom uses three of its outer shell electrons to form three single covalent bonds. The fourth electron in the outer shell of each atom is free to move around throughout the whole of the layer. The electrons that are free to move throughout the layers are called **delocalised electrons**. The movement of these delocalised electrons allows graphite to conduct electricity.

▲ Figure 8.31 Three graphite electrodes glow red hot after their removal from an electric arc furnace used to produce steel.

Some other properties of graphite are:

■ *Graphite is insoluble in all solvents* because it would take too much energy to break all the strong covalent bonds.

■ *Graphite is less dense than diamond* because the layers in graphite are relatively far apart. The distance between the graphite layers is more than twice the distance between atoms in each layer. In a sense, a graphite crystal contains a lot of wasted space, which isn't there in a diamond crystal.

## $C_{60}$ FULLERENE

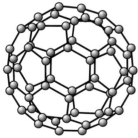

▲ Figure 8.32 A $C_{60}$ fullerene molecule. Remember that molecules consist of fixed number of atoms; each $C_{60}$ fullerene molecule contains 60 carbon atoms joined by covalent bonds.

Diamond and graphite are two **allotropes** of carbon. Allotropes are different forms of the same element. Another allotrope of carbon is $C_{60}$ fullerene. Diamond and graphite both have giant structures but $C_{60}$ fullerene has a simple molecular structure.

In solid or liquid $C_{60}$ fullerene there are $C_{60}$ molecules with weak intermolecular forces between them. The fact that $C_{60}$ has a simple molecular structure has a big influence on its physical properties.

*$C_{60}$ fullerene has lower melting and boiling points than diamond and graphite.* When fullerene is melted, only the relatively weak intermolecular forces of attraction must be broken. This does not require as much energy as breaking all the strong covalent bonds when diamond and graphite are melted.

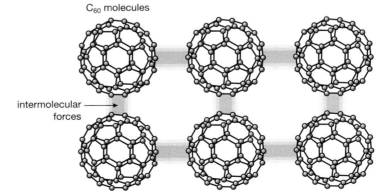

$C_{60}$ molecules

intermolecular forces

▲ Figure 8.33 $C_{60}$ fullerene has a simple molecular structure.

There are different fullerenes, where the molecules contain different numbers of carbon atoms. This is why we include the $C_{60}$ in the name.

$C_{60}$ *fullerene is not as hard as diamond.* It does not take as much energy to break the intermolecular forces of attraction in $C_{60}$ fullerene compared to breaking the strong covalent bonds in diamond.

$C_{60}$ *fullerene does not conduct electricity.* Although all the carbon atoms in $C_{60}$ only form three bonds, the fourth electron on each atom can only move around *within* each $C_{60}$ molecule; the electrons cannot jump from molecule to molecule.

## CHAPTER QUESTIONS

You will need to use the Periodic Table in Appendix A on page 320.

**SKILLS**   CRITICAL THINKING

1   State whether each of the following compounds is ionic or covalent:

   a   MgO          b   $CH_3Br$          c   $H_2O_2$

   d   $FeCl_2$          e   NaF          f   HCN

**SKILLS**   INTERPRETATION

2   a   What is meant by a covalent bond? How does this bond hold two atoms together?

   b   Draw dot-and-cross diagrams (showing outer shell electrons only) to show the covalent bonding in

      i   methane ($CH_4$)

      ii   hydrogen sulfide ($H_2S$)

      iii   phosphine ($PH_3$)

      iv   silicon tetrachloride ($SiCl_4$)

   c   Draw dot-and-cross diagrams (showing outer shell electrons only) to show the covalent bonding in

      i   $O_2$

      ii   $N_2$

3   Draw dot-and-cross diagrams (showing outer shell electrons only) to show the covalent bonding in

   a   ethane ($C_2H_6$)          b   ethene ($C_2H_4$)          c   chloroethane ($CH_3CH_2Cl$)

**SKILLS**   REASONING

4   Explain why carbon dioxide sublimes at −78.5 °C but diamond sublimes at around 4000 °C.

5   Hexane has the formula $C_6H_{14}$. It is a liquid at room temperature.

   a   Explain whether hexane has a simple molecular or giant structure.

   b   Explain whether you would expect pentane ($C_5H_{12}$) to have a higher or lower boiling point than hexane.

   c   Explain whether or not you would expect hexane to conduct electricity.

6   Explain the following in terms of structure and bonding:

   a   diamond is harder than graphite

   b   $C_{60}$ fullerene has a lower melting point than graphite

   c   graphite conducts electricity

   d   diamond does not conduct electricity.

7 The table below gives details of the boiling temperatures of some substances made of covalent molecules. Arrange these substances in increasing order of the strength of their intermolecular forces and explain your order.

| Substance | Boiling point/°C |
|---|---|
| ammonia | −33 |
| ethanamide | 221 |
| ethanol | 78.5 |
| hydrogen | −253 |
| phosphorus trifluoride | −101 |
| water | 100 |

(Don't panic if you don't recognise some of the names. The substances could also have been labelled A, B, C, D, E and F.)

8 The compound $N_2F_2$ has the structure:

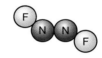

a Explain whether you would expect the bond between the nitrogen atoms to be a single, double or triple bond.

b Draw a dot-and-cross diagram for $N_2F_2$ showing outer electrons only.

9 a Draw a dot-and-cross diagram (showing outer electrons only) to show the covalent bonding in $BCl_3$.

b $BCl_3$ is sometimes described as an *electron-deficient* compound. Explain what you think that means.

# 9 METALLIC BONDING

In this chapter we will discuss the bonding in metals and explain some of the physical properties of metals in terms of structure and bonding.

▲ Figure 9.1 Metals have properties that make them useful as construction materials.

▲ Figure 9.2 Most metals are strong and all metals conduct electricity.

## LEARNING OBJECTIVES

- ■ Know how to represent a metallic lattice by a 2D diagram
- ■ Understand metallic bonding in terms of electrostatic attractions
- ■ Explain typical physical properties of metals, including electrical conductivity and malleability

## METALLIC BONDING

Sodium is a metal. When sodium atoms bond together to form the solid metal, the outer electron on each sodium atom becomes free to move throughout the whole structure. The electrons are said to be *delocalised*. These electrons are no longer attached to particular atoms or pairs of atoms, instead you can think of them as flowing around through the whole metal. When a sodium atom loses its outer electron a sodium ion ($Na^+$) is left behind. A metallic structure consists of *a lattice (regular arrangement) of positive ions in a sea of delocalised electrons* (Figure 9.3).

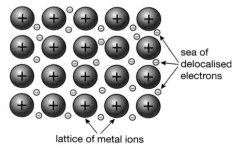

sea of delocalised electrons

lattice of metal ions

▲ Figure 9.3 The metallic structure of a metal such as sodium

**Metallic bonding** is the electrostatic forces of attraction between each
positive ion and the delocalised electrons. This holds the structure together.

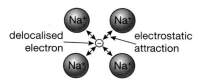

▲ Figure 9.4 The electrostatic attraction between positively charged ions and negatively charged
electrons holds the metallic structure together.

Metals have giant structures. There are no individual molecules and all the
positive ions in the lattice attract all the delocalised electrons.

The ion formed by the metal depends on the number of electrons the original
atom has in its outer shell. Thus all the elements in Group 1 form 1+ ions and
all the elements in Group 2 form 2+ ions.

## PHYSICAL PROPERTIES OF METALS

Most metals are hard and have high melting points. This suggests that the
electrostatic forces of attraction between the positive ions and the delocalised
electrons are strong.

In the case of sodium, only one electron per atom is delocalised, leaving ions
with only one positive charge on them. The bonding in sodium is quite weak,
as metals go, which is why sodium is fairly soft, with a low melting point for
a metal. Magnesium has two outer electrons, both of which are delocalised
into the 'sea', leaving behind ions that carry a charge of 2+. There is a much
stronger electrostatic attraction between the 2+ ions and the delocalised
electrons. This means that the bonding is stronger in magnesium and the
melting point is higher.

Metals conduct electricity because the *delocalised electrons are free to move
throughout the structure.* Imagine what happens if a piece of metal is attached
to an electrical power source.

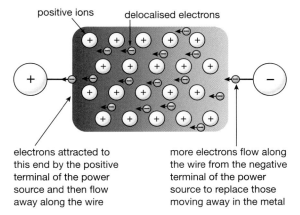

positive ions     delocalised electrons

electrons attracted to
this end by the positive
terminal of the power
source and then flow
away along the wire

more electrons flow along
the wire from the negative
terminal of the power
source to replace those
moving away in the metal

▲ Figure 9.5 How metals conduct electricity.

Metals can be hammered into different shapes; this is what the word
*malleable* means. When we apply a force to a piece of metal *the layers of
positive ions slide over each other.* This does not affect the bonding in the
structure; the positive ions are still attracted to the delocalised electrons.

▲ Figure 9.6 Metals are malleable because the layers of positive ions can slide over each other without changing the bonding.

Metals can also be described as **ductile**. This means that they can be drawn out into wires. The explanation is the same as why they are malleable.

## CHAPTER QUESTIONS

**SKILLS** INTERPRETATION

1 Draw simple 2D diagrams to show the structure of:

  a potassium metal

  b calcium metal

**SKILLS** CRITICAL THINKING ⑧

**SKILLS** REASONING ⑩

2 a A solid metal is often described as having 'a lattice of positive ions in a sea of delocalised electrons'. State the electronic configuration of a magnesium atom and use it to explain what this phrase means.

  b Metallic bonds are not fully broken until the metal has first melted and then boiled. Sodium, magnesium and aluminium are three consecutive elements in the Periodic Table. The boiling points of sodium, magnesium and aluminium are 890 °C, 1110 °C and 2470 °C, respectively. Explain these values in terms of the electronic configurations of the elements and metallic bonding.

⑦

  c Explain why all these elements are good conductors of electricity.

  d Explain why all these metals are malleable.

3 This question uses ideas from Chapters 7, 8 and 9. In these chapters you have met the following types of structure and bonding:

  giant metallic structure
  giant covalent structure
  giant ionic structure
  covalent molecular structure

Some information about some substances is given below. In each case state what type of structure and bonding it has.

  a Substance A melts at 2300 °C. It doesn't conduct electricity even when it is molten. It is insoluble in water.

⑧

  b Substance B is a colourless gas.

⑨

  c Substance C is a yellow solid with a low melting point of 113 °C. It doesn't conduct electricity and it is insoluble in water.

  d Substance D forms brittle orange crystals which melt at 398 °C. It dissolves freely in water to give an orange solution.

  e Substance E is a pink-brown flexible solid. It conducts electricity.

  f Substance F is a liquid with a boiling point of 80 °C. It is insoluble in water.

  g Substance G is a silvery solid which melts at 660 °C. It is used in overhead power cables.

**END OF CHEMISTRY ONLY**

# 10 ELECTROLYSIS

In this chapter we will look at what happens when we pass electricity through various substances.

Some things don't change when we pass electricity through them. For example, if you attach a piece of copper to a battery and make a complete circuit you may notice that the copper gets hot, but apart from that there is no other change. However, when you pass electricity through a solution of potassium iodide, a chemical change occurs, and hydrogen gas and iodine are formed. We will learn about why this happens in this chapter.

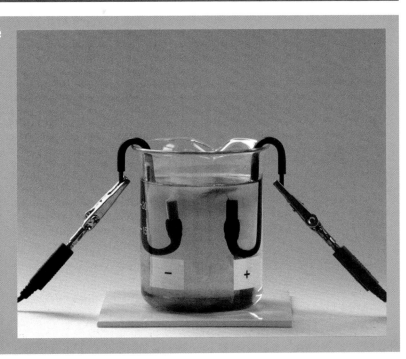

▶ Figure 10.1 This photograph shows what happens if you connect a solution of potassium iodide into a simple electrical circuit. If you look at what is happening in the solution, you can see obvious signs of chemical change. Some coloured substance is being produced at the positive electrode and a gas is being given off at the negative electrode.

## LEARNING OBJECTIVES

- Understand why covalent compounds do not conduct electricity
- Understand why ionic compounds conduct electricity only when molten or in aqueous solution
- Know that anion and cation are terms used to refer to negative and positive ions, respectively
- Describe experiments to investigate electrolysis, using inert electrodes, of molten compounds (including lead(II) bromide) and aqueous solutions (including

sodium chloride, dilute sulfuric acid and copper(II) sulfate) and to predict the products

- Write ionic half-equations representing the reactions at the electrodes during electrolysis and understand why these reactions are classified as oxidation or reduction
- Practical: Investigate the electrolysis of aqueous solutions

## WHY THINGS CONDUCT ELECTRICITY

Before we continue, we need to remind ourselves about why things do or don't conduct electricity. In order for things to conduct electricity, there must be charged particles present and these charged particles must be free to move. The charged particles will be either electrons or ions; it is important that you are clear which one you are talking about.

### METALS

If you remember, the structure of a metal is made up of a lattice of positive ions surrounded by a sea of delocalised electrons. Metals conduct electricity because the delocalised electrons are free to move.

## IONIC COMPOUNDS

These are compounds such as sodium chloride and potassium iodide. Ionic compounds don't conduct electricity when they are solid because the ions are held tightly in position in the lattice – they are not free to move around (they can only vibrate). They do, however, conduct electricity when they are molten (have melted) or if they are dissolved in water (in aqueous solution). This happens because the ions then become free to move around.

Remember that ionic compounds are made up of positive ions and negative ions:

- anions are negative ions
- cations are positive ions.

We will look at where these names come from later in the chapter.

## COVALENT COMPOUNDS

Covalent molecular compounds are substances such as hexane ($C_6H_{14}$), methane ($CH_4$) and carbon dioxide ($CO_2$). These do not conduct electricity in any state or in solution. Covalent molecular compounds consist of individual molecules. These molecules don't have any overall electrical charge, so there are no charged particles to move around. Also, all the electrons are held tightly in the atoms or in covalent bonds and so they are not able to move from molecule to molecule.

### REMINDER

The arrow here shows that this is a reversible reaction. Reversible reactions are discussed in Chapter 21.

There are some exceptions to this, such as covalent compounds that form ions as they react with water, for example ammonia:

$$NH_3(g) + H_2O(l) \rightleftharpoons NH_4^+(aq) + OH^-(aq)$$

Ammonia solution conducts electricity because there are ions which are free to move.

Hydrogen chloride gas dissolves in water to form hydrochloric acid (HCl(aq)). Hydrogen chloride ionises in water:

$$HCl(aq) \rightarrow H^+(aq) + Cl^-(aq)$$

## PASSING ELECTRICITY THROUGH COMPOUNDS: ELECTROLYSIS

When metals conduct electricity you will not notice anything happening, except perhaps that the metal gets hotter. When you pass electricity through an ionic compound, either molten or in solution, a chemical reaction occurs.

**Electrolysis** is a chemical change caused by passing an electric current through a compound which is either molten or in solution.

## SOME OTHER IMPORTANT WORDS

An **electrolyte** is a liquid or solution that undergoes electrolysis. Electrolytes all contain ions. The movement of the ions is responsible for both the conduction of electricity and the chemical changes that take place.

The electricity is passed into and out of the electrolyte through two **electrodes**. Carbon is frequently used for electrodes because it conducts electricity and is chemically fairly **inert** (this means that it does not react with things). Platinum is also fairly inert and can be used instead of carbon. Various other metals are sometimes used as well.

### HINT

Remember **PANiC**: positive anode, negative (is) cathode.

The positive electrode is called the **anode**. The negative electrode is called the **cathode**.

## THE ELECTROLYSIS OF MOLTEN COMPOUNDS

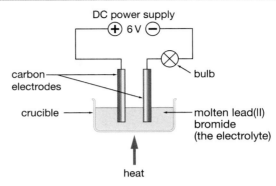

▲ Figure 10.2 Electrolysing molten lead(II) bromide

**KEY POINT**

The power supply can be a 6 volt battery or a power pack. It doesn't matter which. The voltage isn't critical either.

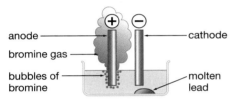

▲ Figure 10.3 What happens when the lead(II) bromide melts and electricity passes through it (the bulb also lights up).

Nothing at all happens until the lead(II) bromide melts. Then:

- the bulb lights up, showing that electrons are flowing through it
- there is bubbling around the electrode (the anode) connected to the positive terminal of the power source as brown bromine gas is given off
- nothing seems to be happening at the electrode (the cathode) connected to the negative terminal of the power source, but afterwards metallic lead is found underneath it
- when you stop heating and the lead(II) bromide solidifies again, everything stops, there is no more bubbling and the bulb goes out.

**EXPLAINING WHAT IS HAPPENING**

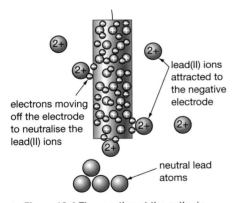

▲ Figure 10.4 The reaction at the cathode

Lead(II) bromide is an ionic compound. The solid consists of a giant structure of lead(II) ions (Pb$^{2+}$) and bromide ions (Br$^-$) packed regularly in a crystal lattice. The ions are locked tightly in the lattice and aren't free to move. The solid lead(II) bromide doesn't conduct electricity. As soon as the solid melts, the ions become free to move around.

As soon as you connect the power source in Figure 10.2, it pumps any mobile electrons away from the left-hand electrode towards the right-hand one. This means that there are extra electrons at the right-hand electrode, so it is negative. The left-hand electrode is positive because some of the electrons have been removed from it. The positive lead(II) ions in the molten lead(II) bromide are attracted to the negative electrode, the cathode. When they get there, each lead(II) ion picks up two electrons from the electrode and forms neutral lead atoms (Figure 10.4). These fall to the bottom of the container as molten lead. This can be represented by a **half-equation**.

$$Pb^{2+} + 2e^- \rightarrow Pb$$

The power source pumps new electrons along the wire to replace the electrons that have been removed from the cathode.

Bromide ions are attracted to the positive electrode, the anode (Figure 10.5). When they get to the positive electrode, the extra electron which makes the bromide ion negatively charged moves onto the electrode. This is because this electrode is short of electrons.

The loss of the extra electron converts each bromide ion into a bromine atom:

$$Br^- \rightarrow Br + e^-$$

**KEY POINT**

Half-equations show either oxidation or reduction reactions (see below). Electrons are shown as e$^-$ in half-equations.

These join in pairs to make bromine molecules:

$$2Br \rightarrow Br_2$$

Overall:

$$2Br^- \rightarrow Br_2 + 2e^-$$

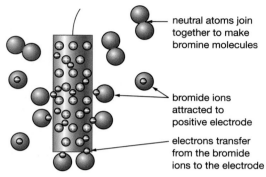

neutral atoms join together to make bromine molecules

bromide ions attracted to positive electrode

electrons transfer from the bromide ions to the electrode

▲ Figure 10.5 The reaction at the anode

The new electrons on the electrode flow back into the power source. Because electrons are flowing in the external circuit, the bulb lights up.

We sometimes talk about ions being **discharged** at the electrodes. Discharging an ion means that it loses its charge. This happens either by giving up electron(s) to the electrode or receiving electron(s) from it. We can therefore say that bromide ions and lead(II) ions were discharged at the electrodes.

## ELECTROLYSIS AND REDOX

You will learn more about **oxidation** and **reduction** reactions in Chapter 14. Oxidation and reduction are words used to describe what is happening to things in certain chemical reactions. We can define oxidation and reduction as follows:

■ *Oxidation occurs when something loses electrons.*

■ *Reduction occurs when something gains electrons.*

We usually simply shorten this to oxidation is loss of electrons and reduction is gain of electrons. A way of remembering this is the mnemonic OILRIG.

If we look again at the electrode equations in the electrolysis of lead(II) bromide, we see that the lead(II) ions gain electrons at the cathode:

$$Pb^{2+} + 2e^- \rightarrow Pb$$

Gain of electrons is reduction. The lead(II) ions are reduced to lead atoms.

The bromide ions lose electrons at the anode:

$$2Br^- \rightarrow Br_2 + 2e^-$$

Loss of electrons is oxidation. Bromide ions are oxidised to bromine molecules.

If something loses electrons something else must gain electrons and so oxidation and reduction always occur at the same time: we talk about **redox** reactions, **red**(uction)**ox**(idation). In the reactions going on at the electrodes each equation only shows one of the processes occurring, either oxidation or reduction, and so we call these *half-equations*. Both these reactions involve ions and so, in the exam, you may be asked to write *ionic half-equations* representing the reactions at the electrodes.

## THE ELECTROLYSIS OF OTHER MOLTEN SUBSTANCES

If we carry out electrolysis of molten sodium chloride, we get sodium at the cathode (negative electrode) and chlorine at the anode (positive electrode). The ionic half-equations are:

cathode:   $Na^+ + e^- \rightarrow Na$          reduction

Sodium ions are reduced to sodium atoms.

anode:     $2Cl^- \rightarrow Cl_2 + 2e^-$          oxidation

Chloride ions are oxidised to chlorine molecules.

These half-equations must always balance in terms of the number of atoms on each side, but also in terms of the charges; the total charge must be the same on both sides. This is why we need 2 electrons in the second half-equation but only 1 in the first.

Let us look at another example, the electrolysis of molten aluminium oxide ($Al_2O_3$). We get aluminium at the cathode and oxygen at the anode. Aluminium is in Group 3 in the Periodic Table and so an aluminium atom has 3 electrons in its outer shell. You may remember from Chapter 7 that this means that Al forms a 3+ ion. This 3+ ion will be attracted to the negative electrode in electrolysis and we will get the reaction:

$Al^{3+} + 3e^- \rightarrow Al$

In order to cancel out the 3+ charge on the $Al^{3+}$ ion we need to add 3 electrons. The total charge on the left-hand side is zero (**3+** + **3−**), which is the same as on the right-hand side. Another way to think about this is that if an aluminium atom loses 3 electrons to form an $Al^{3+}$ ion, we must put those electrons back in order to form the atom again. This is a reduction reaction because the $Al^{3+}$ ion gains electrons.

When we electrolyse molten aluminium oxide, the half-equation for the reaction at the anode is:

$2O^{2-} \rightarrow O_2 + 4e^-$

Each $O^{2-}$ ion has 2 'extra' electrons and oxygen atoms go around in pairs, so we must remove 4 electrons to form $O_2$. These are shown on the right-hand side of the half-equation. There are two Os on each side and the total charge on each side is 4−. This is an oxidation reaction.

When molten zinc(II) chloride is electrolysed, zinc is obtained at the cathode and chlorine at the anode:

cathode:   $Zn^{2+} + 2e^- \rightarrow Zn$   reduction

anode:     $2Cl^- \rightarrow Cl_2 + 2e^-$   oxidation

We can make the following generalisations from the reactions above:

- If you electrolyse a molten ionic compound only containing two elements, you will get the metal at the cathode (because metals form positive ions) and the non-metal at the anode (because non-metals form negative ions).

- Reduction always occurs at the cathode and oxidation always occurs at the anode.

You can probably see now that positive ions are known as cations because they are attracted to the cathode (negative electrode). Negative ions are known as anions because they are attracted to the anode (positive electrode).

### HINT

A common mistake is to put the electrons on the wrong side of the half-equation. Check the charges to make sure you have the same total charge on both sides.

### HINT

A way of remembering this is AN OX RED CAT or AN OILRIG CAT.

Not all ionic compounds can be electrolysed when they are molten. Some break up into simpler substances before their melting point. For example, copper(II) carbonate breaks into copper(II) oxide and carbon dioxide, even on gentle heating. It is impossible to melt it.

## THE ELECTROLYSIS OF AQUEOUS SOLUTIONS

When aqueous solutions are electrolysed the products are not always the same as when molten salts are electrolysed.

**!** Safety Note: Wear eye protection. Do not smell the chlorine, especially if you have asthma. Once chlorine is detected the current must be switched off.

## ACTIVITY 5

### ▼ PRACTICAL: INVESTIGATING THE ELECTROLYSIS OF AQUEOUS SOLUTIONS

We can investigate the electrolysis of an aqueous solution such as sodium chloride solution using the apparatus shown in Figure 10.6.

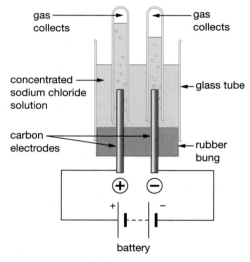

▲ Figure 10.6 Electrolysis of sodium chloride solution

The following procedure is used:

■ Set up the apparatus as shown in Figure 10.6. The glass tube, rubber bung and electrodes together are sometimes called an **electrolytic cell**.

■ Pour concentrated sodium chloride solution into the glass tube.

■ Place a test-tube containing sodium chloride solution over each electrode. The test-tubes must not completely cover the electrodes or ions will be unable to flow and there will be no current.

■ Connect the battery/powerpack to the electrodes.

■ The experiment should be done in a fume cupboard (fume hood) or well-ventilated room because chlorine gas is poisonous.

We can see if something is happening by looking for bubbles of gas or a metal forming at the electrodes. Any gases can be tested.

The tests for gases are covered in Chapter 18.

In this experiment we see bubbles of gas at both electrodes. When the gases are tested we find that hydrogen forms at the negative electrode (cathode) and chlorine forms at the positive electrode (anode).

## KEY POINT

We would expect to get the same volume of hydrogen as chlorine. However, in reality we appear to obtain less chlorine than expected because it is more soluble in water.

## THE ELECTROLYSIS OF SODIUM CHLORIDE SOLUTION

When molten sodium chloride is electrolysed, the products at the electrodes are:

anode:    chlorine

cathode:  sodium

When you electrolyse sodium chloride solution you do not get the same products as when you electrolyse molten sodium chloride. Although chlorine is still formed at the anode, hydrogen is produced at the cathode rather than sodium. The hydrogen at the cathode comes from the water.

Water is called a *weak electrolyte*. It ionises very slightly to give hydrogen ions and hydroxide ions:

$$H_2O(l) \rightleftharpoons H^+(aq) + OH^-(aq)$$

Whenever you have water present, you have to consider these ions *as well as* the ions in the compound you are electrolysing.

**KEY POINT**

The reversible sign shows that as water molecules break up to form hydrogen ions and hydroxide ions, these ions are recombining to make water again.

## AT THE CATHODE

**KEY POINT**

The more reactive something is, the greater tendency it has to form an ion. This means that the more reactive something is, the more difficult it is to turn it back into an atom. For positive ions, the lower the position of an element in the reactivity series, the more easily it will accept an electron. The reactivity series is discussed in Chapter 14.

The solution contains $Na^+(aq)$ and $H^+(aq)$, and these are both attracted to the negative electrode (cathode). However, sodium is a very reactive metal. This means that it is very difficult to add an electron to a sodium ion to convert it back to a sodium atom. Hydrogen is less reactive than sodium so it is easier to add an electron to a hydrogen ion to form a hydrogen atom. Each hydrogen atom formed combines with another one to make a hydrogen molecule:

$$2H^+(aq) + 2e^- \rightarrow H_2(g)$$

Remember that the hydrogen ions come from water molecules splitting up. Each time a water molecule ionises, it also produces a hydroxide ion. There is a build-up of these in the solution around the cathode. These hydroxide ions make the solution alkaline in the region around the cathode.

**KEY POINT**

Because the hydrogen ions are discharged (removed from the solution as hydrogen gas), they can no longer react with the hydroxide ions and reform water. The ionisation of the water becomes a one-way process.

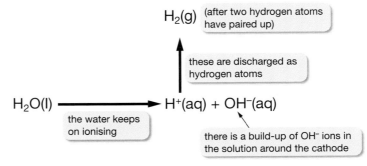

▲ Figure 10.7 The process of electrolysing solutions

**HINT**

You might come across either half-equation in the exam. Either should be accepted in answers.

There is an alternative way of looking at this cathode reaction, starting from neutral water molecules, which can be thought of as taking electrons directly from the cathode:

$$2H_2O(l) + 2e^- \rightarrow H_2(g) + 2OH^-(aq)$$

You can see more easily why the solution becomes alkaline using this half-equation as $OH^-(aq)$ ions are produced.

## AT THE ANODE

$Cl^-(aq)$ and $OH^-(aq)$ are both attracted by the positive anode. It is slightly easier to remove electrons from (oxidise) the hydroxide ion than from the chloride ion, but there isn't much difference. There are, however, many, many more chloride ions present in the solution, and so it is mainly these that are oxidised at the anode:

$$2Cl^-(aq) \rightarrow Cl_2(g) + 2e^-$$

## THE REMAINING SOLUTION

If the electrolysis is carried on for a long time, we can work out what the final solution remaining at the end will be. The ions in the solution were:

from NaCl:    $Na^+$        $\cancel{Cl^-}$

from $H_2O$:    $\cancel{H^+}$        $OH^-$

$Cl^-$ and $H^+$ ions were removed from the solution by being discharged at the electrodes, so we are left with $Na^+$ and $OH^-$, sodium hydroxide solution.

# THE ELECTROLYSIS OF COPPER(II) SULFATE SOLUTION USING INERT ELECTRODES

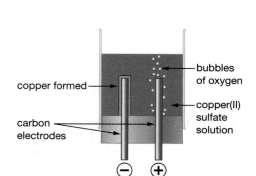

The copper(II) ions and hydrogen ions (from the water) will be attracted to the cathode. Copper is below hydrogen in the reactivity series, which means that it is easier to add electrons to copper ions to form an atom. The cathode will get coated with pink-brown copper:

$$Cu^{2+}(aq) + 2e^- \rightarrow Cu(s) \qquad reduction$$

Sulfate ions and hydroxide ions (from the water) will be attracted to the anode. Sulfate ions aren't easy to oxidise. Instead, you get oxygen from the oxidation of hydroxide ions from the water:

$$4OH^-(aq) \rightarrow 2H_2O(l) + O_2(g) + 4e^- \qquad oxidation$$

There is an alternative way of looking at this anode reaction. The equation this time is:

$$2H_2O(l) \rightarrow O_2(g) + 4H^+(aq) + 4e^-$$

You can see more easily why the solution becomes acidic using this half-equation: $H^+(aq)$ ions are produced. You may come across either half-equation in the exam.

bubbles of oxygen

copper formed

copper(II) sulfate solution

carbon electrodes

$\ominus$  $\oplus$

▲ Figure 10.8 Electrolysis of copper(II) sulfate solution using inert electrodes produces copper at the cathode and oxygen gas at the anode.

If the electrolysis is continued for a long time the copper(II) ions will eventually all be used up, and so the colour of the solution will fade from blue to colourless. What is left in the solution? The ions originally present were:

from $CuSO_4$:    $\cancel{Cu^{2+}}$    $SO_4^{2-}$

from water:    $H^+$    $\cancel{OH^-}$

Copper ions and hydroxide ions are discharged at the electrodes. Hydrogen ions from the water aren't being discharged and neither are the sulfate ions. The solution turns into dilute sulfuric acid ($H_2SO_4$). The electrolysis will then continue as for dilute sulfuric acid (see below).

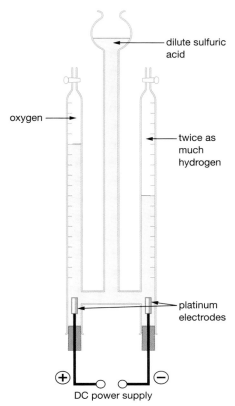

▲ Figure 10.9 Apparatus for electrolysing dilute sulfuric acid and measuring the volume of gases produced.

**DID YOU KNOW?**

This experiment could be used to show that the formula of water is $H_2O$.

## HOW TO WORK OUT WHAT WILL HAPPEN

### EXTENSION WORK

This leaves the problem of what you obtain if you have a moderately reactive metal such as zinc, for example. Reasonably concentrated solutions will give you the metal. Very dilute solutions will give you mainly hydrogen. In between, you will get both. At International GCSE you probably won't have to worry about this. The examples you will see in exams are always clear.

# THE ELECTROLYSIS OF DILUTE SULFURIC ACID USING INERT ELECTRODES

In this case, the only positive ions arriving at the cathode are hydrogen ions (from the acid and the water). These are discharged to give hydrogen gas:

$$2H^+(aq) + 2e^- \rightarrow H_2(g)$$

At the anode – as with copper(II) sulfate solution – sulfate ions and hydroxide ions (from the water) arrive. The sulfate ions are too difficult to oxidise, and so you obtain oxygen from the oxidation of hydroxide ions from the water:

$$4OH^-(aq) \rightarrow 2H_2O(l) + O_2(g) + 4e^-$$

Twice as much hydrogen is produced as oxygen. Look at the half-equations above. For every 4 electrons that flow around the circuit, you would get 1 molecule of oxygen. But 4 electrons would produce 2 molecules of hydrogen. You get twice the number of molecules of hydrogen as of oxygen. Twice the number of molecules occupy twice the volume.

Actually, when we do this experiment the amount of hydrogen we obtain is more than twice as much as the oxygen. This is because oxygen is more soluble in water than hydrogen. What we sometimes do to stop this happening is to carry out the electrolysis experiment for a few minutes first in order to saturate the water with oxygen and then start to collect the gases; this gives much better results.

# THE ELECTROLYSIS OF SOME OTHER SOLUTIONS USING INERT ELECTRODES

- If the metal is high in the reactivity series, you get hydrogen produced at the cathode instead of the metal.

- If the metal is below hydrogen in the reactivity series, you obtain the metal at the cathode.

- If you have solutions of halides (chlorides, bromides or iodides), you obtain the halogen (chlorine, bromine or iodine) at the anode. With other common negative ions (sulfate, nitrate, hydroxide), you obtain oxygen at the anode.

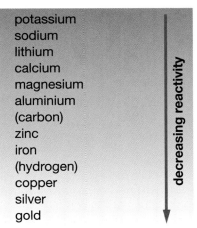

▲ Figure 10.10 The reactivity series shows metals in order of how reactive they are from most reactive to least reactive. This is discussed more in Chapter 14.

The table shows some simple examples of these rules.

Table 10.1 The electrolysis of solutions using inert (carbon or platinum) electrodes

| | Cathode (−) | | | Anode (+) | |
|---|---|---|---|---|---|
| Solution | Product | Half-equation | Product | Half-equation | |
| KI(aq) | hydrogen | $2H^+(aq) + 2e^- \rightarrow H_2(g)$ | iodine | $2I^-(aq) \rightarrow I_2(aq) + 2e^-$ | |
| MgBr$_2$(aq) | hydrogen | $2H^+(aq) + 2e^- \rightarrow H_2(g)$ | bromine | $2Br^-(aq) \rightarrow Br_2(aq) + 2e^-$ | |
| H$_2$SO$_4$(aq) | hydrogen | $2H^+(aq) + 2e^- \rightarrow H_2(g)$ | oxygen | $4OH^-(aq) \rightarrow 2H_2O(l) + O_2(g) + 4e^-$ | |
| CuSO$_4$(aq) | copper | $Cu^{2+}(aq) + 2e^- \rightarrow Cu(s)$ | oxygen | $4OH^-(aq) \rightarrow 2H_2O(l) + O_2(g) + 4e^-$ | |

## WHAT WOULD HAPPEN WITH NON-ELECTROLYTES?

For electrolysis to work, there have to be ions present. The current in the external circuit (with the bulb and power source) can flow only if there are ions which can move and be discharged.

If you tried to electrolyse a covalent compound (either molten or in solution), there wouldn't be a current flow because there aren't any ions. Nothing else would happen either. Sugar, for example, is a non-electrolyte; it doesn't undergo electrolysis. It won't conduct electricity, and won't be decomposed by it, either in solution or when molten.

Simple experiments like those described in this chapter give you an easy way of finding out whether a substance is ionic or not. If it undergoes electrolysis, either molten or in solution, it must contain ions. If it doesn't undergo electrolysis, it doesn't contain ions.

### KEY POINT

There are exceptions to this: covalent compounds that are electrolytes in solution. These include acids and ammonia solution.

**!**

Safety Note: Wear eye protection. Do not smell the chlorine, especially if you have asthma. Once chlorine is detected the current must be switched off.

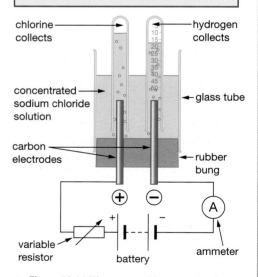

▲ Figure 10.11 We can use this apparatus for a quantitative electrolysis experiment.

## ACTIVITY 6

### ▼ PRACTICAL: QUANTITATIVE ELECTROLYSIS

Quantitative means related to numbers. We can investigate how *much* product we get in an electrolysis experiment. We could, for example, investigate the effect of changing the current on the amount of hydrogen produced at the cathode in the electrolysis of sodium chloride solution. To do this we would use similar apparatus to that in Figure 10.6, but we would also have to put a variable resistor and an ammeter into the circuit to allow us to vary and measure the current. We would also have to use a gas burette or measuring cylinder to measure the volume of gas produced. We would use the following procedure:

■ Set up the apparatus as shown in Figure 10.11.

■ Pour 50 cm³ of concentrated sodium chloride solution into the glass tube.

■ Place a gas burette (or measuring cylinder) filled with sodium chloride solution over the cathode.

■ Turn on the powerpack/connect the battery and set the current to 0.2 A using the variable resistor.

- Take an initial reading on the gas burette. (We have to start the current flowing so that we can see what the current is, therefore we only take a reading on the gas burette after we have turned on the power.)

- Start the timer.

- Stop the timer after 5 minutes and note the final reading on the gas burette.

- Repeat for currents of 0.4 A, 0.6 A, 0.8 A and 1.0 A.

- Repeat each experiment to get more reliable results.

A set of results for this experiment could be:

| Current/A | Volume of hydrogen gas produced/$cm^3$ |
| --- | --- |
| 0.20 | 7.0 |
| 0.40 | 13.9 |
| 0.60 | 15.1 |
| 0.80 | 28.0 |
| 1.0 | 34.9 |

We can plot this data on a graph (Figure 10.12).

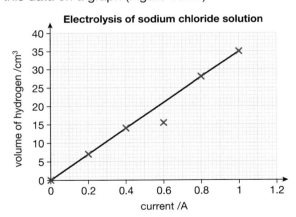

▲ Figure 10.12 How changing the current affects the volume of hydrogen produced at the cathode.

Because the current is continuous data, we draw a line of best fit through the points. The reading at 0.6 A is an anomalous point and we do not include this when drawing our line of best fit. The reading at 0.6 A is too low and could have occurred because the current that we used was too low or some gas escaped.

The line of best fit is a straight line that goes through the origin and so we can say that we have a proportional relationship: the volume of gas produced is directly proportional to the current passed through the solution.

## CHAPTER QUESTIONS

SKILLS ► REASONING   

1 State what is formed at the cathode and at the anode during the electrolysis of the following substances. Assume that carbon electrodes were used each time.

  a   Molten lead(II) bromide

  b   Molten zinc chloride

  c   Sodium iodide solution

  d   Molten sodium iodide

  e   Copper(II) chloride solution

  f   Dilute hydrochloric acid

  g   Magnesium sulfate solution

  h   Sodium hydroxide solution

SKILLS ► PROBLEM SOLVING

2 Copy and complete the following half-equations for reaction at the anode or cathode, and state whether each involves oxidation or reduction. Electrons have been put in for the first four but not for the ones after that.

  a   $Mg^{2+} + e^- \rightarrow Mg$

  b   $Al^{3+} + e^- \rightarrow Al$

  c   $2Br^- \rightarrow Br_2 + e^-$

  d   $O^{2-} \rightarrow O_2 + e^-$

  e   $Cl^- \rightarrow Cl_2$

  f   $Ni^{2+} \rightarrow Ni$

  g   $OH^- \rightarrow O_2 + H_2O$

  h   $H_2O \rightarrow O_2 + H^+$

  i   $H_2O \rightarrow H_2 + OH^-$

SKILLS ► REASONING

3 Some solid potassium iodide was placed in an evaporating basin. Two carbon electrodes were inserted and connected to a 12 volt DC power source and a light bulb. The potassium iodide was heated. As soon as the potassium iodide was molten, the bulb came on. Purple fumes were seen coming from the positive electrode, and lilac flashes were seen around the negative one.

  a   Explain why the bulb didn't come on until the potassium iodide melted.

SKILLS ► CRITICAL THINKING   

  b   State the name of the positive electrode.

  c   Name the purple fumes seen at the positive electrode, and write the ionic half-equation for their formation.

  d   The lilac flashes seen around the negative electrode are caused by the potassium which is formed. The potassium burns with a lilac flame. Write the ionic half-equation for the formation of the potassium.

  e   State the products formed at the electrodes if molten sodium bromide is electrolysed instead of molten potassium iodide.

  f   Write the ionic half-equations for the reactions occurring during the electrolysis of molten sodium bromide.

4  For electrolysis of each of the following
   i   write the ionic half-equation for the reaction occurring at the cathode
   ii  write the ionic half-equation for the reaction occurring at the anode
   iii state what has been oxidised and what has been reduced.

a  Molten lead(II) bromide using carbon electrodes

b  Sodium chloride solution using carbon electrodes

c  Calcium bromide solution using carbon electrodes

d  Copper(II) sulfate solution using platinum electrodes

e  Aluminium nitrate solution using carbon electrodes

f  Molten magnesium iodide using carbon electrodes

g  Dilute hydrochloric acid using platinum electrodes

5  You are asked to find out whether two compounds, S and T, are electrolytes
   or non-electrolytes. S melts at 1261 °C and is soluble in water. T melts at
   265 °C and is insoluble in water. Describe, with the aid of diagrams, how
   you would find out if each of these substances was an electrolyte or a
   non-electrolyte. In each case say what you would look for to help you to
   decide.

6  When copper(II) sulfate solution is electrolysed using copper electrodes the
   reaction at the cathode is the same as with inert electrodes but no oxygen is
   given off at the cathode. Instead the anode gets smaller as copper ions go
   into solution. The half-equation for the reaction at the anode is:

   $$Cu(s) \rightarrow Cu^{2+}(aq) + 2e^-$$

a  Write the half-equation for the reaction at the cathode.

b  When copper(II) sulfate solution is electrolysed using inert electrodes the
   blue colour of the solution fades and the solution becomes more acidic.

   i   Explain these observations.

   ii  Predict and explain what happens to the colour and acidity of the
       solution when copper(II) sulfate solution is electrolysed with copper
       electrodes.

**END OF CHEMISTRY ONLY**

# UNIT QUESTIONS

You may need to refer to the Periodic Table on page 320.

**SKILLS** CRITICAL THINKING ④   **1**   Hydrogen is the most common element in the universe.

The melting point and boiling point of hydrogen are shown in the table:

| Melting point/°C | −259 |
|---|---|
| Boiling point/°C | −253 |

a  Put a cross in the box to show a temperature at which hydrogen is a liquid. **(1)**

−265 °C ☐     −260 °C ☐     −255 °C ☐     −250 °C ☐

**SKILLS** INTERPRETATION ⑥   b  The circle in the diagram represents 1 molecule of hydrogen.
Complete the diagram to show the arrangement of particles in liquid hydrogen. You should add at least 10 more circles to the diagram. **(2)**

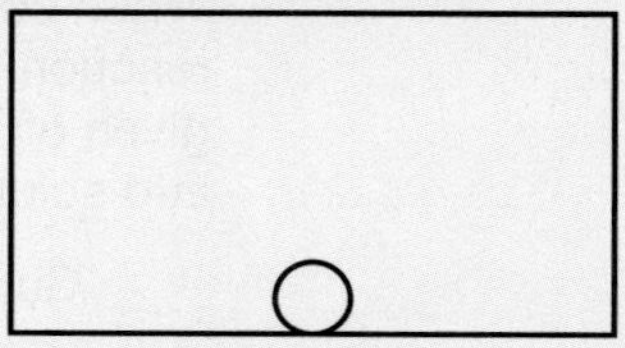

**SKILLS** PROBLEM SOLVING ⑧   c  Hydrogen has three isotopes, the least common of which is tritium, $^3$H.
Use the Periodic Table to work out the number of protons, neutrons and electrons in an atom of tritium. **(3)**

Number of protons:

Number of neutrons:

Number of electrons:

d  Hydrogen reacts with nitrogen to form ammonia ($NH_3$).

⑦   i  The electronic configuration of a nitrogen atom is (put a cross in one box): **(1)**

2, 2, 3 ☐     2, 5 ☐     2, 8, 4 ☐     2, 2, 8, 2 ☐

**SKILLS** CRITICAL THINKING ⑧   ii  Draw a dot-and-cross diagram, showing outer electrons only, to show the bonding in an ammonia molecule. **(2)**

⑥   e  Ammonia forms salts that contain the ammonium ion. Put a cross in the box to show which of the following formulae for ammonium compounds is incorrect.

$NH_4SO_4$ ☐     $NH_4NO_3$ ☐     $NH_4Cl$ ☐     $(NH_4)_2CO_3$ ☐

**5**

f   Two more compounds that contain hydrogen are water and ethanol.
Put a cross in the box to show a method that can be used to separate a
mixture of ethanol and water.      **(1)**

crystallisation ☐      simple distillation ☐

filtration ☐      fractional distillation ☐

**(Total 10 marks)**

**SKILLS** ▸ CRITICAL THINKING   **6**   **2**

a   Complete the following passage by using the words below.
You may use the words once, more than once, or not at all.      **(4)**

mass number      atomic number      groups      periods
protons      electrons      outer shell      nucleus

The elements in the Periodic Table are arranged in order of _____.

The vertical columns are called _____ and contain elements

which have the same number of _____ in their _____.

**5**

b   Put a cross next to the symbol(s) of elements that are non-metals.      **(2)**

H ☐      V ☐      Pb ☐      Ar ☐      W ☐

**SKILLS** ▸ REASONING   **6**

c   Silicon is an element that shows some of the properties of metals and non-
metals. It is sometimes called a *metalloid*.

i   Two properties of silicon are:

▪ silicon conducts electricity

▪ silicon forms an oxide ($SiO_2$) that reacts with hot concentrated sodium
hydroxide.

Explain which of these is not a property of metals.      **(2)**

**SKILLS** ▸ ANALYSIS, REASONING   **8**

ii   The structure of silicon dioxide ($SiO_2$) is shown in the diagram. This
structure continues in all directions.

Explain in terms of electronic configurations and bonding whether the grey
or red circles in this diagram represent silicon atoms.      **(3)**

iii   By referring to the diagram in part **ii** explain whether you would expect
$SiO_2$ to be a solid, liquid or gas at room temperature.      **(2)**

**(Total 13 marks)**

**SKILLS**   INTERPRETATION, REASONING    **7**   **3**

Strontium is an element in Group 2 of the Periodic Table and bromine is an element in Group 7.

a   Explain what happens in terms of electrons when an atom of strontium combines with an atom of bromine to form strontium bromide. You may draw a diagram to illustrate your answer.    **(3)**

b   Explain in terms of structure and bonding whether you would expect strontium bromide to have a high or a low melting point.    **(3)**

c   Explain what you understand by the term 'relative atomic mass' of an element.    **(2)**

**SKILLS**   PROBLEM SOLVING    **10**

d   The natural abundances of the two isotopes of bromine are:

$^{79}Br$    50.69%

$^{81}Br$    49.31%

Calculate the relative atomic mass of bromine. Give your answer to 2 decimal places.    **(2)**

**SKILLS**   CRITICAL THINKING    **6**

e   The formula of strontium nitrate is (put a cross in one box):    **(1)**

$SrNO_3$ ☐     $Sr(NO_3)_2$ ☐     $SrNO_{32}$ ☐     $Sr_3N_2$ ☐

**(Total 11 marks)**

**SKILLS**   REASONING    **9**   **4**

The diagram represents the structure of diamond, a form of carbon.

a   Explain in terms of structure and bonding why diamond is very hard.    **(2)**

b   Two more forms of carbon are graphite and $C_{60}$ fullerene. Part of the structure of graphite and a molecule of $C_{60}$ fullerene are shown in the diagram.

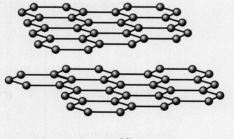 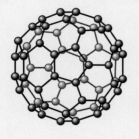

graphite               $C_{60}$ fullerene

i   Explain whether graphite or $C_{60}$ fullerene has the higher boiling point.    **(4)**

ii   Explain why graphite conducts electricity.    **(2)**

iii   Explain why $C_{60}$ fullerene does not conduct electricity.    **(2)**

**(Total 10 marks)**

**SKILLS** ANALYSIS  **5**

a Oxygen gas has the formula $O_2$. Which of the following is a correct dot-and-cross diagram showing the bonding in $O_2$ (outer shells of electrons only are shown). Circle one of the diagrams. **(1)**

**SKILLS** CRITICAL THINKING

b Oxygen reacts with potassium to form potassium oxide. Put a cross in one box to indicate what the formula of potassium oxide is. **(1)**

PO ☐     KO ☐     $K_2O$ ☐     $KO_2$ ☐

**SKILLS** ANALYSIS

c Oxygen reacts with carbon to form carbon dioxide. Which diagram shows the covalent bonds in carbon dioxide? Circle one diagram. **(1)**

O=C—O

O=C=O

O=C≡O

O—C—O

**(Total 3 marks)**

**SKILLS** PROBLEM SOLVING  **6**

In an experiment to find the empirical formula of lead oxide, a small porcelain dish was weighed, filled with lead oxide and weighed again. The dish was placed in a tube and heated in a stream of hydrogen. The hydrogen reduced the lead oxide to a bead of metallic lead. When the apparatus was cool, the dish and its bead of lead were weighed together.

Mass of porcelain dish = 17.95 g

Mass of porcelain dish + lead oxide = 24.80 g

Mass of porcelain dish + lead = 24.16 g

($A_r$: O = 16, Pb = 207)

a Calculate the mass of lead in the lead oxide. **(1)**

b Calculate the mass of oxygen in the lead oxide. **(1)**

c There are three different oxides of lead: $PbO$, $PbO_2$ and $Pb_3O_4$. Use your results from a and b to find the empirical formula of the oxide used in the experiment. **(3)**

**(Total 5 marks)**

**7** Copper reacts with concentrated nitric acid to produce copper(II) nitrate solution and nitrogen dioxide gas.

$$Cu(s) + 4HNO_3(aq) \rightarrow Cu(NO_3)_2(aq) + 2NO_2(g) + 2H_2O(l)$$

A student carried out an experiment to make some copper(II) nitrate crystals.

a The student started with 2.00 g of copper and added excess nitric acid. Calculate the maximum mass of copper(II) nitrate, $Cu(NO_3)_2$, which could be obtained. ($A_r$: N = 14, O = 16, Cu = 63.5) **(3)**

  b Explain the method the student could use to obtain crystals of copper(II) nitrate from the copper nitrate solution. **(3)**

  c The copper nitrate crystallises out of the solution as $Cu(NO_3)_2 \cdot 3H_2O$. The student did some calculations and worked out that he should make 7.61 g of crystals. He actually only made 5.23 g. Calculate the percentage yield of the student's experiment. **(2)**

**(Total 8 marks)**

## CHEMISTRY ONLY

  **8** If pyrite ($FeS_2$) is heated strongly in air it reacts according to the equation:

$$4FeS_2(s) + 11O_2(g) \rightarrow 2Fe_2O_3(s) + 8SO_2(g)$$

Iron can be extracted from the iron(III) oxide produced, and the sulfur dioxide can be converted into sulfuric acid. In this question you can assume that the molar volume of any gas at rtp is $24\,dm^3$.

a Calculate the mass of iron(III) oxide that can be obtained from 480 kg of pure pyrite. ($A_r$: O = 16, S = 32, Fe = 56) **(3)**

b Calculate the volume of sulfur dioxide (measured at rtp) produced from 480 kg of pyrite. **(2)**

c The next stage of the manufacture of sulfuric acid is to convert the sulfur dioxide into sulfur trioxide ($SO_3$) by reacting it with oxygen.

   i Write a balanced equation for this reaction. **(2)**

   ii Calculate the volume of oxygen (measured at rtp) needed for the complete conversion of the sulfur dioxide produced in b into sulfur trioxide. **(2)**

**(Total 9 marks)**

  **9** Strontium hydroxide, $Sr(OH)_2$, is only sparingly soluble in water at room temperature. In an experiment to measure its solubility, a student made a saturated solution of strontium hydroxide. She pipetted $25.0\,cm^3$ of this solution into a conical flask, added a few drops of methyl orange indicator and then titrated it with $0.100\,mol/dm^3$ hydrochloric acid from a burette. She needed to add $32.8\,cm^3$ of the acid to neutralise the strontium hydroxide.

$$Sr(OH)_2(aq) + 2HCl(aq) \rightarrow SrCl_2(aq) + 2H_2O(l)$$

a Explain what is meant by a *saturated solution*. **(1)**

b Calculate the number of moles of HCl in $32.8\,cm^3$ of $0.100\,mol/dm^3$ hydrochloric acid. **(1)**

c Calculate the concentration of the strontium hydroxide in $mol/dm^3$. **(3)**

d The solubility of a substance is usually measured in units of g per 100 g of water. The mass of 1 dm³ of water is 1000 g. Use the value that you obtained in c to calculate the solubility of strontium hydroxide as g of strontium hydroxide per 100 g of water. ($A_r$: H = 1, O = 16, Sr = 88) **(2)**

SKILLS ► EXECUTIVE FUNCTION 7

e The calculation used in d gives only an approximate value for the solubility because the volume changes when a substance is dissolved in water. Describe an experiment that the student could carry out to determine a more accurate value for how much strontium hydroxide dissolves in 100 g of water at this temperature. You can assume that you are given 50 cm³ of saturated strontium hydroxide solution. You should include a description of the apparatus you would use and the measurements you would make. **(5)**

**(Total 12 marks)**

SKILLS ► PROBLEM SOLVING 8 **10** A student carried out some experiments using dilute sulfuric acid.

a He found that 25.0 cm³ of 0.100 mol/dm³ sodium hydroxide solution was neutralised by 20.0 cm³ of the dilute sulfuric acid.

The equation for the reaction is

$$2NaOH(aq) + H_2SO_4(aq) \rightarrow Na_2SO_4(aq) + 2H_2O(l)$$

Calculate the concentration of the sulfuric acid in mol/dm³. **(4)**

 b 100 cm³ of this same sulfuric acid was reacted with magnesium.

$$Mg(s) + H_2SO_4(aq) \rightarrow MgSO_4(aq) + H_2(g)$$

i Magnesium is a metal. Describe the structure and bonding in a piece of magnesium. **(3)**

ii Magnesium is malleable. Explain in terms of structure and bonding why magnesium is malleable. **(2)**

iii Calculate the amount in moles of sulfuric acid that was used. **(1)**

iv The student used 0.100 g of magnesium in the experiment. Work out whether the sulfuric acid or the magnesium was in excess. **(2)**

v Calculate the volume of hydrogen gas that would be produced at rtp in this experiment. (Assume that the molar volume at rtp of hydrogen is 24 000 cm³.) **(2)**

vi When the student did the experiment they collected 94 cm³ of gas. Calculate the percentage yield of the experiment. **(1)**

c The student electrolysed the solution of magnesium sulfate.

i State the name of the product formed at the cathode. **(1)**

ii Oxygen gas is formed at the anode. Write an ionic half-equation for the formation of oxygen at the anode. **(2)**

**(Total 18 marks)**

SKILLS CRITICAL THINKING  **11**

A class of 11 students carried out an experiment to determine how much calcium carbonate ($CaCO_3$) was in a sample of limestone. Each student was given a different mass of limestone, which they added to 50 cm³ of dilute hydrochloric acid and then measured the volume of carbon dioxide produced using the apparatus shown below.

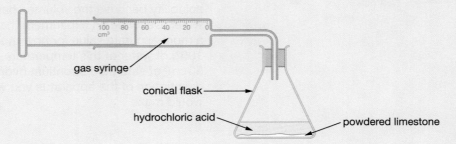

gas syringe

conical flask

hydrochloric acid

powdered limestone

a Name a piece of apparatus that could be used to measure out 50 cm³ of dilute hydrochloric acid. **(1)**

b The diagram shows the reading from the experiment carried out by Student 2.

Complete the table with the volume of gas collected by Student 2. **(1)**

| Student | Mass of limestone/g | Volume of gas produced/cm³ |
|---|---|---|
| 1 | 0.10 | 22 |
| 2 | 0.15 | |
| 3 | 0.20 | 37 |
| 4 | 0.25 | 56 |
| 5 | 0.30 | 66 |
| 6 | 0.35 | 78 |
| 7 | 0.40 | 88 |
| 8 | 0.45 | 91 |
| 9 | 0.50 | 91 |
| 10 | 0.55 | 91 |
| 11 | 0.60 | 91 |

c  Some of the data have been plotted on the graph below. Plot the rest of
  the data on the graph and draw another straight line of best fit.          **(3)**

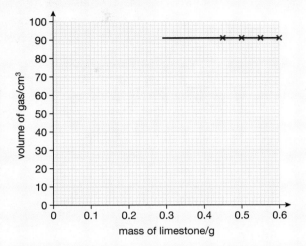

d  The teacher realised that one of the student's results was incorrect.

SKILLS ▶ ANALYSIS

  i  Explain which student's result is incorrect.                            **(2)**

SKILLS ▶ DECISION MAKING

  ii  Suggest one thing that could have gone wrong with the student's
    experiment to produce this result.                                       **(1)**

SKILLS ▶ REASONING

e  Explain why the volume of gas produced in the experiments carried out
  by Students 8–11 is the same.                                             **(2)**

SKILLS ▶ ANALYSIS

f  Use the graph to determine the mass of limestone that reacts exactly with
  the hydrochloric acid.                                                     **(1)**

SKILLS ▶ PROBLEM SOLVING

g  The equation for the reaction between calcium carbonate and hydrochloric
  acid is

$$CaCO_3(s) + 2HCl(...) \rightarrow CaCl_2(aq) + H_2O(...) + CO_2(g)$$

  i  Complete the equation by adding the missing state symbols.              **(2)**

  ii  Calculate the mass of calcium carbonate that would produce 91 cm$^3$
    of carbon dioxide at rtp.                                                **(3)**
    (Assume that the molar volume of a gas at rtp is 24 000 cm$^3$.)

  iii  Calculate the percentage of calcium carbonate in the sample of
    limestone.                                                               **(2)**

**(Total 18 marks)**

**END OF CHEMISTRY ONLY**

# UNIT 2
# INORGANIC CHEMISTRY

Inorganic chemistry is the study of all the elements in the Periodic Table and the compounds they form, except organic compounds formed by carbon. The properties of these elements are very different and they form a huge variety of compounds. The Periodic Table is the great unifying principle used by inorganic chemists as a guide to understanding the behaviour of the elements and their compounds. Most of the elements in the Periodic Table are metals and these are some of the most important materials that we use in everyday life. However, scientists are always searching for new materials with exciting properties. Inorganic chemists are involved in the development of these new materials, for instance high-temperature superconductors that are used in trains that levitate above the tracks.

▲ Figure 11.1 The maglev train levitates above the track due to superconducting materials.

# 11 THE ALKALI METALS

We have already looked at the Periodic Table in Chapter 4. Here we will look at the properties of the elements in Group 1 of the Periodic Table: the alkali metals.

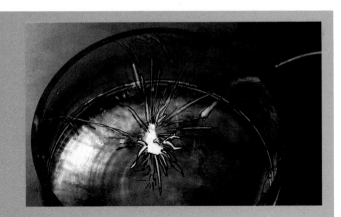

▶ Figure 11.2 Potassium reacting with water. The alkali metals are all reactive metals that react vigorously with water.

## LEARNING OBJECTIVES

■ Understand how the similarities in the reactions of these elements with water provide evidence for their recognition as a family of elements

■ Understand how the differences between the reactions of these elements with air and water provide evidence for the trend in reactivity in Group 1

■ Use knowledge of trends in Group 1 to predict the properties of other alkali metals

### CHEMISTRY ONLY

■ Explain the trend in reactivity in Group 1 in terms of electronic configurations.

| Li | lithium |
| Na | sodium |
| K | potassium |
| Rb | rubidium |
| Cs | caesium |
| Fr | francium |

▲ Figure 11.3 The alkali metals

The elements in Group 1 of the Periodic Table are called the alkali metals. The group contains the elements shown in Figure 11.3.

Francium (pronounced france-ee-um), at the bottom of the group, is radioactive. One of its isotopes is produced during the radioactive decay of uranium-235, but is extremely short-lived. At any one time scientists estimate that there is only about 20–30 g of francium present in the whole of the Earth's crust and no one has ever seen a piece of francium. When you know about the rest of Group 1 you can predict what francium would be like. We will make those predictions later.

## PHYSICAL PROPERTIES

| | Melting point/°C | Boiling point/°C | Density/g/cm³ |
|---|---|---|---|
| Li | 181 | 1342 | 0.53 |
| Na | 98 | 883 | 0.97 |
| K | 63 | 760 | 0.86 |
| Rb | 39 | 686 | 1.53 |
| Cs | 29 | 669 | 1.88 |

The melting and boiling points of the elements are very low for metals, and get lower as you move down the group.

Their densities tend to increase down the group, although not regularly. Lithium, sodium and potassium are all less dense than water, and so will float on it.

The metals are also very soft and are easily cut with a knife, becoming softer as you move down the group. They are shiny and silver when freshly cut, but tarnish very quickly on exposure to air.

## STORAGE AND HANDLING

All these metals are extremely reactive and get more reactive as you go down the group. They all react quickly with oxygen in the air to form oxides, and react rapidly with water to form strongly alkaline solutions of the metal hydroxides. This is why the Group 1 metals are commonly known as the alkali metals.

To stop them reacting with oxygen or water vapour in the air, lithium, sodium and potassium are stored under oil. Rubidium and caesium are so reactive that they have to be stored in sealed glass tubes to stop any possibility of oxygen getting at them.

▲ Figure 11.4 Lithium, sodium and potassium have to be kept in oil to stop them reacting with oxygen in the air.

Great care must be taken not to touch any of these metals with bare fingers. There could be enough sweat on your skin to give a reaction, producing lots of heat and a very **corrosive** metal hydroxide.

## A FAMILY OF ELEMENTS

There are two reasons that we put these elements in Group 1:

1 They all have *one* electron in their outer shell. The electronic configurations are:

lithium        2, 1
sodium        2, 8, 1
potassium    2, 8, 8, 1

2 They have similar chemical properties, for example:

■ they all react with water (this will be discussed below) in the same way to form a hydroxide with the formula MOH (e.g. LiOH, NaOH) and hydrogen

■ they react with oxygen to form an oxide with the formula $M_2O$ ($Na_2O$, $K_2O$)

■ they react with halogens to form compounds with the formula MX (e.g. LiCl, KBr)

■ they form ionic compounds which contain an $M^+$ ion (e.g. $Na^+$, $K^+$).

The chemical properties depend on the number of electrons in the outer shell. The Group 1 elements react in very similar ways because they all have the same number of electrons in the outer shell (one), so reason 2 is really just a consequence of reason 1.

We will discuss the chemical properties in more detail below.

## REACTIONS WITH WATER

All these metals react in the same way with water to produce a metal hydroxide and hydrogen:

alkali metal + water → alkali metal hydroxide + hydrogen

$$2M \quad + 2H_2O \rightarrow \quad 2MOH \quad + \quad H_2$$

The main difference between the reactions is how quickly they happen.

As you go down the group, the metals become more reactive and the reactions occur more rapidly.

The reaction between sodium and water is typical.

### WITH SODIUM

$$2Na(s) + 2H_2O(l) \rightarrow 2NaOH(aq) + H_2(g)$$

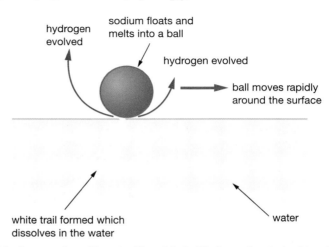

▲ Figure 11.5 Sodium reacting with water. The white trail is the sodium hydroxide, which dissolves in water to form a strongly alkaline solution.

The main observations you can make when this reaction occurs are:

- The sodium floats because it is less dense than water.

- The sodium melts into a ball because its melting point is low and a lot of heat is produced by the reaction.

- There is fizzing because hydrogen gas is produced.

- The sodium moves around on the surface of the water. Because the hydrogen isn't given off symmetrically around the ball, the sodium is pushed around the surface of the water, like a hovercraft.

- The piece of sodium gets smaller and eventually disappears. The sodium is used up in the reaction.

- If you test the solution that is formed with universal indicator solution, you will see that the universal indicator goes blue, indicating an alkaline solution has been formed. The metal hydroxide is alkaline (the solution contains the OH⁻ ion).

▲ Figure 11.6 **A hovercraft**

### LITHIUM

$$2Li(s) + 2H_2O(l) \rightarrow 2LiOH(aq) + H_2(g)$$

The reaction is very similar to sodium's reaction, except that it is slower. Lithium's melting point is higher and the heat isn't produced so quickly, so the lithium doesn't melt.

### POTASSIUM

$$2K(s) + 2H_2O(l) \rightarrow 2KOH(aq) + H_2(g)$$

Potassium's reaction is faster than sodium's. Enough heat is produced to ignite the hydrogen, which burns with a lilac flame. The reaction often ends with the potassium spitting around and exploding.

### RUBIDIUM AND CAESIUM

These react even more violently than potassium, and the reaction can be explosive. Rubidium hydroxide and caesium hydroxide are formed.

## CHEMISTRY ONLY

## EXPLAINING THE INCREASE IN REACTIVITY

As you go down the group, the metals become more reactive.

In all these reactions, the metal atoms are losing electrons and forming metal ions in solution. For example:

$$Na(s) \rightarrow Na^+(aq) + e^-$$

The differences between the reactions depend on how easily the *outer electron* of the metal is lost in each case. That depends on how strongly it is attracted to the nucleus in the original atom. Remember that the nucleus of an atom is positive because it contains protons, and so attracts the negative electrons.

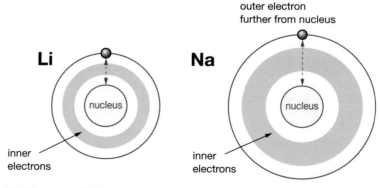

▲ Figure 11.7 Electrons of lithium and sodium

As we move down the group, the atoms have more shells of electrons and get bigger: a sodium atom is bigger than a lithium atom and a potassium atom is bigger than a sodium atom. As the atoms get bigger, the *outer electron*, which is the one lost in the reaction, is further from the nucleus. Because it is further from the nucleus it is less strongly attracted by the nucleus and therefore more easily lost.

**END OF CHEMISTRY ONLY**

▲ Figure 11.8 A piece of sodium. The left-hand edge has been freshly cut, so it is shiny.

## REMINDER

Remember the charge on the oxide ion is $O^{2-}$ and the charge on an alkali metal ion is $M^+$.

## EXTENSION WORK

When the alkali metals react with air, other types of oxide can also be formed, called peroxides ($M_2O_2$) and superoxides ($MO_2$). These all still contain the $M^+$ ion, it is the oxygen bit which is different. Lithium can also form a nitride ($Li_3N$).

## REACTIONS OF THE ALKALI METALS WITH THE AIR

Lithium, sodium and potassium are all stored in oil because they react with the air. If we look at a piece of sodium which has been taken out of the oil, it usually has a crust on the outside. It is not shiny unless it has been freshly cut.

When the piece of sodium is cut, the fresh surface is shiny but it tarnishes rapidly as the freshly exposed sodium reacts with oxygen in the air. If we do the same with a piece of lithium it tarnishes more slowly because lithium reacts more slowly than sodium. A freshly cut piece of potassium tarnishes extremely rapidly, more quickly than sodium. In this way we can see again that potassium is more reactive than sodium, which is more reactive than lithium. In each case the metal reacts with oxygen in the air to form an oxide with the formula $M_2O$.

If we heat each of the metals in the air using a Bunsen burner, we get a much more vigorous reaction and it is more difficult to see which metal is most reactive because all the reactions are so rapid.

Lithium burns with a red flame to form lithium oxide.

Sodium burns with a yellow flame to form sodium oxide.

Potassium burns with a lilac flame to form potassium oxide.

The equation for all these reactions is:

$$4M(s) + O_2(g) \rightarrow 2M_2O(s)$$

In each case the product formed is a white powder – the alkali metal oxide.

## COMPOUNDS OF THE ALKALI METALS

All Group 1 metal ions are colourless. That means that their compounds will be colourless or white unless they are combined with a coloured negative ion. Potassium dichromate(VI) is orange, for example, because the dichromate(VI) ion is orange, and potassium manganate(VII) is purple because the manganate(VII) ion is purple. Group 1 compounds are typical ionic solids and are mostly soluble in water.

## SUMMARISING THE MAIN FEATURES OF THE GROUP 1 ELEMENTS

Group 1 elements:

- are metals
- are soft with melting points and densities which are very low for metals
- have to be stored out of contact with air or water
- react rapidly with air to form coatings of the metal oxide
- react with water to produce an alkaline solution of the metal hydroxide and hydrogen gas
- increase in reactivity as you go down the group
- form compounds in which the metal has a 1+ ion
- have mainly white/colourless compounds which dissolve to produce colourless solutions.

## PREDICTING THE PROPERTIES OF FRANCIUM

As we move down a group in the Periodic Table the properties of the elements change gradually. So, if we know the properties of most of the elements in a group, we should be able to predict the properties of elements we don't know. Francium is extremely radioactive and at any time, anywhere in the world, there is only a tiny amount present; nobody has actually seen a piece of francium. We can, however, predict the properties of francium using the properties of the other alkali metals.

We can predict that francium:

■ is very soft

■ will have a melting point around room temperature

■ has density which is probably just over $2\,g/cm^3$

■ will be a silvery metal, but will tarnish almost instantly in air

■ will react violently with water to give francium hydroxide and hydrogen

■ will be more reactive than caesium

■ will have a hydroxide, francium hydroxide, with the formula FrOH, which will be soluble in water and form a strongly alkaline solution

■ will form compounds that are white/colourless and dissolve in water to give colourless solutions.

We could use a graphical method to predict the melting point of francium. If we plot the melting point of the alkali metals against atomic number then draw a line of best fit we get:

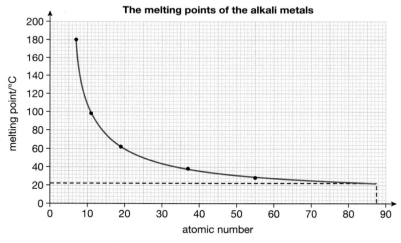

▲ Figure 11.9 This graph allows us to predict the melting point of francium.

If we carry on the line to atomic number 87 we can predict a melting point of about 22 °C.

**KEY POINT**

Various other predictions give a melting point for francium between 21 °C and 27 °C.

**CHAPTER QUESTIONS**

1  Explain why lithium, sodium and potassium are put in the same group in the Periodic Table.

2  This question concerns the chemistry of the elements Li, Na, K, Rb and Cs. In a and b, you should name the substances represented by letters.

a  **A** is the least dense of all the metals.

b  When metal **B** is dropped onto water it melts into a small ball and moves rapidly around the surface. A gas, **C**, is given off and this burns with a lilac flame. A solution of **D** is formed.

c  Write an equation for the reaction of **B** with water.

d  What would you expect to see if solution **D** was tested with universal indicator paper?

e  Explain why **B** melts into a small ball when it is dropped onto water.

f  **E** burns in air with a yellow flame to form compound **F**. Write a word equation and balanced symbol equation for the reaction that occurs.

3  Explain whether each of the following statements is true or false.

a  Sodium forms mostly covalent compounds.

b  A rubidium atom is larger than a potassium atom.

c  All the alkali metals react with air to form oxides.

d  Lithium reacts with chlorine to form lithium chloride, which has the formula $Li_2Cl$.

4  Imagine that a new alkali metal has recently been discovered and that it fits into the Periodic Table below francium. We will call this new element edexcelium.

a  Explain whether you would expect edexcelium to be more or less dense than francium.

b  State how many electrons edexcelium will have in its outer shell.

c  State the names of the products that will be formed when edexcelium reacts with water.

d  Explain whether edexcelium will be more or less reactive than francium.

e  If the symbol for edexcelium is Ed, write a balanced chemical equation for the reaction of edexcelium with water.

f  When edexcelium reacts with water, will the solution formed be acidic, alkaline or neutral?

g  Write the formula for the compound formed when edexcelium reacts with air.

# 12 THE HALOGENS

The elements in Group 7 of the Periodic Table are called the halogens. In this chapter we will look at some of the properties of these elements. When we first look at them, it is hard to see that they form a family of elements because they look so different, but they react in very similar ways.

▶ Figure 12.1 Chlorine, bromine and iodine are three of the halogens.

## LEARNING OBJECTIVES

- Know the colours, physical states (at room temperature) and trends in physical properties of the halogens

- Use knowledge of trends in Group 7 to predict the properties of other halogens

- Understand how displacement reactions involving halogens and halides provide evidence for the trend in reactivity in Group 7

**CHEMISTRY ONLY**

- Explain the trend in reactivity in Group 7 in terms of electronic configurations.

| F | fluorine |
| Cl | chlorine |
| Br | bromine |
| I | iodine |
| At | astatine |

▲ Figure 12.2 The halogens

▲ Figure 12.3 Iodine is a grey solid but has a purple vapour.

## THE HALOGENS

The atoms of fluorine, chlorine, bromine, iodine and astatine all have 7 electrons in their outer shell, therefore these elements are put in Group 7 in the Periodic Table. The number of electrons in the outer shell of an atom determines how something reacts, so all the elements in Group 7 react in a similar way.

The name 'halogen' means 'salt-producing'. When they react with metals, these elements produce a wide range of salts, including calcium fluoride, sodium chloride, silver bromide and potassium iodide. All the salts contain the X$^-$ ion (where X stands for any halogen atom).

The halogens are non-metallic elements with diatomic molecules: $F_2$, $Cl_2$, etc.

Table 12.1 The melting and boiling points of the halogens increase and the colour becomes darker down the group

| | Physical state at room temperature | Colour |
|---|---|---|
| $F_2$ | gas | yellow |
| $Cl_2$ | gas | green |
| $Br_2$ | liquid | red-brown liquid, orange/brown vapour |
| $I_2$ | solid | grey solid, purple vapour |

**REMINDER**

Remember that intermolecular forces are the forces of attraction between molecules.

*The melting and boiling points increase down the group.* The halogens are all covalent molecular substances and the melting and boiling points increase as the relative molecular mass increases. As the relative molecular mass increases, the intermolecular forces of attraction become stronger and therefore more energy must be put in to overcome these stronger forces of attraction. Remember, no covalent bonds are broken when these melt/boil.

Because the halogens are non-metals, they are poor conductors of heat and electricity.

Astatine is radioactive and is formed during the radioactive decay of other elements, such as uranium and thorium. Most of its isotopes are so unstable that their lives can be measured in seconds or fractions of a second.

## SAFETY

Fluorine is so dangerously reactive that you would never expect to find it in a school lab. Apart from any safety problems, due to the reactivity of the elements (especially fluorine and chlorine), all the halogens have extremely poisonous **vapours** and have to be handled in a fume cupboard. Liquid bromine is also very corrosive, and great care has to be taken to keep it off the skin.

## REACTIONS OF THE HALOGENS

The halogens react with hydrogen to form **hydrogen halides**: hydrogen fluoride, hydrogen chloride, hydrogen bromide and hydrogen iodide. For example:

$$H_2(g) + Br_2(g) \rightarrow 2HBr(g)$$

The hydrogen halides are all acidic, poisonous gases. In common with all the compounds formed between the halogens and non-metals, the gases are covalently bonded. They are very soluble in water, reacting with it to produce solutions of acids. For example, hydrochloric acid is a solution of hydrogen chloride in water:

$$HCl(g) \xrightarrow{\text{dissolve in water}} HCl(aq)$$

hydrogen chloride     hydrochloric acid

**KEY POINT**

This is something that can be confusing. Hydrogen chloride is a gas and when it dissolves in water it becomes hydrochloric acid. Hydrogen chloride is a gas consisting of HCl molecules and hydrochloric acid is a solution containing $H^+$ and $Cl^-$ ions. The acids formed by other hydrogen halides are:

HF(aq)    hydrofluoric acid
HBr(aq)    hydrobromic acid
HI(aq)    hydroiodic acid

The halogens react with alkali metals to form salts. For instance, sodium burns in chlorine with its typical yellow flame to produce white, solid sodium chloride:

$$2Na(s) + Cl_2(g) \rightarrow 2NaCl(s)$$

Sodium chloride is an ionic solid. Typically, when the halogens react with metals from Groups 1 and 2, they form ions.

▲ Figure 12.4 Sodium burning in chlorine to produce sodium chloride.

# DISPLACEMENT REACTIONS INVOLVING THE HALOGENS

In this section we will look at what happens when we react a solution of a halogen with a solution containing halide ions. These are called **displacement reactions** – you will see why below. We can use these displacement reactions to show that chlorine is more reactive than bromine, which is more reactive than iodine.

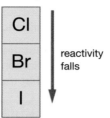

reactivity falls

▲ Figure 12.5 Reactivity decreases down the group.

**KEY POINT**

We will concentrate on the three commonly used halogens, but the trend continues for the rest of the group.

## REACTING CHLORINE WITH POTASSIUM BROMIDE OR POTASSIUM IODIDE SOLUTIONS

**KEY POINT**

Although pure liquid bromine is red-brown, in solution it is orange. If the solution is very dilute it may even look yellow.

**HINT**

You have to be careful with the words you use here. In the exam you must write 'chlorine displaces bromine' not 'chlorine displaces bromide'.

▲ Figure 12.6 Bromine and iodine displaced from potassium bromide and potassium iodide solutions.

If you add chlorine solution ('chlorine water') to colourless potassium bromide solution, the solution becomes orange as bromine is formed:

$$2KBr(aq) + Cl_2(aq) \rightarrow 2KCl(aq) + Br_2(aq)$$

potassium + chlorine → potassium + bromine
bromide                         chloride

Chlorine is more reactive than bromine and has displaced the bromine from solution. You can think about this in the following way: if something is more reactive, it has a greater tendency to *react* to form a compound. Something that is less reactive is more likely to go back to being the element (the unreacted form).

Similarly, adding chlorine solution to potassium iodide solution gives a brown (orange if it is very dilute) solution of iodine.

$$2KI(aq) + Cl_2(aq) \rightarrow 2KCl(aq) + I_2(aq)$$

In exactly the same way, the more reactive bromine displaces the less reactive iodine from potassium iodide solution. Adding bromine solution ('bromine water') to colourless potassium iodide solution gives a brown solution of iodine:

$$2KI(aq) + Br_2(aq) \rightarrow 2KBr(aq) + I_2(aq)$$

▲ Figure 12.7 Bromine displaces iodine from potassium iodide solution.

All of the following experiments show no reaction:

■ bromine + potassium chloride solution: the less reactive bromine cannot displace the more reactive chlorine

■ iodine + potassium chloride solution: the less reactive iodine cannot displace the more reactive chlorine

■ iodine + potassium bromide solution: the less reactive iodine cannot displace the more reactive bromine.

## IONIC EQUATIONS FOR DISPLACEMENT REACTIONS

In all of the above reactions it would not have mattered if we had reacted chlorine with *sodium* bromide or *magnesium* bromide, for example. The reaction occurs between the chlorine molecule and the bromide ion; the metal ion does not do anything, we call it a **spectator ion**. If we write the reaction between chlorine and potassium iodide showing all the charges, we can see this more clearly:

$$2K^+(aq) + 2Br^-(aq) + Cl_2(aq) \rightarrow 2K^+(aq) + 2Cl^-(aq) + Br_2(aq)$$

You can see that the potassium ions are the same on each side. They have not changed and so we can write a new equation simply concentrating on the things that have changed:

$$2Br^-(aq) + Cl_2(aq) \rightarrow 2Cl^-(aq) + Br_2(aq)$$

This is called an *ionic equation* and does not include the spectator ions.

## DISPLACEMENT REACTIONS AS REDOX REACTIONS

The reactions above that do show a colour change are often discussed in terms of oxidation and reduction, they are *redox reactions*. If you have not already covered redox reactions, you should look at Chapter 14 before reading the next section.

If we look at the reaction between chlorine and either bromide ions or iodide ions, we can see why they are described as redox reactions.

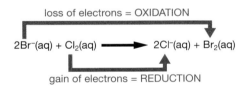

▲ Figure 12.8 The reaction of chlorine with potassium bromide (top) or potassium iodide (below).

In each case, the chlorine is acting as an oxidising agent. In the first equation it oxidises the $Br^-$ ions by taking electrons away from them. Because the oxidising agent takes electrons away from something else, it is reduced itself: the $Cl_2$ is reduced to $2Cl^-$ ions.

In the reaction between bromine and iodide ions, the bromine is the oxidising agent:

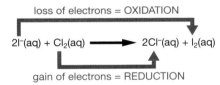

▲ Figure 12.9 The reaction between bromine solution and potassium iodide solution.

## CHEMISTRY ONLY

## EXPLAINING THE TREND IN THE REACTIVITY OF THE HALOGENS

We can explain the reactivity of the halogens in terms of how readily they form negative ions. When the halogens react (in these reactions) they form 1– ions by taking electrons away from something else. A chlorine atom is smaller than a bromine atom, so when we add an electron to the outer shell of a chlorine atom, we are adding it to a shell closer to the nucleus. If it is in a shell closer to the nucleus it is more strongly attracted to the nucleus. The bromine atom is bigger and so the outer electron is added to a shell further away from the nucleus, where it is not as strongly attracted. Chlorine therefore has a stronger tendency to form a 1– ion than bromine, and a chlorine atom will take an electron away from a bromide ion.

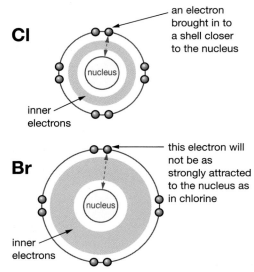

▲ Figure 12.10 Chlorine has a stronger attraction for an electron than bromine.

## END OF CHEMISTRY ONLY

## SUMMARISING THE MAIN FEATURES OF THE GROUP 7 ELEMENTS

Group 7 elements:

- have diatomic molecules, $X_2$ (e.g. $F_2$, $Cl_2$, $Br_2$)

- go from gases to liquid to solid as you move down the group: the melting points and boiling points increase down the group

- have coloured poisonous vapours: the colours of the elements get darker down the group

- form compounds with the formula HX (e.g. HF, HCl, HBr) when reacted with hydrogen and these dissolve in water to form acids

- form ionic salts with metals and covalent compounds with non-metals

- form $X^-$ (e.g. $F^-$, $Cl^-$, $Br^-$) ions in ionic compounds

- become less reactive towards the bottom of the group

- will displace elements lower down the group from their salts.

We can use these properties and trends to predict the properties of astatine, the element below iodine in Group 7. Astatine is an extremely rare radioactive element and no one has ever seen a sample of it, but we can predict that it:

- will be a darker colour than iodine, very dark grey or black

- will be a solid at room temperature and have a higher melting point than iodine

- will be diatomic and contain $At_2$ molecules

- will react with hydrogen to form HAt, which will dissolve in water to form an acid

- will form the salt NaAt with sodium or KAt with potassium

- will contain the astatide ion ($At^-$) in its ionic salts

- will be less reactive than iodine

- will be displaced from solution by iodine

## CHAPTER QUESTIONS

SKILLS  INTERPRETATION        1 a Draw a diagram to show the arrangement of electrons in a fluorine atom.

SKILLS  REASONING        b Explain why fluorine is placed in Group 7 in the Periodic Table.

SKILLS  INTERPRETATION    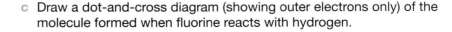    c Draw a dot-and-cross diagram (showing outer electrons only) of the molecule formed when fluorine reacts with hydrogen.

SKILLS  REASONING    d Explain why fluorine has a lower melting point than chlorine.

e Explain what you would observe when a chlorine solution is added to a solution of potassium fluoride.

SKILLS ⟩ INTERPRETATION    ⑧

2 This question is about astatine, At, at the bottom of Group 7 of the Periodic Table. You are asked to make some predictions about astatine and its chemistry.

  a Showing only the outer electrons, draw diagrams to show the arrangement of electrons in an astatine *atom*, an astatide *ion* and an astatine *molecule*.

SKILLS ⟩ REASONING    ⑦

  b Explain what physical state you would expect astatine to be in at room temperature.

  c Explain whether you would expect astatine to be more or less reactive than iodine.

⑥

  d Suggest a likely pH for a reasonably concentrated solution of hydrogen astatide in water.

⑧

  e Explain what you would expect caesium astatide to look like. Will it be soluble in water? Explain your reasoning.

SKILLS ⟩ PROBLEM SOLVING, REASONING    ⑩

  f Write an ionic equation for the reaction that will occur if you add chlorine water to a solution of sodium astatide. Assume that astatine is insoluble in water. Explain clearly why this reaction would be counted as a redox reaction.

SKILLS ⟩ PROBLEM SOLVING    ⑨

3 Chlorine reacts with fluorine to form chlorine monofluoride (ClF).

  a Write an equation for this reaction, including state symbols. All the substances are gases at room temperature.

SKILLS ⟩ INTERPRETATION    ⑧

  b Draw a dot-and-cross diagram for ClF.

SKILLS ⟩ REASONING

  c Arrange the substances $Cl_2$, $F_2$ and ClF in order of increasing boiling point and explain your order.

SKILLS ⟩ PROBLEM SOLVING

  d Chlorine and fluorine can also react together to form chlorine trifluoride, which is a gas at room temperature and pressure.

    i State the formula of chlorine trifluoride.

⑨

    ii Chlorine trifluoride is an extremely reactive substance and, for instance, reacts explosively with water. The products of the reaction are hydrofluoric acid, hydrochloric acid and oxygen gas. Write an equation, including state symbols, for the reaction of chlorine trifluoride with water.

# 13 GASES IN THE ATMOSPHERE

This chapter looks at the most common gases in the atmosphere, concentrating on oxygen. We will also look at carbon dioxide, which is a greenhouse gas and may contribute to climate change.

▶ Figure 13.1 Despite the gases added by industry, the air around us is mostly nitrogen and oxygen.

## LEARNING OBJECTIVES

- Know the approximate percentages by volume of the four most abundant gases in dry air

- Understand how to determine the percentage by volume of oxygen in air using experiments involving the reactions of metals (e.g. iron) and non-metals (e.g. phosphorus) with air

- Describe the combustion of elements in oxygen, including magnesium, hydrogen and sulfur

- Describe the formation of carbon dioxide from the thermal decomposition of metal carbonates, including copper(II) carbonate

- Know that carbon dioxide is a greenhouse gas and that increasing amounts in the atmosphere may contribute to climate change

- Practical: Determine the approximate percentage by volume of oxygen in air using a metal or a non-metal

## THE COMPOSITION OF THE AIR

The approximate percentages (by volume) of the four most abundant gases present in *unpolluted, dry* air are shown in Table 13.1.

Table 13.1 Approximate percentages (by volume) of the main gases in unpolluted, dry air

| Gas | Amount in air (%) | Amount in air (fraction) |
| --- | --- | --- |
| nitrogen | 78.1 | about 4/5 |
| oxygen | 21.0 | about 1/5 |
| argon | 0.9 | |
| carbon dioxide | 0.04 | |

There are also very small amounts of the other noble gases in the air.

**KEY POINT**

It is important to realise that these figures apply only to dry, unpolluted air. Air can have anywhere between 0 and 4% water vapour. The percentage of carbon dioxide in the air, although very small, is rising steadily because of human activity.

**SHOWING THAT AIR CONTAINS ABOUT ONE-FIFTH OXYGEN**

We will look at a few different methods for measuring the percentage of oxygen in the air. All the methods rely on the same basic principle: we react something with the oxygen in the air and look at how much the volume decreases as the oxygen is removed.

Safety Note: Wear eye protection and heat carefully to avoid cracking the glass syringe with the flame. Take care not to push the opposite plunger completely out!.

## ACTIVITY 1

## ▼ PRACTICAL: USING COPPER

The apparatus shown in Figure 13.2 can be used to find the percentage of oxygen in the air.

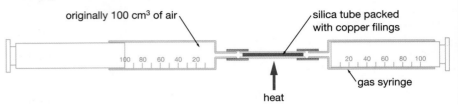

originally 100 cm³ of air     silica tube packed with copper filings

heat     gas syringe

▲ Figure 13.2 Using copper to measure the percentage of oxygen in air.

- The plunger on one of the gas syringes is pushed all the way in and the other moved out to 100 cm³. We now know that the apparatus contains 100 cm³ air.
- The silica tube is heated strongly (roaring Bunsen flame).
- The plunger in the left-hand gas syringe is pushed in. This causes the air to pass over the heated copper. This pushes out the plunger on the right-hand gas syringe.
- The plungers are pushed in sequence so that the air in the system keeps passing over the heated copper. The pink-brown copper turns black as copper(II) oxide is formed.
- As the copper reacts, the Bunsen burner is moved along the tube so that it is always heating fresh copper.
- The volume of gas in the syringes falls as the oxygen is consumed.
- We keep pushing the plungers in and out until there is no change in volume.
- The apparatus is then allowed to cool to room temperature again before taking the final volume of gas (because gases expand as they are heated).

## RESULTS

| initial volume of air In the system/cm³ | 100 |
|---|---|
| final volume of 'air' in the system/cm³ | 79 |

The volume of the air has decreased because the oxygen has been removed as it reacts with the copper. The reaction that occurs is

$$2Cu(s) + O_2(g) \rightarrow 2CuO(s)$$

copper + oxygen → copper(II) oxide

We can work out from these data that the volume of oxygen that reacted was $100 - 79 = 21$ cm³.

The original volume of air was 100 cm³ so we can work out the percentage of oxygen in the air:

percentage oxygen = $21/100 \times 100 = 21\%$

**KEY POINT**

## KEY POINT

Iron needs oxygen and water to rust. The rusting of iron is discussed in Chapter 14.

Safety Note: Iron filings can irritate the skin and are a particular hazard if they get in the eyes.

### ACTIVITY 2

## ▼ PRACTICAL: USING THE RUSTING OF IRON

Iron rusts in damp air, using up oxygen as it does so. We can use this reaction to determine how much oxygen there is in the air.

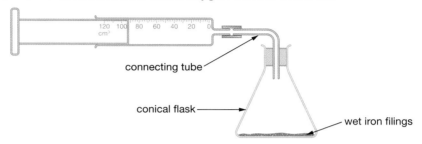

connecting tube

conical flask

wet iron filings

▲ Figure 13.3 This apparatus can be used to find the percentage of oxygen in the air.

We will use the apparatus from Figure 13.3, but before we start we need to know the volume of air present in the apparatus. We can find this by filling up the conical flask and connecting tube with water and then transferring the water to a measuring cylinder. On the conical flask we mark the position of the bung and only fill with water to that point. We will assume that the small volume occupied by the iron filings is negligible (very small compared to the overall volume).

The procedure for the experiment is as follows:

■ Set up the apparatus as shown in Figure 13.3.

■ Put wet iron filings into the conical flask.

■ Record the initial reading on the gas syringe.

■ Leave the apparatus in place for about a week, until the reading on the gas syringe stops changing.

■ Record the final reading on the gas syringe.

| | |
|---|---|
| volume of air in conical flask/cm$^3$ | 130 |
| volume of air in connecting tube/cm$^3$ | 12 |
| initial reading on gas syringe/cm$^3$ | 92 |
| final reading on gas syringe/cm$^3$ | 43 |

So, the total volume of air inside the apparatus at the beginning of the experiment is $130 + 12 + 92 = 234\,cm^3$.

The total volume of 'air' in the apparatus at the end = 130 + 12 + 43
$$= 185\,cm^3$$

Volume of oxygen used up = $234 - 185 = 49\,cm^3$

The percentage of oxygen in the air is $49/234 \times 100 = 21\%$.

Sometimes when you do this experiment, the answer comes out as less than 21%. Possible reasons for this could be:

■ The experiment was not left set up for long enough. The iron has not had enough chance to react with all the oxygen in the apparatus.

■ Not enough iron was added at the beginning. The iron must be in excess, that is, there must be enough iron to react with all the oxygen present.

**!**

Safety Note: The teacher demonstrating needs to wear goggles or a face shield, and have the room well ventilated. After igniting the phosphorous the bell-jar needs to be held down as pressure rises initially. After all the fumes have gone the evaporating basin needs to be sunk using the glass rod to prevent any unreacted phosphorous re-igniting.

## ACTIVITY 3

## ▼ PRACTICAL: USING PHOSPHORUS

Phosphorus is a very reactive element that reacts with the oxygen in the air to form a phosphorus oxide. This oxide is very soluble in water.

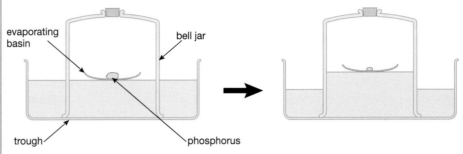

▲ Figure 13.4 Finding the percentage of oxygen in air using phosphorus.

■ The apparatus shown in Figure 13.4 is set up with the piece of phosphorus on an evaporating basin, which is floating in the water.

■ The initial level of water is marked on the side of the bell jar with a waterproof pen or a sticker.

■ The bung is removed from the bell jar and the phosphorus is touched with a hot metal wire in order to ignite it.

■ The bung is quickly put back into the bell jar.

■ The phosphorus burns, the bell jar becomes filled with a white smoke (phosphorus oxide) and the level of water rises inside the bell jar. The smoke eventually clears as the phosphorus oxide dissolves in the water.

■ When the level of water inside the bell jar stops rising, the final level is marked.

■ To find how much the water level has changed, the bell jar is turned upside down, filled with water to each mark in turn and the water is poured into a large measuring cylinder.

It is important that there is still some phosphorus left on the evaporating basin at the end of the experiment. We have used an excess of phosphorus so that there is more than enough to react with all the oxygen. If there was no phosphorus left, then we would probably get a lower value for the percentage of oxygen in the air because not all the oxygen might have been used up.

## THE COMBUSTION OF ELEMENTS IN OXYGEN

Some elements burn in oxygen, these reactions are called *combustion reactions*. Elements burn more brightly and rapidly in pure oxygen than in air because air only contains 21% oxygen.

## BURNING MAGNESIUM

Magnesium burns in oxygen with an extremely bright white flame to give a white, powdery ash of magnesium oxide:

$$2Mg(s) + O_2(g) \rightarrow 2MgO(s)$$

The white powder formed is not very soluble in water but a very small amount does dissolve to form an alkaline solution:

$$MgO(s) + H_2O(l) \rightarrow Mg(OH)_2(aq)$$

▲ Figure 13.5 Magnesium ribbon burning in air

## BURNING SULFUR

### KEY POINT

Sulfur dioxide is also called sulfur(IV) oxide.

Sulfur burns in oxygen with a blue flame. Poisonous, colourless sulfur dioxide gas is produced.

$$S(s) + O_2(g) \rightarrow SO_2(g)$$

▲ Figure 13.6 Sulfur burning in oxygen

### KEY POINT

Sulfurous acid ($H_2SO_3$), not sulfuric acid ($H_2SO_4$), is formed.

The sulfur dioxide dissolves in water to form an acidic solution of *sulfurous acid*:

$$SO_2(g) + H_2O(l) \rightarrow H_2SO_3(aq)$$

## BURNING HYDROGEN

Hydrogen burns in oxygen with a pale blue flame. The product is water:

$$2H_2(g) + O_2(g) \rightarrow 2H_2O(l)$$

If you ignite a mixture of hydrogen and oxygen it will explode. This is the basis of the 'squeaky pop' test for hydrogen (see Chapter 18).

## THE PROPERTIES OF OXIDES

We can make some generalisations about the properties of oxides formed when elements burn in oxygen.

## METAL OXIDES

- Metal oxides are ionic compounds containing $O^{2-}$ ions.
- Metal oxides are usually basic oxides, which means that they react with acids to form salts.
- Metal oxides are usually insoluble in water. Those metal oxides that are soluble in water react with it to form alkaline solutions containing hydroxide ($OH^-$) ions.

- Non-metal oxides are covalent compounds.

- Non-metal oxides are usually acidic oxides, which react with alkalis/bases to form salts.

- Non-metal oxides are often soluble in water and react with it to form acidic solutions containing hydrogen ($H^+$) ions.

## CARBON DIOXIDE

▲ Figure 13.7 Copper(II) carbonate decomposes to form carbon dioxide when heated.

Carbon dioxide is a colourless gas that is most easily made in the laboratory by the reaction between dilute hydrochloric acid and calcium carbonate in the form of marble chips:

$$CaCO_3(s) + 2HCl(aq) \rightarrow CaCl_2(aq) + CO_2(g) + H_2O(l)$$

Carbon dioxide can also be obtained when metal carbonates are heated strongly.

Most carbonates split to give the metal oxide and carbon dioxide when you heat them. This is an example of **thermal decomposition**, breaking up something by heating it.

For example, copper(II) carbonate is a *green* powder which decomposes on heating to produce *black* copper(II) oxide:

$$CuCO_3(s) \rightarrow CuO(s) + CO_2(g)$$

Calcium carbonate doesn't decompose unless it is heated at quite high temperatures. This is a commercially important reaction because it is used to convert limestone (calcium carbonate) into quicklime (calcium oxide):

$$CaCO_3(s) \rightarrow CaO(s) + CO_2(g)$$

## CARBON DIOXIDE AND GLOBAL WARMING: THE GREENHOUSE EFFECT

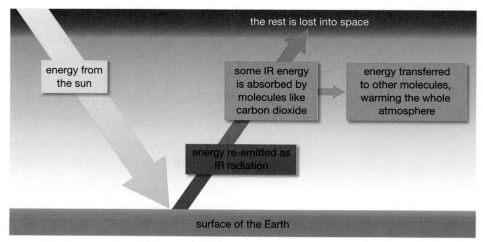

the rest is lost into space

energy from the sun

some IR energy is absorbed by molecules like carbon dioxide

energy transferred to other molecules, warming the whole atmosphere

energy re-emitted as IR radiation

surface of the Earth

▲ Figure 13.8  The greenhouse effect

Carbon dioxide is produced when fossil fuels (coal, oil and gas) burn. For instance, when coal, which is mostly carbon, burns in excess oxygen:

$$C(s) + O_2(g) \rightarrow CO_2(g)$$

Petrol is a mixture containing many different hydrocarbons. An example of a reaction that occurs when petrol burns is:

$$2C_8H_{18}(l) + 25O_2(g) \rightarrow 16CO_2(g) + 18H_2O(l)$$

Carbon dioxide is a greenhouse gas. The greenhouse effect occurs when high-energy UV and visible light from the sun pass through the atmosphere and warm up the surface of the Earth. The surface of the Earth (like any other warm surface) radiates infrared (IR) radiation. This IR radiation is absorbed by molecules such as $CO_2$ in the atmosphere. These then give out this energy again in all directions, heating the atmosphere.

For approximately the last 200 years the level of $CO_2$ in the atmosphere has been increasing. This has occurred since the industrial revolution and is due to the burning of fossil fuels and deforestation (cutting down trees to create more land for agriculture). Many people believe that the increase in the level of carbon dioxide in the atmosphere may contribute to climate change. The exact nature of this climate change is difficult to know, but some of the things that scientists believe could happen are:

- polar ice caps could melt
- sea levels could rise
- there could be more extreme weather (such as floods, droughts and heat waves).

**LOOKING AHEAD**

### Why is $CO_2$ a greenhouse gas and oxygen and nitrogen are not?

The presence of greenhouse gases in the atmosphere is essential to the maintenance of life on Earth as they keep the planet at a temperature at which water is mostly a liquid. However, it is very important that we do not have too much of these greenhouse gases. If nitrogen and oxygen, the main constituents of the atmosphere, were greenhouse gases the temperature of the Earth would be too high to support life as we know it. So, what makes something a greenhouse gas? A greenhouse gas must be able to absorb IR radiation emitted by the surface of the Earth and then emit it again in all directions, warming the atmosphere. All molecules vibrate and if these vibrations involve a change in the polarity of the molecule then the molecule will be able to absorb IR radiation and be a greenhouse gas.

What is polarity? A molecule is polar if one end of the molecule is slightly negative and the other end is slightly positive. Polarity arises because different atoms attract the electrons in a covalent bond to different extents (this is a property called *electronegativity*). The polarity also depends on the shape of the molecule. Water is a good example of a polar molecule.

The polarity of a water molecule changes as it vibrates and water vapour is an important greenhouse gas.

Carbon dioxide is a non-polar molecule because it is linear (the atoms are in a straight line) and symmetrical, and therefore there is no positive end and no negative end of the molecule.

However, as a $CO_2$ molecule vibrates it becomes non-symmetrical and polar. Because the polarity changes as a $CO_2$ molecule vibrates it can absorb IR radiation and is a greenhouse gas.

$O_2$ and $N_2$ are non-polar because both the atoms are the same and therefore attract the electrons in the covalent bonds equally, so there are no small charges. The molecules can only vibrate in one way and the molecules remain non-polar as they vibrate. This means that there is no change in polarity as they vibrate, therefore $N_2$ and $O_2$ cannot absorb IR radiation and are not greenhouse gases.

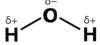

▲ Figure 13.9 Water is a polar molecule. The oxygen atoms attract the electrons in the covalent bonds more than the hydrogen atoms. The δ+ (delta positive) and δ− (delta negative) represent small charges on the atoms.

$$\overset{\delta-}{O} = \overset{\delta+}{C} = \overset{\delta-}{O}$$

▲ Figure 13.10 The oxygen atoms attract the electrons in the covalent bonds more than the carbon atoms do (oxygen is more electronegative than carbon) but the charges cancel due to the shape (a better way of saying this is that the dipoles cancel).

## CHAPTER QUESTIONS

1 State the approximate percentage of each of the following gases in dry air:

  a nitrogen     b oxygen     c carbon dioxide     d argon

2 A student carried out an experiment to measure the percentage oxygen in the air by using the rusting of iron. The experimental set-up is shown below. The initial reading on the measuring cylinder was $95\,cm^3$. They left the apparatus for 2 days and the new reading on the measuring cylinder was $80\,cm^3$.

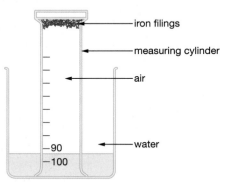

iron filings

measuring cylinder

air

—90
—100

water

a Calculate the percentage oxygen in the air using this data.

b Explain why the answer is not the same as you had expected and how the experiment could be improved.

3 A student carries out an experiment to investigate how much gas is produced when various metal carbonates are heated. Their data are shown in the table.

| Metal carbonate | Mass of carbonate used/g | Volume of gas collected at rtp/cm³ | Mass of solid left at the end/g |
|---|---|---|---|
| lead(II) carbonate | 1.00 | 89 | 0.84 |
| copper(II) carbonate | 1.00 | 194 | 0.64 |
| sodium carbonate | 1.00 | 0 | 1.00 |

a Describe an experiment that would allow the student to collect these data.

b Name the gas collected.

c Write an equation for the reaction occurring with lead(II) carbonate.

d Name the type of reaction occurring with lead(II) carbonate and copper(II) carbonate.

e Explain what these results indicate about the thermal stability of sodium carbonate compared to the other two substances.

(7)

f Explain why much more gas is given off in the copper(II) carbonate experiment than in the lead(II) carbonate experiment.

g Use the data from the lead(II) carbonate experiment to calculate the mass of $100\,cm^3$ of the gas.

h The student repeated the experiment for copper(II) carbonate and only obtained $152\,cm^3$ of gas this time. Explain two things that could have gone wrong with the second experiment to give this result.

4 Carbon dioxide is described as a *greenhouse gas*. Explain what this means.

# 14 REACTIVITY SERIES

The reactivity series lists elements (mainly metals) in order of decreasing reactivity. It is likely that you will have come across some of this chemistry already in earlier years. This chapter looks at some of the reactions of metals and introduces the idea of redox reactions.

▲ Figure 14.1 Rails are welded together using molten iron, produced by a reaction between aluminium and iron(III) oxide.

▲ Figure 14.2 Gold is so unreactive that it will remain chemically unchanged in contact with air or water basically forever.

## LEARNING OBJECTIVES

- Understand how metals can be arranged in a reactivity series based on their reactions with:
  - water
  - dilute hydrochloric or sulfuric acid

- Understand how metals can be arranged in a reactivity series based on their displacement reactions between:
  - metals and metal oxides
  - metals and aqueous solutions of metal salts

- Know the order of reactivity of these metals: potassium, sodium, lithium, calcium, magnesium, aluminium, zinc, iron, copper, silver, gold

- Know the conditions under which iron rusts

- Understand how the rusting of iron may be prevented by
  - barrier methods
  - galvanising
  - sacrificial protection

- Understand the terms:
  - oxidation
  - reduction
  - redox
  - oxidising agent
  - reducing agent

  in terms of gain or loss of oxygen and loss or gain of electrons

- Practical: Investigate reactions between dilute hydrochloric and sulfuric acids and metals (e.g. magnesium, zinc and iron)

# DISPLACEMENT REACTIONS INVOLVING METAL OXIDES

## THE REACTION BETWEEN MAGNESIUM AND COPPER(II) OXIDE

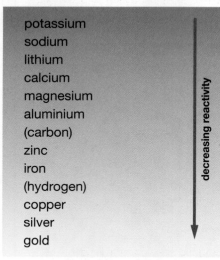

potassium
sodium
lithium
calcium
magnesium
aluminium
(carbon)
zinc
iron
(hydrogen)
copper
silver
gold

decreasing reactivity

▲ Figure 14.4 The reactivity series. Although carbon and hydrogen are not metals they are usually included in the series. You will learn why in this chapter.

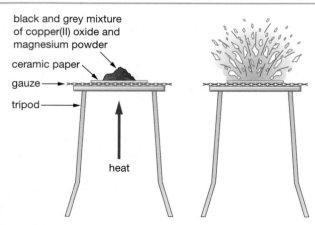

black and grey mixture of copper(II) oxide and magnesium powder
ceramic paper
gauze →
tripod →

heat

▲ Figure 14.3 Magnesium reacts with copper(II) oxide.

**Safety Note:** For this demonstration goggles or a face shield are needed, together with safety screens in a well-ventilated room. The bench should be protected by a large sheet of hardboard. The pupils should be at the back of the room. One gram of the mixture is strongly heated in a metal crown bottle cap before retiring to the back of the room to await a violent reaction. If nothing happens after a minute or so turn off the gas and do not touch the mixture but leave it to cool and dispose of it in water.

Magnesium powder and copper(II) oxide are mixed together and heated very strongly. At the end, traces of white magnesium oxide are left on the ceramic paper.

magnesium + copper(II) oxide → magnesium oxide + copper

$$Mg(s) + CuO(s) \rightarrow MgO(s) + Cu(s)$$

This is an example of a displacement reaction. The less reactive metal, copper, has been displaced from its compound by the more reactive magnesium. Any metal higher in the reactivity series will displace one lower down from a compound.

If you heated copper with magnesium oxide, nothing would happen because copper is less reactive than magnesium. Copper isn't capable of displacing magnesium from magnesium oxide.

## THE REACTION BETWEEN MAGNESIUM AND ZINC OXIDE

Heating magnesium with zinc oxide produces zinc metal:

magnesium + zinc oxide → magnesium oxide + zinc

$$Mg(s) + ZnO(s) \rightarrow MgO(s) + Zn(s)$$

## THE REACTION BETWEEN CARBON AND COPPER(II) OXIDE

**Safety Note:** Wear eye protection. The tube will get very hot and remain so for some time.

A black mixture of carbon and copper(II) oxide is heated in a test-tube. The mixture glows red hot because of the heat given out during the reaction, and you are left with pink-brown copper in the tube.

$$C(s) + 2CuO(s) \rightarrow CO_2(g) + 2Cu(s)$$

Carbon is above copper in the reactivity series and displaces the copper from copper(II) oxide.

## CHEMISTRY ONLY

Carbon is included in the reactivity series because it is important in extracting several metals (including iron) from metal oxides. If the metal is less reactive than carbon (below carbon in the reactivity series), then heating with carbon can be a cheap way of removing oxygen from the oxide to leave the metal. Copper isn't, in fact, extracted like this. This reaction is simply a lab illustration that carbon is above copper in the reactivity series.

### END OF CHEMISTRY ONLY

## OXIDATION AND REDUCTION

**DID YOU KNOW?**

Many metals are found as metal oxides in nature. If they are found as other compounds these are often converted to metal oxides before the metal is extracted.

**OXYGEN TRANSFER**

A substance has been oxidised if it gains oxygen. **Oxidation** is gain of oxygen.

A substance has been reduced if it loses oxygen. **Reduction** is loss of oxygen.

Consider the reaction between magnesium and copper(II) oxide again. Figure 14.5 shows what happens in terms of oxidation and reduction.

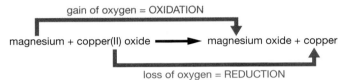

▲ Figure 14.5 Magnesium reacting with copper(II) oxide

A **redox** reaction is one in which both **red**uction and **ox**idation are occurring. Oxidation and reduction always occur together because if something loses oxygen, something else must gain it.

■ A **reducing agent** is a substance that reduces something else. In this case, the magnesium is the reducing agent because it takes the oxygen away from the copper oxide, reducing it.

■ An **oxidising agent** is a substance that oxidises something else. The copper(II) oxide is the oxidising agent in this reaction because it gives oxygen to the magnesium, oxidising it.

An oxidising agent always gets reduced in a chemical reaction because it oxidises something else by giving away its oxygen.

A reducing agent always gets oxidised in a chemical reaction because it takes the oxygen away from something else and therefore gains oxygen itself.

In the reaction between magnesium and zinc oxide on page 146, the magnesium is the reducing agent and the zinc oxide is the oxidising agent. In the reaction between copper(II) oxide and carbon, the carbon is the reducing agent and the copper(II) oxide is the oxidising agent.

**ELECTRON TRANSFER**

We are now going to look very closely at what happens in the reaction between magnesium and copper(II) oxide in terms of the various particles involved. Here is the equation again:

$$Mg(s) + CuO(s) \rightarrow MgO(s) + Cu(s)$$

The magnesium and the copper are metals, and are made of metal atoms, but the copper(II) oxide and the magnesium oxide are both ionic compounds.

The copper(II) oxide contains $Cu^{2+}$ and $O^{2-}$ ions, and the magnesium oxide contains $Mg^{2+}$ and $O^{2-}$ ions. Writing these into the equation gives:

$$Mg(s) + Cu^{2+}O^{2-}(s) \rightarrow Mg^{2+}O^{2-}(s) + Cu(s)$$

The oxide ion ($O^{2-}$) does not change in this reaction. It ends up with a different partner, but is totally unchanged itself.

What is happening in this reaction is that magnesium *atoms* are turning into magnesium *ions*. The magnesium atoms lose electrons to form magnesium ions. These electrons are gained by the copper ions to form copper atoms. We have described this as a redox reaction above in terms of loss/gain of oxygen, but now we can see that it was not the oxygen that was actually the important part, it has not changed! To understand what is happening better, we need a more fundamental definition of oxidation and reduction:

- *oxidation is loss of electrons*
- *reduction is gain of electrons.*

The mnemonic OILRIG can be helpful in remembering this:

**O**xidation

**I**s

**L**oss of electrons

**R**eduction

**I**s

**G**ain of electrons

## DISPLACEMENT REACTIONS INVOLVING SOLUTIONS OF SALTS

Salts are compounds such as copper(II) sulfate, silver nitrate or sodium chloride. You will find a definition of what a salt is on page 173. This section explores some reactions between metals and solutions of salts in water.

### THE REACTION BETWEEN ZINC AND COPPER(II) SULFATE SOLUTION

The copper is displaced by the more reactive zinc. The blue colour of the copper(II) sulfate solution fades as colourless zinc sulfate solution is formed:

$$Zn(s) + CuSO_4(aq) \rightarrow ZnSO_4(aq) + Cu(s)$$

The zinc and the copper are metals consisting simply of atoms, but the copper(II) sulfate and the zinc sulfate are metal compounds and so are ionic.

The equation can be rewritten showing the ions:

$$Zn(s) + Cu^{2+}(aq) + SO_4^{2-}(aq) \rightarrow Zn^{2+}(aq) + SO_4^{2-}(aq) + Cu(s)$$

We can see that the sulfate ions are exactly the same on both sides of the equation – they have not changed at all in the reaction. We call the sulfate ions here *spectator ions*. Removing the spectator ions (because they aren't changed during the reaction) leaves:

$$Zn(s) + Cu^{2+}(aq) \rightarrow Zn^{2+}(aq) + Cu(s)$$

This is called an ionic equation and just shows the things that change in the reaction.

This is another redox reaction.

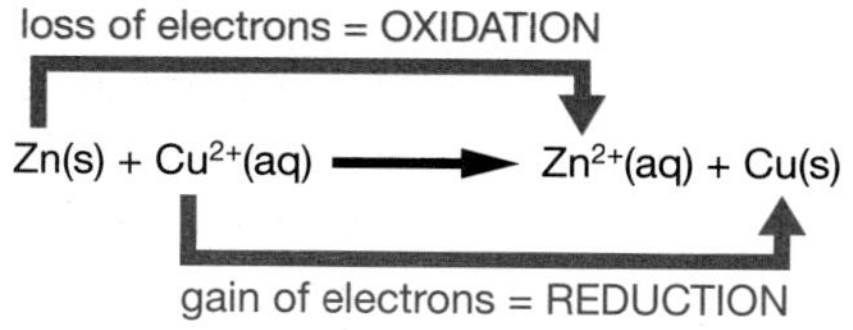

**KEY POINT**

When we looked at the reaction between Mg and CuO above, we also wrote ions in the equation but we did not separate them completely because the ions are in a solid. We only separate everything like this when things are in solution. In a solution the ions are free to move around separately, in a solid they are not.

**KEY POINT**

Which copper(II) salt you started with would not matter, as long as it was soluble in water. Copper(II) chloride or copper(II) nitrate would react in exactly the same way with zinc because the chloride ions or the nitrate ions would again be spectator ions, not participating in the reaction.

loss of electrons = OXIDATION

$$Zn(s) + Cu^{2+}(aq) \longrightarrow Zn^{2+}(aq) + Cu(s)$$

gain of electrons = REDUCTION

▲ Figure 14.6 The reaction between zinc and copper(II) sulfate solution

The Zn atoms are oxidised to $Zn^{2+}$ ions because they lose electrons. The $Cu^{2+}$ ions gain electrons and are reduced to Cu atoms. We can split up the ionic equation to show the individual oxidation and reduction processes:

$$Zn(s) \rightarrow Zn^{2+}(aq) + 2e^- \quad \text{oxidation}$$

$$Cu^{2+}(aq) + 2e^- \rightarrow Cu(s) \quad \text{reduction}$$

These are called *ionic half-equations* as they each only show one of the processes (either oxidation or reduction) occurring in the reaction. In reality, these processes cannot occur one without the other: if something gains electrons, it has to get them from somewhere, so something else must lose electrons.

## OXIDISING AND REDUCING AGENTS IN TERMS OF ELECTRONS

A couple more definitions (this time in terms of electrons):

- *An oxidising agent is something that oxidises something else by taking electrons away from it. Oxidising agents accept electrons and therefore are reduced in a reaction.*

- *A reducing agent is something that reduces something else by giving electrons to it. Reducing agents give away electrons and therefore are oxidised in a reaction.*

Let us look again at the overall ionic equation for the displacement reaction between zinc and copper ions:

$$Zn(s) + Cu^{2+}(aq) \rightarrow Zn^{2+}(aq) + Cu(s)$$

The Zn reduces the $Cu^{2+}$ by giving electrons to it and therefore the Zn is the reducing agent. In the process of the reaction the Zn is oxidised because it has given electrons away (lost them).

The $Cu^{2+}$ oxidises the Zn by taking electrons away from it, therefore the $Cu^{2+}$ is the oxidising agent. In the process the $Cu^{2+}$ is reduced because it has taken electrons (gained them).

**HINT**

You now have two definitions of oxidation (and its reverse, reduction). Which one should you use? Use whichever is simpler in the case you are asked about! Both definitions are true. Don't worry too much about this at the moment. With a little experience, you will find it is obvious which one you need to use in various cases.

## THE REACTION BETWEEN COPPER AND SILVER NITRATE SOLUTION

Silver is below copper in the reactivity series, so a coil of copper wire in silver nitrate solution will produce metallic silver. Figure 14.7 shows the silver being produced as a mixture of grey 'fur' and delicate crystals. Notice the solution becoming blue, as copper(II) nitrate is produced:

$$Cu(s) + 2AgNO_3(aq) \rightarrow Cu(NO_3)_2(aq) + 2Ag(s)$$

This time the nitrate ions are spectator ions, and the final version of the ionic equation looks like this:

$$Cu(s) + 2Ag^+(aq) \rightarrow Cu^{2+}(aq) + 2Ag(s)$$

This is another redox reaction.

The Cu is the reducing agent (it is oxidised) because it gives electrons to the $Ag^+$ ions to reduce them to Ag. The $Ag^+$ ion is the oxidising agent (it is reduced) because it takes electrons away from the Cu atoms.

▲ Figure 14.7 Displacing silver from silver nitrate solution

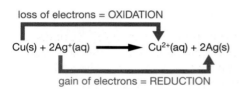

loss of electrons = OXIDATION

$$Cu(s) + 2Ag^+(aq) \longrightarrow Cu^{2+}(aq) + 2Ag(s)$$

gain of electrons = REDUCTION

▲ Figure 14.8 Copper reacting with silver nitrate solution

The ionic half-equations are:

$$Cu(s) \rightarrow Cu^{2+}(aq) + 2e^- \quad \text{oxidation}$$
$$Ag^+(aq) + e^- \rightarrow Ag(s) \quad \text{reduction}$$

# REACTIONS OF METALS WITH WATER

## A GENERAL SUMMARY

### METALS ABOVE HYDROGEN IN THE REACTIVITY SERIES

Metals above hydrogen in the reactivity series react with water (or steam) to produce hydrogen.

If the metal reacts with *cold water*, the metal *hydroxide* and hydrogen are formed.

metal + cold water → metal hydroxide + hydrogen

If the metal reacts with *steam*, the metal *oxide* and hydrogen are formed.

metal + steam → metal oxide + hydrogen

As you move down the reactivity series, the reactions become less and less vigorous.

### METALS BELOW HYDROGEN IN THE REACTIVITY SERIES

Metals below hydrogen in the reactivity series (such as copper) don't react with water or steam. This is why copper can be used for both hot and cold water pipes.

## REACTIONS OF POTASSIUM, SODIUM OR LITHIUM WITH COLD WATER

These reactions are described in detail on pages 125–126. They are very vigorous reactions, but become less violent in the following order: potassium > sodium > lithium. The equations all look like this:

$$2M(s) + 2H_2O(l) \rightarrow 2MOH(aq) + H_2(g)$$

Replace M by K, Na or Li, depending on which metal you want.

## REACTION OF CALCIUM WITH COLD WATER

Calcium reacts gently with cold water. The grey granules sink, but are carried back to the surface again as bubbles of hydrogen are formed around them. The mixture becomes warm as heat is produced.

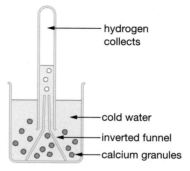

hydrogen collects

cold water

inverted funnel

calcium granules

▲ Figure 14.9 Calcium reacting with cold water

Calcium hydroxide is formed. This isn't very soluble in water. Some of it dissolves to give a colourless solution, but most of it is left as a white, insoluble solid.

$$Ca(s) + 2H_2O(l) \rightarrow Ca(OH)_2(aq \text{ or } s) + H_2(g)$$

### REACTION OF MAGNESIUM WITH COLD WATER

There is almost no reaction. If the magnesium is very clean, a few bubbles of hydrogen form on it, but the reaction soon stops again. This is because the magnesium becomes coated with insoluble magnesium hydroxide, which prevents any more water coming into contact with the magnesium.

### REACTION OF MAGNESIUM WITH STEAM

Magnesium ribbon can be heated in steam using the apparatus shown in Figure 14.10.

#### KEY POINT

You might wonder why there is no description for aluminium and steam. The reactivity of aluminium is supposed to be between that of magnesium and that of zinc. However, aluminium has only a very slow reaction with steam because it is covered in a very thin, but very strong, layer of aluminium oxide. It only really shows its true reactivity if that layer can be penetrated in some way. Water or steam don't do that very well. We will talk about this again when we look at reactions between metals and acids.

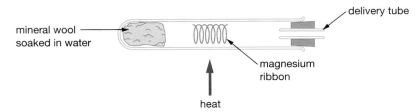

▲ Figure 14.10 Magnesium reacting with steam

The mineral wool isn't heated directly. Enough heat moves back along the test-tube to turn the water to steam.

The magnesium burns with a bright white flame in the steam, producing hydrogen, which can be ignited at the end of the delivery tube. White magnesium oxide is formed:

$$Mg(s) + H_2O(g) \rightarrow MgO(s) + H_2(g)$$

### REACTIONS OF ZINC OR IRON WITH STEAM

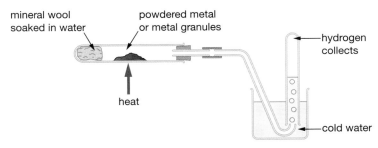

▲ Figure 14.11 Zinc or iron reacting with steam

#### KEY POINT

Care has to be taken during this experiment to avoid 'suck-back'. If you stop heating while the delivery tube is still under the surface of the water, water is sucked back into the hot tube, which often results in it cracking.

With both zinc and iron, the hydrogen comes off slowly enough to be collected. Neither metal burns.

#### WITH ZINC

Zinc oxide is formed. This is yellow when it is hot, but white on cooling.

$$Zn(s) + H_2O(g) \rightarrow ZnO(s) + H_2(g)$$

#### WITH IRON

The iron becomes slightly darker grey. A complicated oxide is formed, called tri-iron tetroxide, $Fe_3O_4$:

$$3Fe(s) + 4H_2O(g) \rightarrow Fe_3O_4(s) + 4H_2(g)$$

#### KEY POINT

Notice that in these equations, water now has a state symbol (g) because we are talking about it as steam.

#### REMINDER

Remember that *metals below hydrogen* in the reactivity series, such as copper, *don't react* with water or steam.

# REACTIONS OF METALS WITH DILUTE ACIDS

## ACTIVITY 4

### ▼ PRACTICAL: INVESTIGATION INTO THE REACTIONS BETWEEN METALS AND DILUTE ACIDS

- Set up four test-tubes and put about 2 cm³ of dilute hydrochloric acid into each one.
- Put a small piece of magnesium, zinc, iron or copper into each test-tube and observe any reaction that occurs.
- If there is fizzing, collect or trap the gas and test with a lighted splint – a squeaky pop indicates the presence of hydrogen gas.
- Repeat the experiments with dilute sulfuric acid.

The results that we could obtain from the experiments with dilute hydrochloric acid are shown in Figure 14.12 and Table 14.1.

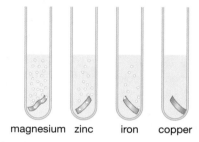

magnesium    zinc    iron    copper

▲ Figure 14.12 The reactions between metals and hydrochloric acid

Table 14.1 The results of experiments on the reaction between hydrochloric acid and some metals

| Metal | Reaction with dilute hydrochloric acid |
|---|---|
| Magnesium | Reacts vigorously with lots of fizzing. The gas produced gave a squeaky pop with a lighted splint. A colourless solution is formed. The test-tube gets hot. |
| Zinc | Steady reaction. Fizzing. Enough gas eventually collected to produce a squeaky pop with a lighted splint. A colourless solution formed. The test-tube gets warmer. |
| Iron | Slow fizzing. Very little gas was collected in the time available. A very pale green solution formed. The test-tube got slightly warmer. |
| Copper | No change. |

These reactions are all **exothermic** (they give out heat). See Chapter 19 for more about exothermic reactions.

Now that we have noticed that there is a temperature change in the reaction we could measure this temperature change to compare the reactivity of the metals. However, in order to do this we will have to think more carefully about the things that we have to keep the same in order to make this a valid (fair) test.

The volume and concentration of the hydrochloric acid are perhaps the most obvious things that we have to keep the same, but what about the mass of the metal? This is a more difficult question. If we use 0.24 g of metal in each experiment, we can work out that this is 0.24/24 = 0.010 mol magnesium but 0.24/56 = 0.0040 mol iron.

If the reaction occurs very slowly, the heat will be given out over a long period of time. As the heat is being given out it will also be lost to the surroundings and the temperature change will be quite low. If the reaction occurs more quickly, the heat is given out more quickly to the solution, and there is less time for heat to be lost to the surroundings – we obtain a larger temperature change. A large surface area causes a faster rate of reaction, so we should use metal powders in these reactions. These things are discussed more in Chapters 19 and 20.

Safety Note: Wear eye protection and avoid splashing the acid when stirring.

Now, chemical reactions depend on how many particles are present: more moles and therefore more particles are present in the Mg experiment and more heat would be given out because of this. It is therefore important that we use the same number of moles in each experiment rather than the same number of grams.

The other variable that we should control is the surface area of the metal. Although this does not affect the overall amount of heat given out in the reaction, it will affect the speed at which the heat is given out, which can affect the temperature change.

We will use the following procedure to investigate this reaction quantitatively (involving numbers):

■ Measure out 50 cm$^3$ of 1 mol/dm$^3$ hydrochloric acid using a 50 cm$^3$ measuring cylinder (this amount was chosen so that the hydrochloric acid is in excess – there is more than enough to react with all the metal in each case).

■ Pour the hydrochloric acid into a polystyrene cup (this is an insulator).

■ Measure the initial temperature of the hydrochloric acid.

■ Weigh out 0.010 mol of magnesium (0.010 × 24 = 0.24 g) powder.

■ Add the magnesium powder to the polystyrene cup, stirring rapidly, and measure the maximum temperature reached.

■ Repeat with the other metals using 0.010 mol in each case.

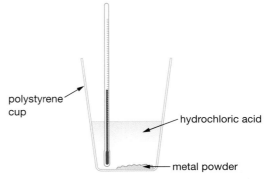

▲ Figure 14.13 Measuring the temperature change when a metal reacts with dilute hydrochloric acid.

The data for this experiment could be:

| Metal | Mass/g | Initial temperature/°C | Maximum temperature/°C |
|---|---|---|---|
| magnesium | 0.24 | 18 | 40 |
| zinc | 0.65 | 19 | 26 |
| iron | 0.56 | 18 | 22 |
| copper | 0.64 | 18 | 18 |

We can work out the temperature change for each experiment and plot this as a bar chart. We use a bar chart because the type of metal is not continuous data: the metal can be either magnesium or zinc, it cannot be anything in between.

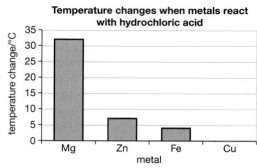

**Temperature changes when metals react with hydrochloric acid**

▲ Figure 14.14 Magnesium gives out the most heat when reacted with hydrochloric acid.

We can see from this data that magnesium caused the greatest temperature change and so is the most reactive metal of the four; copper caused no change and is the least reactive.

## A GENERAL SUMMARY

The pattern for the reaction of metals with acids is the same as for the reaction between the metals and water, but in each case the reaction is much more vigorous.

### METALS ABOVE HYDROGEN IN THE REACTIVITY SERIES

Metals above hydrogen react with acids to form a salt (e.g. magnesium sulfate or zinc chloride) and hydrogen. The higher the metal in the series, the more violent the reaction.

metal + acid → salt + hydrogen

metal + dilute sulfuric acid → metal sulfate + hydrogen

metal + dilute hydrochloric acid → metal chloride + hydrogen

### METALS BELOW HYDROGEN IN THE REACTIVITY SERIES

Metals such as copper, silver and gold do not react with simple dilute acids such as sulfuric or hydrochloric acid.

### POTASSIUM, SODIUM, LITHIUM AND CALCIUM WITH DILUTE ACIDS

These are too reactive to add safely to acids, the reaction is too violent. Calcium can be used if the acid is very dilute.

### METALS FROM MAGNESIUM TO IRON IN THE REACTIVITY SERIES

**Magnesium** reacts vigorously with cold dilute acids, and the mixture becomes hot. A colourless solution of magnesium sulfate or chloride is formed. With dilute sulfuric acid:

$$Mg(s) + H_2SO_4(aq) → MgSO_4(aq) + H_2(g)$$

**Aluminium** is slow to start reacting, but after warming it reacts very vigorously. There is a very thin, but very strong, layer of aluminium oxide on the surface of the aluminium, which stops the acid from getting to it. On heating, the acid removes this layer, and the aluminium can show its true reactivity. With dilute hydrochloric acid:

$$2Al(s) + 6HCl(aq) → 2AlCl_3(aq) + 3H_2(g)$$

*Zinc* and *iron* react slowly in the cold, but more rapidly on heating. The vigour of the reactions is less than that of aluminium. The zinc forms zinc sulfate or zinc chloride and hydrogen. The iron forms iron(II) sulfate or iron(II) chloride and hydrogen. For example:

$$Zn(s) + H_2SO_4(aq) \rightarrow ZnSO_4(aq) + H_2(g)$$
$$Fe(s) + 2HCl(aq) \rightarrow FeCl_2(aq) + H_2(g)$$

▲ Figure 14.15 Iron reacting with dilute hydrochloric acid

## FINDING THE APPROXIMATE POSITION OF A METAL IN THE REACTIVITY SERIES USING WATER AND DILUTE ACIDS

Add a very small piece of metal to some cold water. If there is any rapid reaction, then the metal must be above magnesium in the reactivity series.

If there isn't any reaction, add a small amount of metal to some dilute hydrochloric acid (or dilute sulfuric acid). If there isn't any reaction in the cold acid, warm it carefully.

If there is still no reaction, the metal is probably below hydrogen in the reactivity series. If there is a reaction, then it is somewhere between magnesium and hydrogen.

## MAKING PREDICTIONS USING THE REACTIVITY SERIES

You can make predictions about the reactions of unfamiliar metals if you know their position in the reactivity series.

**A PROBLEM INVOLVING MANGANESE**

Suppose you have a question as follows.

Manganese (Mn) lies between aluminium and zinc in the reactivity series, and forms a 2+ ion. Solutions of manganese(II) salts are very pale pink (almost colourless).

a Use the reactivity series to predict whether manganese will react with copper(II) sulfate solution. If it will react, describe what you would see, name the products and write an equation for the reaction.

b Explain why you would expect manganese to react with steam. Name the products of the reaction and write the equation.

potassium
sodium
lithium
calcium
magnesium
aluminium
**manganese**
zinc
iron
(hydrogen)
copper

▲ Figure 14.16 Where is manganese in the reactivity series?

a **The reaction between manganese and copper(II) sulfate solution**

Manganese is above copper in the reactivity series and so will displace it from the copper(II) sulfate:

A pink-brown deposit of copper will be formed. The colour of the solution will fade from blue and leave a very pale pink (virtually colourless) solution of manganese(II) sulfate.

$$Mn(s) + CuSO_4(aq) \rightarrow MnSO_4(aq) + Cu(s)$$

b **The reaction between manganese and steam**

Manganese is above hydrogen in the reactivity series and so reacts with steam to give hydrogen and the metal oxide, in this case manganese(II) oxide.

You couldn't predict the colour of the manganese(II) oxide, and the question doesn't ask you to do this.

$$Mn(s) + H_2O(g) \rightarrow MnO(s) + H_2(g)$$

## RUSTING OF IRON

Iron **rusts** in the presence of *oxygen* and *water*. Rusting occurs with iron and the most common alloy of iron, mild steel (see below).

The formula of rust is $Fe_2O_3 \cdot xH_2O$, where *x* is a variable number. It can be called 'hydrated iron(III) oxide'. Rusting can be described as a redox reaction – the iron is oxidised.

**HINT**

**Warning!** Many metals corrode, but it is only the corrosion of iron that is referred to as *rusting*.

**KEY POINT**

Forming rust from iron is a surprisingly complicated process. The iron loses electrons to form iron(II) ions, $Fe^{2+}$, which are then oxidised by the air to iron(III) ions, $Fe^{3+}$. Reactions involving the water produce the actual rust.

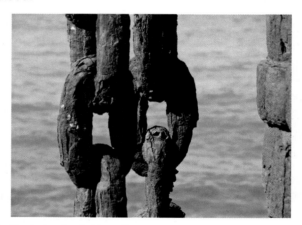

▲ Figure 14.17 Rusting is accelerated by salty water.

### PREVENTING RUSTING BY USING BARRIERS

The most obvious way of preventing rusting is to keep water and oxygen away from the iron. You can do this by painting it, coating it in oil or grease, or covering it in plastic. Coating the iron with a metal *below* it in the reactivity series (e.g. coating steel with tin for tin cans) is also a **barrier method**.

Barrier methods are usually quite cheap ways of preventing rusting. A problem with barrier methods is that once the coating is broken, the iron underneath is exposed to oxygen and water and the iron will rust (even the bits that are not directly exposed to the air/water).

### PREVENTING RUSTING BY GALVANISING

**Galvanised** iron is iron that is coated with a layer of zinc. As long as the zinc layer is unscratched, it serves as a barrier to air and water. However, the iron still doesn't rust, even when some of the zinc on the surface is scratched away

## EXTENSION WORK

During the process, the zinc loses electrons (is oxidised) to form zinc ions.

$$Zn(s) \rightarrow Zn^{2+}(aq) + 2e^-$$

These electrons flow into the iron. Any iron atom which has lost electrons to form an ion immediately regains them. If the iron can't form ions, it can't rust.

to expose the iron. This is because the zinc is more reactive than iron, and so reacts with oxygen/water more readily than the iron does. The zinc corrodes instead of the iron.

▲ Figure 14.18 Galvanised iron doesn't rust even in constant contact with air and water.

## PREVENTING RUSTING BY USING SACRIFICIAL PROTECTION

▲ Figure 14.19 A sacrificial anode on a ship

Zinc, magnesium or aluminium blocks are attached to metal hulls or keels of ships to prevent the iron/steel from rusting. For this to work, you have to use a metal that is *more reactive than iron;* the more reactive metal reacts (is oxidised) more readily in the presence of oxygen/water than the iron. The corrosion of the more reactive metal prevents the iron from rusting. Such blocks are called sacrificial anodes.

These sacrificial anodes have to be replaced occasionally when all the more reactive metal has been oxidised. This type of protection is used on large structures where it would be very difficult to use a barrier method effectively.

Galvanising is a combination of a barrier method and sacrificial protection.

Underground pipelines are also protected using sacrificial anodes. In this case, lumps of magnesium are attached at intervals along the pipe. The very reactive magnesium corrodes in preference to the iron. The electrons produced as the magnesium forms its ions prevent the ionisation of the iron.

## CHAPTER QUESTIONS

SKILLS  CRITICAL THINKING

1  a  List the following metals in order of decreasing reactivity: aluminium, copper, iron, sodium.

b  Some magnesium powder was mixed with some copper(II) oxide and heated strongly. There was a vigorous reaction, producing a lot of sparks and a bright flash of light.

i  State the names of the products of the reaction.

SKILLS  PROBLEM SOLVING

ii  Write a balanced symbol equation for the reaction.

SKILLS  REASONING

iii  Explain which substance in the reaction has been reduced.

iv  Explain which substance is the oxidising agent.

c  If a mixture of zinc powder and cobalt(II) oxide is heated, the following reaction occurs:

$$Zn(s) + CoO(s) \rightarrow ZnO(s) + Co(s)$$

i  Explain which metal is higher in the reactivity series.

SKILLS  CRITICAL THINKING

ii  The zinc can be described as a reducing agent. Using this example, explain what is meant by the term 'reducing agent'.

iii  Explain which substance in this reaction has been oxidised.

d Aluminium, chromium and manganese are all moderately reactive metals. Use the following information to arrange them in the correct reactivity series order. Start with the most reactive one.

- Chromium is manufactured by heating chromium(III) oxide with aluminium.
- If manganese is heated with aluminium oxide there is no reaction.
- If manganese is heated with chromium(III) oxide, chromium is produced.

2 Study the following equations and in each case decide whether the substance in bold type is oxidised or reduced. Explain your choice in terms of either oxygen transfer or electron transfer as appropriate.

a **Zn(s)** + CuO(s) → ZnO(s) + Cu(s)

b **$Fe_2O_3$(s)** + 3C(s) → 2Fe(s) + 3CO(g)

c **Mg(s)** + $Zn^{2+}$(aq) → $Mg^{2+}$(aq) + Zn(s)

d Zn(s) + **$Cu^{2+}$(aq)** → $Zn^{2+}$(aq) + Cu(s)

3 The equation for the reaction when solid magnesium and solid lead(II) oxide are heated together is:

Mg(s)+ PbO(s) → MgO(s) + Pb(s)

Explain what this tells you about the position of lead in the reactivity series.

4 Some iron filings were shaken with some copper(II) sulfate solution. The ionic equation for the reaction is:

Fe(s) + $Cu^{2+}$(aq) → $Fe^{2+}$(aq) + Cu(s)

a Explain any one change that you would observe during this reaction.

b Explain which substance has been oxidised in this reaction.

c Write down the full (not ionic) equation for this reaction.

5 Some experiments were carried out to place the metals copper, nickel and silver in reactivity series order.

*Experiment 1*: A piece of copper was placed in some green nickel(II) sulfate solution. There was no change to either the copper or the solution.

*Experiment 2*: A coil of copper wire was suspended in some silver nitrate solution. A furry grey growth appeared on the copper wire, out of which grew spiky silvery crystals. The solution gradually turned from colourless to blue.

a Use this information to place copper, nickel and silver in reactivity series order, starting with the most reactive one.

b In another experiment, a piece of nickel was placed in some copper(II) sulfate solution.

i State one change that you would observe during this reaction.

ii Write the full balanced equation for this reaction.

iii Write the ionic equation for this reaction and use it to explain which substance has been oxidised during the reaction.

6 a Look carefully at the following equations and then explain what you can say about the position of the metal X in the reactivity series.

X(s) + 2HCl(aq) → $XCl_2$(aq) + $H_2$(g)

X(s) + $CuSO_4$(aq) → $XSO_4$(aq) + Cu(s)

X(s) + $FeSO_4$(aq) → no reaction

b Decide whether X will react with the following substances. If it will react, state the names of the products, if not state 'no reaction'.

    i Silver nitrate solution
    ii Zinc oxide
    iii Cold water
    iv Copper(II) chloride solution
    v Dilute sulfuric acid

7 If you add some powdered aluminium to a small amount of cold dilute hydrochloric acid in a boiling tube, very little happens. If you warm this gently, it starts to fizz very rapidly.

a State the name of the gas given off to produce the fizzing.

b If you used an excess of hydrochloric acid, the result would be a colourless solution. State the name of the solution.

c Write the full balanced equation for the reaction.

d Explain why the aluminium hardly reacts at all with the dilute acid in the cold, but reacts vigorously after even gentle heating.

8 If you have some small pieces of the metal titanium and any simple apparatus that you might need, describe how you would find out the approximate position of titanium in the reactivity series, using only water and dilute hydrochloric acid. You only need to find out that the reactivity is 'similar to iron' or 'similar to magnesium', for example. Your experiments should be done in an order that guarantees maximum safety. For example, if titanium's reactivity turned out to be similar to potassium, dropping it into dilute hydrochloric acid wouldn't be a good idea!

9 In the past, cars were made from mild steel, which was then painted. In more modern cars, the mild steel is galvanised before it is painted.

a State the conditions under which iron/steel rusts.

b Explain how painting prevents iron/steel from rusting.

c What is meant by *galvanised steel*?

d Describe and explain the effect that galvanised steel has on the life of the car.

10 A student tested four unknown metals (the names are all imaginary) with different metal salt solutions. She put a tick (✔) if there was a reaction and a cross (✖) if there was no change. Some of her results are shown in the table.

| | Pearsonium chloride solution | Mollium chloride solution | Rosium chloride solution | Amelium chloride solution |
|---|---|---|---|---|
| pearsonium | ✖ | | ✖ | |
| mollium | ✔ | ✖ | ✔ | ✔ |
| rosium | | | ✖ | ✖ |
| amelium | | | | ✖ |

a Copy and complete the table.

b Arrange the metals in an order of reactivity (most reactive first).

c Copy and complete the equation:

    mollium + pearsonium chloride →

**CHEMISTRY ONLY**

# 15 EXTRACTION AND USES OF METALS

Metals are some of the most important materials that we use in everyday life. This chapter explores the principles behind the extraction of metals from their ores. We will also look at the properties of alloys and the uses of some metals.

## LEARNING OBJECTIVES

- Know that most metals are extracted from ores found in the Earth's crust and that unreactive metals are often found as the uncombined element

- Explain how the method of extraction of a metal is related to its position in the reactivity series, illustrated by carbon extraction for iron and electrolysis for aluminium

- Be able to comment on a metal extraction process, given appropriate information *(detailed knowledge of the processes used in the extraction of a specific metal is not required)*

- Explain the uses of aluminium, copper, iron and steel in terms of their properties *(the types of steel will be limited to low-carbon (mild), high-carbon and stainless)*

- Know that an alloy is a mixture of a metal and one or more elements, usually other metals or carbon

- Explain why alloys are harder than pure metals

## EXTRACTING METALS FROM THEIR ORES

### MINERALS AND ORES

Most metals are found in the Earth's crust combined with other elements. The individual compounds are called **minerals**.

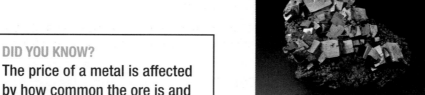

▲ Figure 15.1 Pyrite (iron pyrites), $FeS_2$     ▲ Figure 15.2 Magnetite, $Fe_3O_4$

Figures 15.1 and 15.2 show samples of some iron-containing minerals; they are normally found mixed with other unwanted minerals in rocks. An **ore** is a sample of rock that contains enough of a mineral for it to be worthwhile to extract the metal. Most metals are extracted from ores found in the Earth's crust.

A few very unreactive metals, such as gold, are found **native**. That means that they exist naturally as the uncombined element. Silver and copper are also sometimes found native, although much more rarely.

### EXTRACTING THE METAL

Many ores contain either oxides or compounds that are easily converted to oxides. Sulfides such as sphalerite (zinc blende), ZnS, can be easily converted into an oxide by heating in air, a process known as **roasting**.

$$2ZnS(s) + 3O_2(g) \rightarrow 2ZnO(s) + 2SO_2(g)$$

**REMINDER**

If you have forgotten about oxidation and reduction, you might find it useful to re-read Chapter 14.

To obtain the metal from the oxide, you have to remove the oxygen. Removal of oxygen is called reduction. Metals exist as positive ions in their ionic compounds, and to produce the metal you would have to add electrons to the positive ion. Addition of electrons is also called reduction.

## METHODS OF EXTRACTION AND THE REACTIVITY SERIES

How a metal is extracted depends to a large extent on its position in the reactivity series. A manufacturer obviously wants to use the cheapest possible method of extracting a metal from an ore. There are two main economic factors to take into account:

- the cost of energy
- the cost of the reducing agent.

potassium

sodium

calcium

magnesium

aluminium

**(carbon)**

zinc

iron

copper

▲ Figure 15.3 A part of the reactivity series

### METALS BELOW CARBON IN THE REACTIVITY SERIES

For a metal below carbon in the reactivity series, the cheapest method of reducing the ore is often to heat it with carbon.

The extraction of iron is a good example of this. One of the main ores of iron contains a high percentage of iron(III) oxide. The iron can be extracted from this by heating with carbon:

$$Fe_2O_3(s) + 3C(s) \rightarrow 2Fe(l) + 3CO(g)$$

Carbon is higher in the reactivity series than iron and will take the oxygen away from the iron oxide. This is a redox reaction; the $Fe_2O_3$ is reduced to Fe in the reaction and the C is oxidised to CO. In this reaction the carbon is the reducing agent, it reduces the iron(III) oxide.

The extraction of iron is carried out in a blast furnace. The process is a bit more complicated than we have shown here. Other reactions also occur and the main reducing agent is actually carbon monoxide:

$$Fe_2O_3(s) + 3CO(g) \rightarrow 2Fe(l) + 3CO_2(g)$$

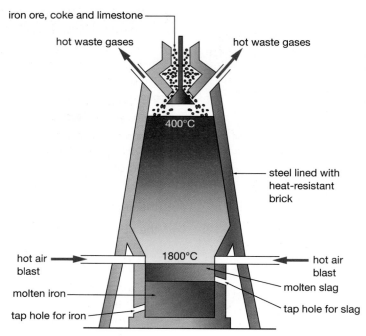

iron ore, coke and limestone

hot waste gases

hot waste gases

400°C

steel lined with
heat-resistant
brick

hot air
blast

1800°C

hot air
blast

molten slag

molten iron

tap hole for iron

tap hole for slag

▲ Figure 15.4 Iron is extracted from its ore in a blast furnace.

## METALS ABOVE CARBON IN THE REACTIVITY SERIES

Ores of metals higher in the reactivity series than zinc can't be reduced using carbon at reasonable temperatures. This is because the metals are more reactive than carbon and therefore carbon cannot take the oxygen away from the metal oxide. Metals above zinc are usually produced by *electrolysis*.

Aluminium is extracted by the electrolysis of aluminium oxide ($Al_2O_3$) dissolved in a molten salt called cryolite.

**KEY POINT**

This is essentially the same as the electrolysis of molten aluminium oxide. The cryolite is simply used to make the process more economical. Aluminium oxide melts at over 2000 °C but if it is dissolved in molten cryolite the whole process can be carried out at around 1000 °C. This saves a lot of energy and therefore also saves a lot of money.

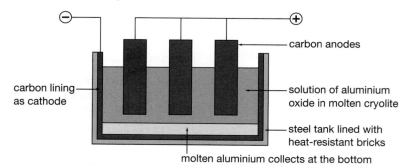

carbon anodes

carbon lining
as cathode

solution of aluminium
oxide in molten cryolite

steel tank lined with
heat-resistant bricks

molten aluminium collects at the bottom

▲ Figure 15.5 Aluminium is produced by electrolysis.

At the cathode the reaction is $Al^{3+} + 3e^- \rightarrow Al$     reduction

At the anode the reaction is $2O^{2-} \rightarrow O_2 + 4e^-$     oxidation

This process requires huge amounts of electricity, which makes it expensive. A metal such as aluminium, which has to be extracted by electrolysis, is much more expensive than one like iron, which can be extracted by reduction with carbon, despite the fact that aluminium is more abundant in the Earth's crust than iron. Iron (and all other metals) could also be extracted by electrolysis but because the process is so expensive we do not use it unless we have to.

Some metals, such as titanium, are extracted by heating the compound with a more reactive metal. This is also an expensive method because the more reactive metal itself will have had to be extracted by an expensive process first.

## ALLOYS

*An* **alloy** *is a mixture of a metal with, usually, other metals or carbon.* For example, brass is a mixture of copper and zinc, and steel is an alloy of iron with carbon.

*Alloys are harder than the individual pure metals from which they are made.* In an alloy, the different metals/elements have slightly differently sized atoms. This breaks up the regular lattice arrangement and makes it more difficult for the layers of ions to slide over each other.

▲ Figure 15.6 A brass propeller. Brass is an alloy.

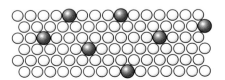

▲ Figure 15.7 Atoms in an alloy

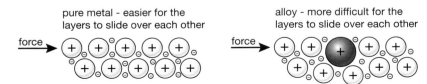

▲ Figure 15.8 The layers cannot slide over each other as easily in an alloy as in a pure metal.

Other common alloys include bronze (a mixture of copper and tin), stainless steel (an alloy of iron with chromium and usually nickel), and the mixture of copper and nickel (cupronickel) which is usually used to make 'silver' coins.

## PROPERTIES AND USES OF SOME METALS

### USES OF ALUMINIUM

Pure aluminium isn't very strong, so aluminium alloys are normally used instead. The aluminium can be strengthened by adding other elements, such as silicon, copper or magnesium.

Figure 15.9 shows one of the uses of aluminium.

▲ Figure 15.9 Aluminium resists corrosion, it has a low density and is a good conductor of electricity. The aluminium in the cables in the photo is strengthened by a core of steel.

Other uses of aluminium include planes, electricity cables, and pots and pans. Its uses depend on its low density and strength (when alloyed), its ability to conduct electricity and heat, and its ability to resist corrosion.

Aluminium resists corrosion because it has a very thin, but very strong, layer of aluminium oxide on the surface. This prevents anything else reaching the surface and reacting with it.

# PROPERTIES AND USES OF DIFFERENT KINDS OF STEEL

There are lots of different alloys of iron that contain various proportions of carbon and sometimes other metals. An alloy of iron and carbon is called steel.

## MILD STEEL

Mild steel is the name given to an alloy of iron containing up to about 0.25% of carbon. This small amount of carbon increases the hardness and strength of the iron. Mild steel is a strong material that can be easily hammered into various shapes (malleable) and drawn into wires (ductile). It is used for (among other things) nails, car bodies, ship building, girders and bridges.

▲ Figure 15.10 Mild steel is used for car bodies . . .

A disadvantage of mild steel is that it rusts when exposed to oxygen and water. It is also about three times denser than aluminium. Some car bodies are made from aluminium. This has the advantage that the car body will not rust and, because the car is lighter, less fuel will have to be used.

▲ Figure 15.11 Some cars have a body made of aluminium. In this photo the car body has been painted. The paint is mainly for decoration; it is not required to prevent rusting.

## HIGH-CARBON STEEL

High-carbon steel is iron containing about 0.6–1.2% carbon (these amounts are variable). High-carbon steel is harder and more resistant to wear than mild steel but more brittle (not as malleable and ductile). It is used for cutting tools. High-carbon steel also usually contains small amounts of manganese.

▲ Figure 15.12 High-carbon steel is used to make cutting tools.

## STAINLESS STEEL

Figure 15.13 Some cars have a stainless steel body. They do not need to be painted.

Stainless steel is an alloy of iron with chromium and often nickel. Chromium forms a strong oxide layer in the same way as aluminium, and this oxide layer protects the iron as well. Stainless steel is therefore very resistant to corrosion.

Uses include kitchen sinks, saucepans, knives and forks, and gardening tools, but there are also major uses for it in the brewing (making beer), dairy (milk and cheese production) and chemical industries, where corrosion-resistant vessels are essential. Stainless steel is significantly more expensive than mild steel.

Table 15.1 A summary of types of steel

| Type of steel | Iron mixed with | Some uses |
|---|---|---|
| mild steel | up to 0.25% carbon | nails, car bodies, ship building, girders |
| high-carbon steel | 0.6–1.2% carbon | cutting tools, masonry nails |
| stainless steel | chromium (and nickel) | cutlery, cooking utensils, kitchen sinks |

## COPPER

Figure 15.14 Copper pots and pans

Some properties and uses of copper and its alloys are summarised in Table 15.2.

Table 15.2 Copper and its alloys have a variety of uses

| Use | Property |
|---|---|
| electrical wires | very good conductor of electricity and ductile |
| pots and pans | very good conductor of heat (thermal conductor), very unreactive and malleable |
| water pipes | unreactive – does not react with hot or cold water and malleable |
| surfaces in hospitals | antimicrobial properties and malleable |

## CHAPTER QUESTIONS

SKILLS  REASONING, PROBLEM SOLVING

1  Explain why different methods are used to extract aluminium and iron from their ores and write equations for the reactions occurring.

SKILLS  CREATIVITY

2  Sodium is the sixth most abundant element in the Earth's crust, occurring in large quantities as common salt, NaCl, and yet sodium metal wasn't first produced until the early 19th century.

   a  From your knowledge of the position of sodium in the reactivity series, suggest a method for manufacturing sodium from sodium chloride. You aren't expected to give details of the manufacturing process, but should describe and explain (including equation(s) where relevant) how sodium is formed in your process.

SKILLS  REASONING

   b  Suggest why sodium wasn't produced until the early 19th century.

   c  Suggest three other metals which might have been first isolated from their compounds at the same sort of time.

3 Lead is between iron and copper in the reactivity series. Explain a method that could be used to extract lead from its ore.

4 The extraction of chromium from its ore is quite complicated, but in the final stage chromium(III) oxide is reacted with aluminium to form chromium and aluminium oxide.

a Write a balanced equation for this reaction.

b Explain what the reducing agent is in this reaction.

c Explain whether chromium is more or less reactive than aluminium.

5 The first stage in the extraction of titanium from its ore is heating the ore with chlorine and carbon. The equation for the reaction is:

$$2FeTiO_3 + 7Cl_2 + C \rightarrow 2TiCl_4 + 2FeCl_3 + CO$$

This equation is not completely balanced.

The $TiCl_4$ is separated from the $FeCl_3$ and then, in the second stage, heated with magnesium.

a Balance the equation for the first stage in this process.

b Write a balanced chemical equation for the second stage in the process.

c Explain whether magnesium is more or less reactive than titanium.

d Explain a method that could be used to extract magnesium from its ore.

e Explain whether you would expect titanium to be more or less expensive than magnesium.

6 This question is about the properties and uses of some metals.

a Aluminium alloys are used in aircraft construction.

  i Explain which property of aluminium makes it particularly suitable for this purpose.

  ii Explain, in terms of structure and bonding, why aluminium alloys are used in preference to pure aluminium.

  iii Some cars have aluminium bodies rather than mild steel bodies. Explain two advantages and one disadvantage of using aluminium rather than mild steel.

b Explain two properties of copper that make it suitable for use in:

  i electrical wiring

  ii water pipes

c Explain why car bodies are made from mild steel rather than high-carbon steel.

**END OF CHEMISTRY ONLY**

# 16 ACIDS, ALKALIS AND TITRATIONS

This chapter explores what indicators, acids and alkalis are, and how to carry out an acid–alkali titration. The reactions of acids are also discussed in Chapter 17.

▲ Figure 16.1 Acids range from the extremely dangerous, needing protective clothing to clean up spills . . .

▲ Figure 16.2 . . . to a natural part of our diet: oranges contain citric acid.

## LEARNING OBJECTIVES

- Describe the use of litmus, phenolphthalein and methyl orange to distinguish between acidic and alkaline solutions

- Understand how the pH scale, from 0–14, can be used to classify solutions as strongly acidic (0–3), weakly acidic (4–6), neutral (7), weakly alkaline (8–10) and strongly alkaline (11–14)

- Describe the use of universal indicator to measure the approximate pH value of an aqueous solution

- Know that acids in aqueous solution are a source of hydrogen ions, and alkalis in an aqueous solution are a source of hydroxide ions

- Know that alkalis can neutralise acids

CHEMISTRY ONLY

- Describe how to carry out an acid–alkali titration.

## pH AND INDICATORS

### THE pH SCALE

The pH scale ranges from about 0 to about 14, and tells you how acidic or how alkaline a solution is.

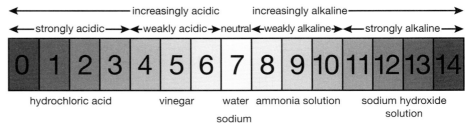

▲ Figure 16.3 The pH scale. Vinegar is a solution of ethanoic acid.

We can classify substances as strongly acidic/alkaline or weakly acidic/alkaline based on their pH. Remember, when writing pH, it is a small p and a big H (the symbol for hydrogen).

Table 16.1 We can classify solutions according to their pH

|  | pH | Solution |
|---|---|---|
| strongly acidic | 0–3 | hydrochloric acid |
| weakly acidic | 4–6 | ethanoic acid (vinegar) |
| neutral | 7 | sodium chloride |
| weakly alkaline | 8–10 | ammonia |
| strongly alkaline | 11–14 | sodium hydroxide |

### EXTENSION WORK

These are actually only approximate ranges and depend on the concentration of the acid/alkali solution. A 0.10 mol/dm³ solution of ethanoic acid actually has a pH of 2.88.

## MEASURING pH

### USING UNIVERSAL INDICATOR

#### HINT

You can measure pH much more accurately using a pH meter.

Universal indicator is made from a mixture of indicators, which change colour in a gradual way over a range of pH values. It can be used as a solution or as paper. The most common form is known as *full-range* universal indicator. It changes through a variety of colours from pH 1 right up to pH 14, but it isn't very accurate.

▲ Figure 16.4 Using universal indicator solution to measure the pH of various solutions.

The colour of the paper or solution is checked against a chart to find the pH.

### ACID–ALKALI INDICATORS

Any substance that has different colours depending on the pH can be used as an indicator. Some common indicators and their colours in acidic and alkaline solutions are shown in Table 16.2.

Table 16.2 The colours of various indicators in acidic and alkaline solutions

|  | Acid | Alkali |
|---|---|---|
| litmus | red | blue |
| methyl orange | red | yellow |
| phenolphthalein | colourless | pink |
| universal indicator | red | blue |

#### KEY POINT

Litmus can be used as a solution or as litmus paper.

Litmus is red in acidic solutions and blue in alkaline ones. In neutral solutions, the colour is purple, which is an equal mixture of the red and blue forms. Universal indicator is green in neutral solutions.

Table 16.3 Some acids showing the replaceable hydrogen

| Acid | Formula |
|---|---|
| hydrochloric acid | HCl |
| nitric acid | $HNO_3$ |
| sulfuric acid | $H_2SO_4$ |
| ethanoic acid | $CH_3COOH$ |
| phosphoric acid | $H_3PO_4$ |

## ACIDS

The formulae of some acids are given in Table 16.3.

All acids contain hydrogen and when acids react the hydrogen shown in red in Table 16.3 is replaced by something else; all acids have replaceable H. For example, when hydrochloric acid reacts with sodium hydroxide we obtain:

$$HCl(aq) + NaOH(aq) \rightarrow NaCl(aq) + H_2O(l)$$

The H of the HCl has been replaced by an Na.

Not all of the hydrogens in acids are replaceable, for example in ethanoic acid only the H attached to the O is replaceable, not the ones joined to the C. The reactions of carboxylic acids such as ethanoic acid are discussed further in Chapter 27.

When acids are in water they dissociate (break apart) to form hydrogen ions ($H^+$), for example:

$$HCl(aq) \rightarrow H^+(aq) + Cl^-(aq)$$
$$HNO_3(aq) \rightarrow H^+(aq) + NO_3^-(aq)$$
$$H_2SO_4(aq) \rightarrow 2H^+(aq) + SO_4^{2-}(aq)$$

When we are measuring pH we are actually measuring the concentration of these $H^+$ ions in the solution. This is why the H in pH is written with a capital letter.

We can define acids as *substances that act as a source of hydrogen ions ($H^+$) in solution*.

## BASES

Bases are substances that neutralise acids by combining with the hydrogen ions in them. When we are referring to a base at this level we usually mean a metal oxide, a metal hydroxide or ammonia ($NH_3$).

## ALKALIS

Figure 16.5 Ammonia solution is sometimes called ammonium hydroxide. It is an alkali and is used in some cleaning products. Some universal indicator solution has been added to it in the beaker.

Some bases dissolve in water to form solutions containing hydroxide ions. These are alkalis.

*Alkalis are a source of hydroxide ($OH^-$) ions in solution.*

Examples of alkalis are sodium hydroxide and potassium hydroxide (all the Group 1 hydroxides). When sodium hydroxide is in water it breaks apart to form sodium and hydroxide ions:

$$NaOH(aq) \rightarrow Na^+(aq) + OH^-(aq)$$

The other alkali you will meet is a solution of ammonia ($NH_3$). The ammonia reacts with the water to form ammonium ions and hydroxide ions:

$$NH_3(aq) + H_2O(l) \rightleftharpoons NH_4^+(aq) + OH^-(aq)$$

These alkalis all have a pH greater than 7.

## OTHER ALKALINE SOLUTIONS

There are some other substances, such as soluble metal carbonates, that react with water to form hydroxide ions:

$$Na_2CO_3(aq) + H_2O(l) \rightleftharpoons NaOH(aq) + NaHCO_3(aq)$$

There are not many soluble carbonates, but sodium carbonate and potassium carbonate are both alkalis with a pH greater than 7. This is due to the $OH^-$ ions in the solution. Only some of the carbonate ions react with water, so these solutions are only weakly alkaline.

## REACTING ACIDS WITH BASES AND ALKALIS

*Acids react with bases or alkalis in a neutralisation reaction.*

### REACTING ACIDS WITH BASES

▲ Figure 16.6 Copper(II) oxide reacts with hot dilute sulfuric acid to form blue copper(II) sulfate solution.

*Metal oxides, such as copper(II) oxide and magnesium oxide, are bases.*

Copper(II) oxide reacts with hot dilute sulfuric acid, in a neutralisation reaction, to produce a solution of copper(II) sulfate and water:

$$CuO(s) + H_2SO_4(aq) \rightarrow CuSO_4(aq) + H_2O(l)$$

Copper(II) oxide is an ionic compound containing the $O^{2-}$ ion and what has happened in this reaction is that the $H^+$ ions from the acid have combined with the $O^{2-}$ ions from the base to form water ($H_2O$).

### REACTING ACIDS WITH ALKALIS

Sodium hydroxide solution (an alkali) reacts with dilute hydrochloric acid to form sodium chloride and water:

$$NaOH(aq) + HCl(aq) \rightarrow NaCl(aq) + H_2O(l)$$

This is a neutralisation reaction. Sodium hydroxide and sodium chloride are both ionic compounds and so will be present as ions in solution. HCl(aq) is an acid and so will dissociate into $H^+(aq)$ and $Cl^-(aq)$. We can therefore re-write this equation showing all the ions.

$$\boxed{Na^+(aq)} + OH^-(aq) + H^+(aq) + \boxed{Cl^-(aq)} \rightarrow \boxed{Na^+(aq)} + \boxed{Cl^-(aq)} + H_2O(l)$$

The $Na^+(aq)$ and $Cl^-(aq)$ ions are the same on both sides of the equation and so are spectator ions. We can leave the spectator ions out of this equation to just show the things that have changed:

$$OH^-(aq) + H^+(aq) \rightarrow H_2O(l)$$

This is an ionic equation. All neutralisation reactions for an acid reacting with an alkali will have the same ionic equation. This is because they all involve the $OH^-$ ions from the alkali reacting with the $H^+$ ions from the acid to form water.

## CHEMISTRY ONLY

### TITRATION

We can use titration to follow the course of a neutralisation reaction between an acid and alkali. Titration can be used to find out how much of the acid/alkali reacts with a certain volume of the alkali/acid.

The technique for carrying out a titration using phenolphthalein as indicator is shown in Figure 16.7.

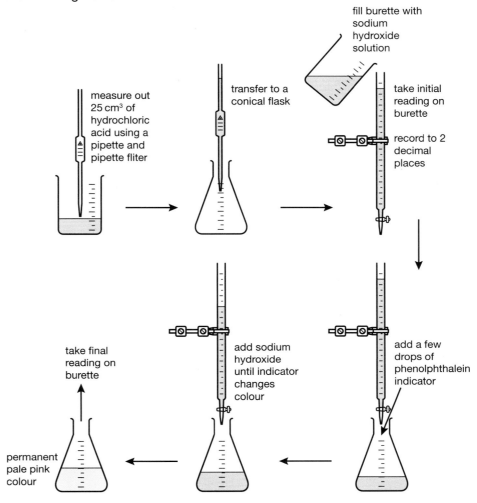

▲ Figure 16.7 How to carry out an acid–alkali titration

Calculations involving titrations are discussed in Chapter 6.

The alkali is added to the acid from the burette until the indicator changes colour. We usually do a rough titration first in order to find out approximately how much alkali is required to neutralise the acid. Imagine we found out that we required 22.50 cm³ of alkali in the rough titration. We would then do the experiment again but this time running in about 20 cm³ of the alkali quite quickly, swirling the conical flask all the time to mix the solutions. We would then add the alkali very slowly (dropwise) until one drop of alkali causes the indicator to change from its acid to its alkali colour. We then know the amount needed for neutralisation to within 1 drop; this is about as precise as we can be. In the above titration we could also have added the acid to the alkali but it is easier to see this indicator turning pink rather than going colourless.

When reading a burette it is important to remember that the numbers increase from the top to the bottom: 0 is at the top and 50 is at the bottom. We normally record readings from the burette to the nearest 0.05 cm³, therefore all readings should be written down to 2 decimal places. The second decimal place is given as '0' if the level of solution is on the line and '5' if it is between the lines. The readings from the burette in Figure 16.8 should be recorded as shown in Table 16.4.

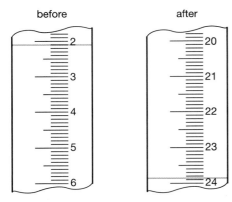

▲ Figure 16.8 Reading a burette

**KEY POINT**

The endpoint of a titration is the point at which the indicator changes colour.

The point at which the indicator changes colour does not necessarily indicate the point at which the solution is neutral. The pH changes very rapidly near the endpoint of the titration and in some titrations 1 drop of an alkali can change the pH from about 3 to 11. So the most precise we can be is to know the neutralisation point to within 1 drop.

Table 16.4 Results table from a titration experiment

| final reading on burette/cm$^3$ | 23.85 |
|---|---|
| initial reading on burette/cm$^3$ | 2.10 |
| volume of alkali added/cm$^3$ | 21.75 |

The volume of alkali added is obtained by subtracting the initial reading from the final reading:

volume of alkali = 23.85 − 2.10 = 21.75 cm$^3$

Various other indicators can also be used, such as methyl orange, but not universal indicator. We do not use universal indicator because it has a range of colours and changes gradually between them; this means we would not be able to see a clear endpoint for the titration.

**END OF CHEMISTRY ONLY**

## CHAPTER QUESTIONS

**SKILLS** CRITICAL THINKING

1 State the colours of methyl orange, phenolphthalein and litmus in each of the following solutions:
   a sodium hydroxide solution     b hydrochloric acid solution.

**SKILLS** ANALYSIS

2 Table 16.5 gives the pH of some solutions. Classify each as strongly acidic, strongly alkaline, weakly acidic, weakly alkaline or neutral by copying the table and putting ticks in the appropriate boxes.

Table 16.5

| Solution | pH | Strongly acidic | Weakly acidic | Neutral | Weakly alkaline | Strongly alkaline |
|---|---|---|---|---|---|---|
| potassium iodide | 7 | | | | | |
| propanoic acid | 4.2 | | | | | |
| sodium carbonate | 9.5 | | | | | |
| potassium hydroxide | 13 | | | | | |
| iron(III) chloride | 2.4 | | | | | |
| nitric acid | 1.3 | | | | | |

**SKILLS** PROBLEM SOLVING

3 Write equations for the reactions between:
   a sodium oxide and nitric acid     b potassium hydroxide and sulfuric acid.

**SKILLS** CRITICAL THINKING

4 Explain what type of reaction is occurring in 3.

**CHEMISTRY ONLY**

5 A student carried out an acid–alkali titration. They used the following set of instructions. Copy out the paragraph and fill in the blanks.

Measure out 25.0 cm$^3$ of potassium hydroxide solution using a _____.
Transfer the potassium hydroxide solution to a _____ _____. Add a few drops of an _____. Put sulfuric acid into the _____.
Add the acid to the alkali until the _____ _____ _____.

**END OF CHEMISTRY ONLY**

# 17 ACIDS, BASES AND SALT PREPARATIONS

Acids were introduced in Chapter 16. In this chapter we will look more closely at some of the reactions of acids and how to make salts.

▲ Figure 17.1 Sodium chloride is common salt. You don't normally make it. It occurs naturally.

## LEARNING OBJECTIVES

■ Know the general rules for predicting the solubility of ionic compounds in water:
  ■ common sodium, potassium and ammonium compounds are soluble
  ■ all nitrates are soluble
  ■ common chlorides are soluble, except those of silver and lead(II)
  ■ common hydroxides are insoluble except for those of sodium, potassium and calcium (calcium hydroxide is slightly soluble)

■ Understand acids and bases in terms of proton transfer

■ Understand that an acid is a proton donor and a base is a proton acceptor

■ Describe the reactions of hydrochloric acid, sulfuric acid and nitric acid with metals, bases and metal carbonates (excluding the reactions between nitric acid and metals) to form salts

■ Know that metal oxides, metal hydroxides and ammonia can act as bases, and that alkalis are bases that are soluble in water

■ Describe an experiment to prepare a pure, dry sample of a soluble salt, starting from an insoluble reactant

■ Practical: Prepare a sample of pure, dry hydrated copper(II) sulfate crystals, starting from copper(II) oxide

### CHEMISTRY ONLY

■ Describe an experiment to prepare a pure, dry sample of a soluble salt, starting from an acid and alkali.

■ Describe an experiment to prepare a pure, dry sample of an insoluble salt, starting from two soluble reactants.

■ Practical: Prepare a sample of pure, dry lead(II) sulfate

## SALTS

All acids contain hydrogen. When that hydrogen is replaced by a metal, the compound formed is called a salt. Magnesium sulfate is a salt, and so is zinc chloride, and so is potassium nitrate.

The formulae of some acids and salts are given in Table 17.1.

Table 17.1 Some acids and salts

| Acid | Formula | Example of salt | Name of salts |
|------|---------|-----------------|---------------|
| hydrochloric acid | $HCl$ | $NaCl$ | chlorides |
| nitric acid | $HNO_3$ | $KNO_3$ | nitrates |
| sulfuric acid | $H_2SO_4$ | $CuSO_4$ | sulfates |
| ethanoic acid | $CH_3COOH$ | $CH_3COONa$ | ethanoates |
| phosphoric acid | $H_3PO_4$ | $K_3PO_4$ | phosphates |

Sulfuric acid can be thought of as the **parent acid** of all the sulfates (the salts formed from sulfuric acid are all called sulfates). It doesn't matter if the replacement of the hydrogen can't be done directly. For example, you can't make copper(II) sulfate from copper and dilute sulfuric acid because they don't react. There are, however, other ways of making it from sulfuric acid – copper(II) sulfate is still a salt.

Salts are also formed when the hydrogen in an acid is replaced with $NH_4$. These are ammonium salts, for example $NH_4Cl$ (ammonium chloride) and $(NH_4)_2SO_4$ (ammonium sulfate).

**REMINDER**

These salts contain the $NH_4^+$ ion.

## REACTIONS OF ACIDS

Before we look at some practical methods to make salts, we need to study the reactions of acids.

### REACTING ACIDS WITH METALS

**KEY POINT**

Of the common acids in the lab, nitric acid has much more complex reactions with metals. You won't be asked about this at International GCSE.

**REMINDER**

Remember: MASH

Simple dilute acids react with metals depending on their positions in the reactivity series.

- Metals below hydrogen in the series don't react with dilute acids.
- Metals above hydrogen in the series react to produce hydrogen gas.
- The higher the metal is in the reactivity series, the more vigorous the reaction. You would never mix metals such as sodium or potassium with acids because their reactions are too violent.

A summary equation for metals above hydrogen in the reactivity series:

metal + acid → salt + hydrogen

## REACTIONS INVOLVING MAGNESIUM AND ACIDS

### WITH DILUTE SULFURIC ACID

There is rapid fizzing and a colourless gas is evolved, which pops with a lighted splint (the test for hydrogen). The reaction mixture becomes very warm as heat is produced. The magnesium gradually disappears to leave a colourless solution of magnesium sulfate.

$$Mg(s) + H_2SO_4(aq) \rightarrow MgSO_4(aq) + H_2(g)$$

This is a displacement reaction. The more reactive magnesium has displaced the less reactive hydrogen.

▲ Figure 17.2 Magnesium reacting with dilute sulfuric acid

## WITH DILUTE HYDROCHLORIC ACID

The reaction looks exactly the same. The only difference is that this time a solution of magnesium chloride is formed.

$$Mg(s) + 2HCl(aq) \rightarrow MgCl_2(aq) + H_2(g)$$

### EXTENSION WORK

Acids in solution form ions. Dilute sulfuric acid contains hydrogen ions and sulfate ions. Dilute hydrochloric acid contains hydrogen ions and chloride ions.

You can rewrite the equations above as ionic equations. In the case of sulfuric acid:

$$Mg(s) + 2H^+(aq) + \boxed{SO_4^{2-}(aq)} \rightarrow Mg^{2+}(aq) + \boxed{SO_4^{2-}(aq)} + H_2(g)$$

The sulfate ion hasn't been changed by the reaction. It is a spectator ion, and so we don't include it in the ionic equation:

$$Mg(s) + 2H^+(aq) \rightarrow Mg^{2+}(aq) + H_2(g)$$

Repeating this with hydrochloric acid, we find that the chloride ions are also spectator ions:

$$Mg(s) + 2H^+(aq) + \boxed{2Cl^-(aq)} \rightarrow Mg^{2+}(aq) + \boxed{2Cl^-(aq)} + H_2(g)$$

Not including the spectator ions produces the ionic equation:

$$Mg(s) + 2H^+(aq) \rightarrow Mg^{2+}(aq) + H_2(g)$$

The reactions look the same because they are the same. All solutions of acids contain hydrogen ions. That means that magnesium will react with any simple dilute acid in the same way.

### REMINDER

Ionic compounds are present as separate ions in solution, so in $MgSO_4(aq)$ the $Mg^{2+}(aq)$ and $SO_4^{2-}(aq)$ ions are separated from each other.

## REACTIONS INVOLVING ZINC AND ACIDS

Again, the reactions between zinc and the two acids look exactly the same. The reactions are slower because zinc is lower down the reactivity series than magnesium.

The equations are:

$$Zn(s) + H_2SO_4(aq) \rightarrow ZnSO_4(aq) + H_2(g)$$
$$Zn(s) + 2HCl(aq) \rightarrow ZnCl_2(aq) + H_2(g)$$

### EXTENSION WORK

The ionic equations are both the same:

$Zn(s) + 2H^+(aq) \rightarrow Zn^{2+}(aq) + H_2(g)$

## BASES

We saw in Chapter 16 that bases are substances that neutralise acids by combining with the hydrogen ions in them to produce water. When we are referring to a base at this level, we usually mean a metal oxide, a metal hydroxide or ammonia.

**REACTING ACIDS WITH BASES**

Metal oxides, such as copper(II) oxide and magnesium oxide, are bases.

**REACTING DILUTE SULFURIC ACID WITH COPPER(II) OXIDE**

The copper(II) oxide (black powder) reacts with hot dilute sulfuric acid to produce a blue solution of copper(II) sulfate.

$$CuO(s) + H_2SO_4(aq) \rightarrow CuSO_4(aq) + H_2O(l)$$

All the metal oxide and acid combinations that you will meet at International GCSE behave in exactly the same way as the reaction between copper(II) oxide and dilute sulfuric acid, that is, they produce a salt and water. Most need to be heated for the reaction to start.

Copper(II) oxide is an ionic compound containing the $O^{2-}$ ion. What has happened in this reaction is that the $H^+$ ions from the acid have combined with the $O^{2-}$ ions to form water ($H_2O$).

The general equation for the reaction of a metal oxide (base) with an acid is

$$metal\ oxide + acid \rightarrow salt + water$$

This is a neutralisation reaction – the base neutralises the acid (water is formed).

## BASES AND ALKALIS

Some metal oxides are soluble in water and react with it to form solutions of metal hydroxides, for example:

$$Na_2O(s) + H_2O(l) \rightarrow 2NaOH(aq)$$

All the Group 1 oxides do this reaction, so for potassium oxide we would get:

$$K_2O(s) + H_2O(l) \rightarrow 2KOH(aq)$$

Most other metal oxides are not soluble in water.

Calcium oxide dissolves slightly to form calcium hydroxide:

$$CaO(s) + H_2O(l) \rightarrow Ca(OH)_2(aq)$$

Another alkali you will meet is a solution of ammonia ($NH_3$). The ammonia reacts with the water to form ammonium ions and hydroxide ions:

$$NH_3(aq) + H_2O(l) \rightleftharpoons NH_4^+(aq) + OH^-(aq)$$

All these solutions contain hydroxide ions. We learned in the last chapter that we can define alkalis as solutions that are a source of hydroxide ($OH^-$) ions, so all of these solutions are alkalis.

**REACTING ACIDS WITH METAL HYDROXIDES**

All metal hydroxides react with acids in a neutralisation reaction:

$$metal\ hydroxide + acid \rightarrow salt + water$$

## REACTING DILUTE HYDROCHLORIC ACID WITH SODIUM HYDROXIDE SOLUTION

Mixing sodium hydroxide solution and dilute hydrochloric acid produces a colourless solution so not much seems to have happened. But if you repeat the reaction with a thermometer in the beaker, the temperature rises several degrees, showing that there has been a chemical change. Sodium chloride solution has been formed:

$$NaOH(aq) + HCl(aq) \rightarrow NaCl(aq) + H_2O(l)$$

We saw in Chapter 16 that the ionic equation for this reaction is:

$$OH^-(aq) + H^+(aq) \rightarrow H_2O(l)$$

All neutralisation reactions for an acid reacting with an alkali have the same ionic equation: they all involve the hydroxide ions from the alkali reacting with the $H^+$ ions from the acid to form water.

## REACTING ACIDS WITH CARBONATES

Carbonates react with cold dilute acids to produce carbon dioxide gas.

A summary equation for acids and carbonates:

carbonate + acid → salt + carbon dioxide + water

## THE REACTION BETWEEN COPPER(II) CARBONATE AND DILUTE ACIDS

Green copper(II) carbonate reacts with the common dilute acids to give a blue or blue-green solution of copper(II) sulfate, copper(II) nitrate or copper(II) chloride. Carbon dioxide gas is given off. You can recognise this because it turns limewater milky.

All the equations have the same form:

$$CuCO_3(s) + H_2SO_4(aq) \rightarrow CuSO_4(aq) + CO_2(g) + H_2O(l)$$

$$CuCO_3(s) + 2HNO_3(aq) \rightarrow Cu(NO_3)_2(aq) + CO_2(g) + H_2O(l)$$

$$CuCO_3(s) + 2HCl(aq) \rightarrow CuCl_2(aq) + CO_2(g) + H_2O(l)$$

▲ Figure 17.3 The reaction between copper(II) carbonate and dilute sulfuric acid.

## THE REACTION BETWEEN SODIUM CARBONATE AND DILUTE ACIDS

Sodium carbonate is soluble in water. The equation for the reaction of sodium carbonate with hydrochloric acid is:

$$Na_2CO_3(aq) + 2HCl(aq) \rightarrow 2NaCl(aq) + CO_2(g) + H_2O(l)$$

### HINT

Carbonates are not mentioned as bases on the syllabus; only metal oxides, metal hydroxides and ammonia.

### EXTENSION WORK

If we show the ions (remembering that HCl splits apart into its ions in solution):

$$2Na^+(aq) + CO_3^{2-}(aq) + 2H^+(aq) + 2Cl^-(aq) \rightarrow 2Na^+(aq) + 2Cl^-(aq) + CO_2(g) + H_2O(l)$$

By not including the spectator ions we obtain the ionic equation:

$$CO_3^{2-}(aq) + 2H^+(aq) \rightarrow CO_2(g) + H_2O(l)$$

Because carbonate ions react by accepting hydrogen ions, we can classify carbonate ions as bases.

# SALT PREPARATIONS

In the next sections we will learn about some practical methods to make salts. Before we can do this, we need to know which substances are soluble and which are insoluble in water because the method we use depends on this.

## THE SOLUBILITY OF IONIC COMPOUNDS IN WATER

### NOTES ON THE TABLE

To keep the table simple, it includes one or two compounds (aluminium carbonate, for example) which don't actually exist. Don't worry about these. The problem won't arise at International GCSE.

There is no clear line between 'insoluble' and 'almost insoluble' compounds. The ones picked out as 'almost insoluble' include the more common ones that you might need to know about elsewhere in the course.

|            | nitrate | chloride  | sulfate          | carbonate | hydroxide        |
|------------|---------|-----------|------------------|-----------|------------------|
| ammonium   | soluble | soluble   | soluble          | soluble   | soluble          |
| potassium  | soluble | soluble   | soluble          | soluble   | soluble          |
| sodium     | soluble | soluble   | soluble          | soluble   | soluble          |
| barium     | soluble | soluble   | insoluble        | insoluble | soluble          |
| calcium    | soluble | soluble   | almost insoluble | insoluble | almost insoluble |
| magnesium  | soluble | soluble   | soluble          | insoluble | insoluble        |
| aluminium  | soluble | soluble   | soluble          | insoluble | insoluble        |
| zinc       | soluble | soluble   | soluble          | insoluble | insoluble        |
| iron       | soluble | soluble   | soluble          | insoluble | insoluble        |
| lead       | soluble | insoluble | insoluble        | insoluble | insoluble        |
| copper     | soluble | soluble   | soluble          | insoluble | insoluble        |
| silver     | soluble | insoluble | almost insoluble | insoluble | insoluble        |

key

soluble    insoluble    almost insoluble (slightly soluble)

▲ Figure 17.4 **Solubility patterns**

### HINT

It can seem a bit frightening to have to remember all this, but it isn't as difficult as it looks at first sight.

Except for the carbonates and hydroxides, most of these compounds are soluble. Learn the exceptions in the sulfates and chlorides.

The reason for the exceptions in the carbonates and hydroxides is that all sodium, potassium and ammonium compounds are soluble.

Note that:

■ all sodium, potassium and ammonium compounds are *soluble*

■ all nitrates are *soluble*

■ most common chlorides are *soluble*, except lead(II) chloride and silver chloride

■ most common sulfates are *soluble*, except lead(II) sulfate, barium sulfate, silver sulfate and calcium sulfate

■ most common carbonates are *insoluble*, except sodium, potassium and ammonium carbonates

■ most metal hydroxides are *insoluble* (or *almost insoluble*), except sodium, potassium and ammonium hydroxides. Calcium hydroxide is slightly soluble in water.

## MAKING SOLUBLE SALTS (EXCEPT SODIUM, POTASSIUM AND AMMONIUM SALTS)

These all involve reacting a solid with an acid. You can use any of the following mixtures:

- acid + metal (but only for the moderately reactive metals from magnesium to iron in the reactivity series)
- acid + metal oxide or hydroxide
- acid + carbonate.

Whatever mixture you use, the method is basically the same.

### ACTIVITY 5

#### ▼ PRACTICAL: MAKING COPPER(II) SULFATE CRYSTALS

The practical procedure for this is as follows:

- Measure 50 cm³ of dilute sulfuric acid into a beaker and heat it on a tripod and gauze using a Bunsen burner.
- Add a spatula full of black copper(II) oxide and continue heating. If all the copper(II) oxide disappears add more copper(II) oxide until there is some left in the beaker. Stir the mixture well to make sure that no more will react. At this stage, we have added *excess* copper(II) oxide – there is more than enough to react with all the acid present. When there is copper(II) oxide left, we know that all the acid has been neutralised.
- Filter off the excess copper(II) oxide and transfer the filtrate (solution), which is blue, to an evaporating basin. The solution we have now is copper(II) sulfate.

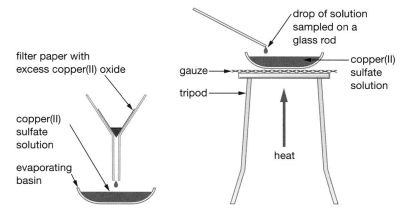

filter paper with excess copper(II) oxide

copper(II) sulfate solution

evaporating basin

drop of solution sampled on a glass rod

copper(II) sulfate solution

gauze

tripod

heat

▲ Figure 17.5 Making copper(II) sulfate crystals

The equation for the formation of the solution is:

$$CuO(s) + H_2SO_4(aq) \rightarrow CuSO_4(aq) + H_2O(l)$$

- Heat the solution of copper(II) sulfate over a Bunsen burner to boil off some of the water and concentrate the solution.
- Keep heating until a saturated solution is formed. We can test this by dipping a glass rod into the solution. If crystals form on the glass rod when we remove it we know that the solution is very close to saturated and crystals will also begin to form in the solution.

> **Safety Note:** Direct heating of the acid is not recommended. Wear eye protection and do not evaporate the solution to dryness. The acid is best heated in a boiling tube placed in a water bath. Avoid contact with the crystals as they are harmful and irritating to both the skin and eyes.

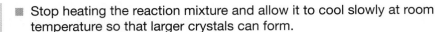

■ Stop heating the reaction mixture and allow it to cool slowly at room temperature so that larger crystals can form.

■ Remove the blue crystals from the reaction mixture by filtration, or by just pouring off the remaining solution.

■ The crystals can be dried by blotting them with a paper towel, or they can be left to dry in a warm place.

### WHY NOT SIMPLY EVAPORATE THE SOLUTION TO DRYNESS?

It would seem much easier to just evaporate off all the water by boiling rather than crystallising the solution slowly, but evaporating to dryness wouldn't give you blue copper(II) sulfate crystals. Instead, you would produce a white powder of anhydrous copper(II) sulfate.

When many salts form their crystals water from the solution becomes chemically bound up with the salt. This is called *water of crystallisation*. A salt which contains water of crystallisation is said to be *hydrated*. We can show the crystallisation part of the reaction as:

$$CuSO_4(aq) + 5H_2O(l) \rightarrow CuSO_4{\cdot}5H_2O(s)$$

▲ Figure 17.6 Copper(II) sulfate crystals

> **KEY POINT**
>
> 'Anhydrous' means 'without water'.

> **!**
>
> Safety Note: Wear eye protection. Heating must be very gentle to avoid 'spitting' of hot particles.

## MAKING MAGNESIUM SULFATE CRYSTALS

We can add excess magnesium to sulfuric acid. This time the acid does not have to be heated.

When we add the magnesium the reaction mixture will fizz (hydrogen is given off). We keep adding magnesium until the fizzing stops and there is magnesium left in the beaker. This means that all the acid has reacted. The equation for the formation of the solution of magnesium sulfate is:

$$Mg(s) + H_2SO_4(aq) \rightarrow MgSO_4(aq) + H_2(g)$$

Again the solution is concentrated by heating it and allowed to crystallise. The crystallisation reaction is:

$$MgSO_4(aq) + 7H_2O(l) \rightarrow MgSO_4{\cdot}7H_2O(s)$$

▲ Figure 17.7 Dilute sulfuric acid with magnesium

## HOW DO YOU KNOW WHETHER YOU NEED TO HEAT THE MIXTURE?

Carbonates react with dilute acids in the cold, and so does magnesium. Most other things that you are likely to come across need to be heated.

| cold | hot |
|------|-----|
| carbonates magnesium | most other substances |

▲ Figure 17.8 Do you need to heat the mixture?

## CHEMISTRY ONLY

## MAKING SODIUM, POTASSIUM AND AMMONIUM SALTS

### THE NEED FOR A DIFFERENT METHOD

In the method we've just been looking at, you add an excess of a solid to an acid, and then filter off the unreacted solid. You do this to make sure all the acid is used up. This method does not work for making sodium, potassium and ammonium salts.

The problem is that all sodium, potassium and ammonium compounds are soluble in water. The solid you added to the acid would not only react with the acid, but any excess would just dissolve in the water present. You wouldn't have any visible excess to filter off. There's no simple way of seeing when you have added just enough of the solid to neutralise the acid.

### SOLVING THE PROBLEM BY DOING A TITRATION

You normally make these salts from sodium or potassium hydroxide or ammonia solution, but you can also use the carbonates. Fortunately, the solutions of all these are alkaline. That means you can find out when you have a neutral solution by using an indicator.

The method of finding out exactly how much of two solutions you need for them to neutralise each other is called titration (see Chapter 16). You carry out a titration using an indicator to tell you when you have added exactly enough of the acid to neutralise the alkali. Having found out how much acid and alkali are needed, you can make a pure solution of the salt by mixing those same volumes again, but without the indicator.

### MAKING SODIUM SULFATE CRYSTALS

#### KEY POINT

Methyl orange is red in acidic solutions and yellow in alkaline ones. However, when it is used as an indicator in titrations (see Chapter 16) involving adding an acid to an alkali, an orange colour is usually obtained, which indicates the endpoint. Orange is a mixture of the red and yellow forms.

- 25 cm³ of sodium hydroxide solution is transferred to a conical flask using a pipette, and a few drops of methyl orange are added as the indicator.
- Dilute sulfuric acid is run in from the burette until the indicator just turns from yellow to orange.
- The volume of acid needed is noted, and the same volumes of acid and alkali are mixed together in a clean flask *without any indicator*.

The equation for the formation of the solution is:

$$2NaOH(aq) + H_2SO_4(aq) \rightarrow Na_2SO_4(aq) + 2H_2O(l)$$

- The solution is heated to evaporate off some of the water until a saturated solution is formed. It is then left to cool so that crystals form.
- The crystals are finally separated from any remaining solution by filtration (or the remaining solution can be poured away).
- The crystals are dried by patting them dry with a paper towel or by leaving them in a warm place.

The equation for the crystallisation process is:

$$Na_2SO_4(aq) + 10H_2O(l) \rightarrow Na_2SO_4 \cdot 10H_2O(s)$$

**EXTENSION WORK**

Methyl orange actually changes colour between pH 3.1 and pH 4.4, so below pH 3.1 methyl orange is red and above pH 4.4 it is yellow.

Between these pHs methyl orange will be a mixture of red and yellow, that is, orange. This means that it is actually yellow in a solution with pH 7. This technique does not, therefore, give us the exact neutralisation point, but because the pH changes very rapidly near the endpoint of a titration, if we do the titration carefully we will be able to find the neutralisation point to the nearest drop of acid, which is very good!

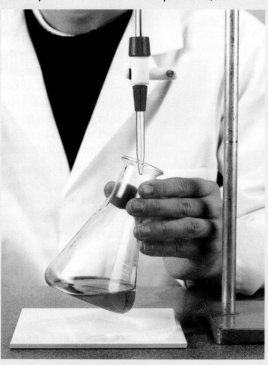

▶ Figure 17.9 **Titration using methyl orange indicator**

**MAKING SODIUM CHLORIDE CRYSTALS**

$$NaOH(aq) + HCl(aq) \rightarrow NaCl(aq) + H_2O(l)$$

You would need to do the titration using dilute hydrochloric acid rather than dilute sulfuric acid. You would then follow the same procedure as for sodium sulfate.

**HINT**

Once you have re-mixed the acid and the alkali without the indicator, you could evaporate the sodium chloride solution to dryness rather than crystallising it slowly. Sodium chloride crystals don't contain any water of crystallisation, so you could save time by evaporating all the water in one go. The disadvantage is that you obtain either a powder or very tiny crystals. In the exam you will not necessarily know whether something contains water of crystallisation or not, so it is always better to concentrate the solution and let crystals form slowly.

**MAKING AMMONIUM SULFATE CRYSTALS**

$$2NH_3(aq) + H_2SO_4(aq) \rightarrow (NH_4)_2SO_4(aq)$$

Using ammonia solution rather than sodium hydroxide solution makes no difference to the method. Although simple ammonium salts don't have water of crystallisation, you would still crystallise them slowly rather than evaporating them to dryness. Heating dry ammonium salts tends to break them up.

## MAKING INSOLUBLE SALTS

**PRECIPITATION REACTIONS**

▲ Figure 17.10 A precipitate of silver chloride

The basic procedure is to mix solutions of two soluble salts to form an insoluble salt and a solution of a soluble one. For instance, to make insoluble silver chloride we would mix together solutions of silver nitrate and sodium chloride.

$$AgNO_3(aq) + NaCl(aq) \rightarrow AgCl(s) + NaNO_3(aq)$$

This is called a **precipitation reaction**. A **precipitate** is a solid that is formed by a chemical reaction involving liquids or gases. A precipitation reaction is simply a reaction that produces a precipitate. In this reaction, a white precipitate of silver chloride is produced. A white precipitate is formed because silver chloride won't dissolve in water, and so it is seen as a fine white solid. Figure 17.10 shows the results of this reaction.

**EXPLAINING WHAT'S HAPPENING**

Silver nitrate solution contains silver ions and nitrate ions in solution. The positive and negative ions are attracted to each other, but the attractions aren't strong enough to make them stick together. Similarly, sodium chloride solution contains sodium ions and chloride ions; again, the attractions aren't strong enough for them to stick together.

When you mix the two solutions, the various ions meet each other. When silver ions meet chloride ions, the attractions are so strong that the ions clump together and form a solid. The sodium and nitrate ions remain in solution because they aren't sufficiently attracted to each other.

**KEY POINT**

The water molecules in the solutions have been left out to avoid cluttering the diagram.

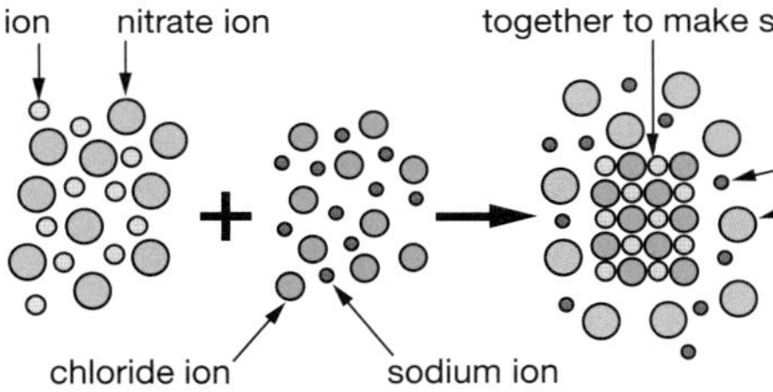

▶ Figure 17.11 **Precipitation of silver chloride**

**EXTENSION WORK**

The ionic equation for a precipitation reaction is much easier to write than the full equation. All that is happening is that the ions of the insoluble salt are coming together to form the solid. The ionic equation simply shows that happening.

To work out the ionic equation for the precipitation reaction:

■ Write down the formula for the precipitate/insoluble salt on the right-hand side of the equation.

■ Write down the formulae for the ions that have come together to produce it on the left-hand side.

■ Don't forget the state symbols.

So, for the formation of silver chloride we have:

$$Ag^+(aq) + Cl^-(aq) \rightarrow AgCl(s)$$

or for the formation of barium sulfate:

$$Ba^{2+}(aq) + SO_4^{2-}(aq) \rightarrow BaSO_4(s)$$

You don't need to worry about the spectator ions because they aren't doing anything.

▲ Figure 17.12 A precipitate of lead(II) iodide

**EXTENSION WORK**

The ionic equation is:
$$Pb^{2+}(aq) + 2I^-(aq) \rightarrow PbI_2(s)$$

**TO MAKE BARIUM SULFATE**

**EXTENSION WORK**

The ionic equation is:
$$Ba^{2+}(aq) + SO_4^{2-}(aq) \rightarrow BaSO_4(s)$$

Safety Note: Wear eye protection. The lead nitrate and sulfate are toxic and all skin contact must be avoided.

# WHAT DO WE MIX TOGETHER TO MAKE INSOLUBLE SALTS?

Our procedure for making insoluble salts is to mix together two solutions containing soluble salts, but how do we know what to mix together? To determine this, two really useful facts are:

■ all nitrates are soluble

■ all sodium and potassium salts are soluble.

So all we have to do is mix the nitrate of the metal part of our insoluble salt with the sodium or potassium salt of the non-metal part.

For instance, if we have to make lead(II) iodide:

■ the metal part is lead, so we use a solution of lead(II) nitrate

■ the non-metal part is iodide, so we use a solution of sodium iodide.

The equation for the reaction is:

$$Pb(NO_3)_2(aq) + 2NaI(aq) \rightarrow PbI_2(s) + 2NaNO_3(aq)$$

Figure 17.12 shows the yellow precipitate of lead(II) iodide.

We can use barium nitrate solution and potassium sulfate solution:

$$Ba(NO_3)_2(aq) + K_2SO_4(aq) \rightarrow BaSO_4(s) + 2KNO_3(aq)$$

The barium sulfate is a white precipitate.

There are other ways we could make barium sulfate. We could also use barium chloride, which is soluble in water. The sulfate part doesn't necessarily have to come from a salt; dilute sulfuric acid contains sulfate ions, so we can use that as well.

## ACTIVITY 6

## ▼ PRACTICAL: PREPARING A PURE, DRY SAMPLE OF LEAD(II) SULFATE

The following procedure is used:

■ Take 25 cm$^3$ of a solution of lead(II) nitrate in a beaker and add 25 cm$^3$ of a solution of sodium sulfate. The amounts don't really matter as we will get rid of any excess later.

■ A white precipitate of lead(II) sulfate will form.

■ The equation for the reaction is:

$$Pb(NO_3)_2(aq) + Na_2SO_4(aq) \rightarrow PbSO_4(s) + 2NaNO_3(aq)$$

■ The reaction mixture is filtered.

■ A white residue of lead(II) sulfate is left in the filter paper and a colourless solution of sodium nitrate and any excess reactants (either lead(II) nitrate or sodium sulfate) passes through into the beaker or conical flask. The white residue on the filter paper is the compound we want, but it is contaminated with solutions of sodium nitrate and the reactant that was in excess.

- Wash the residue with *distilled water* by pouring a $20\,cm^3$ portion (the amount is not important) into the filter paper and allowing it to filter through the paper. This washes away everything apart from the insoluble lead(II) sulfate. This should be repeated several times to make sure that our sample is not contaminated with anything.
- We then transfer the filter paper and lead(II) sulfate to a warm oven to dry (the water evaporates).

We must wash the lead(II) sulfate with distilled (pure) water because tap water contains dissolved substances. If we wash with tap water and then transfer our sample to the oven, the water will evaporate, leaving behind the dissolved solids from the water, which will contaminate our salt.

## SUMMARISING THE METHODS OF MAKING SALTS

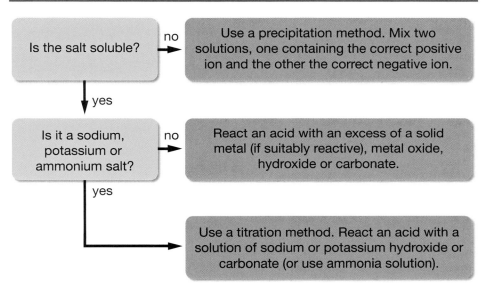

▲ Figure 17.13 Methods of making salts

### END OF CHEMISTRY ONLY

## THEORIES OF ACIDS AND BASES

In Chapter 16, we met the idea that an acid is something that produces $H^+(aq)$ ions in solution and an alkali is something that produces $OH^-(aq)$ ions in solution. This is called the Arrhenius theory of acids and bases. For example, when hydrogen chloride gas is dissolved in water to form hydrochloric acid the molecules dissociate to form $H^+$ ions:

$$HCl(aq) \rightarrow H^+(aq) + Cl^-(aq)$$

A neutralisation reaction can be shown as $H^+$ from an acid reacting with $OH^-$ from an alkali to form water.

$$H^+(aq) + OH^-(aq) \rightarrow H_2O(l)$$

The problem with this definition of acids and bases is that it only applies to reactions that occur in aqueous solution and is not more widely applicable.

**HINT**

You do not have to remember the names of the theories.

**EXTENSION WORK**

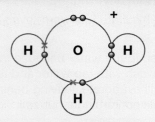

In the dot-and-cross diagram for the $H_3O^+$ ion you can see that the covalent bond that formed between the $H_2O$ and the $H^+$ is unusual in that both the shared electrons come from the O. This is called a dative covalent bond or a co-ordinate covalent bond. Once it has been formed, it is exactly the same as a 'normal' covalent bond.

**KEY POINT**

You have already met the ammonium ($NH_4^+$) ion, which is obtained when $H^+$ is added to ammonia ($NH_3$). The hydroxonium ion is the equivalent ion for water. You do not have to remember the name *hydroxonium ion*.

**KEY POINT**

To understand the reaction between ammonia and hydrochloric acid in relation to the Arrhenius theory we have to realise that ammonia reacts with water to form $OH^-$ ions:

$NH_3(aq) + H_2O(l) \rightleftharpoons NH_4^+(aq) + OH^-(aq)$

and that hydrochloric acid dissociates to form $H^+$ ions:

$HCl(aq) \rightarrow H^+(aq) + Cl^-(aq)$

and that the overall reaction involves $H^+$ ions reacting with $OH^-$ ions to form water:

$H^+(aq) + OH^-(aq) \rightarrow H_2O(l)$

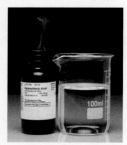

▲ Figure 17.17 Ammonia and hydrogen chloride gases also react to form ammonium chloride (the white smoke): $NH_3(g) + HCl(g) \rightleftharpoons NH_4^+Cl^-(s)$

There is a more general theory of acids and bases that can be used in more situations:

■ *An acid is a proton (hydrogen ion) donor.*

■ *A base is a proton (hydrogen ion) acceptor.*

This is called the Brønsted–Lowry theory.

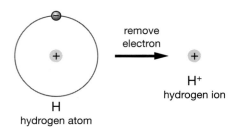

▲ Figure 17.14 A hydrogen ion ($H^+$) is a proton. If an electron is removed from a hydrogen atom, only a proton is left.

A better way of writing the reaction that occurs when hydrogen chloride dissolves in water is:

$H_2O(l) + HCl(aq) \rightarrow H_3O^+(aq) + Cl^-(aq)$

The HCl donates (gives) a proton ($H^+$) to an $H_2O$ molecule. So, according to our definition, the HCl is an acid and $H_2O$ is a base because it accepts the proton.

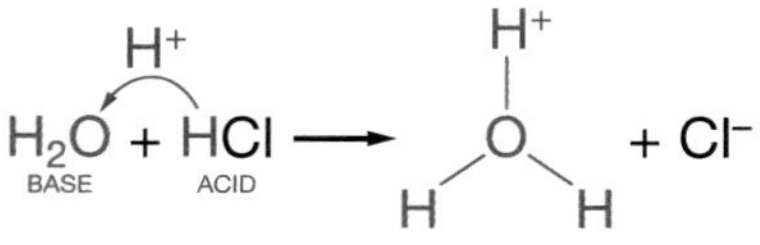

▲ Figure 17.15 HCl donates a proton and $H_2O$ accepts it.

The $H_3O^+(aq)$ ion is called the **hydroxonium ion**. This is the ion that we usually write simply as $H^+(aq)$. You can think of it as a hydrogen ion joined to a water molecule. $Cl^-$ has a negative charge because the $H^+$ left an electron behind when it transferred to the $H_2O$.

## THE REACTION BETWEEN NH₃ AND HCl

Dilute hydrochloric acid reacts with ammonia solution in a neutralisation reaction to form a salt, ammonium chloride:

$NH_3(aq) + HCl(aq) \rightarrow NH_4Cl(aq)$

Ammonium chloride is an ionic salt so we can show the ions that are present:

$NH_3(aq) + HCl(aq) \rightarrow NH_4^+(aq) + Cl^-(aq)$

We can see now that, in order to form ammonium chloride, the HCl must donate a proton ($H^+$) to the $NH_3$. The HCl is an acid because it donates a proton and the $NH_3$ is a base because it accepts the proton.

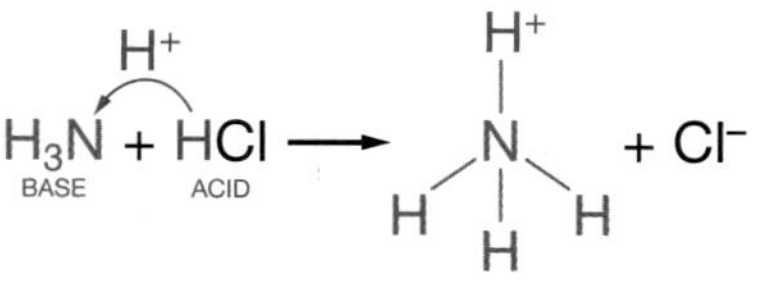

▲ Figure 17.16 HCl is an acid and donates a proton to $NH_3$ (a base).

## MORE ACID-BASE REACTIONS

Acids in solution are acidic because they donate a proton ($H^+$) to water to form the hydroxonium ion. If we take a general acid (HA), this is what happens when it is put into water:

$$HA(aq) + H_2O(l) \rightarrow H_3O^+(aq) + A^-(aq)$$

HA donates a proton to $H_2O$, which acts as a base. For nitric acid this would be:

$$HNO_3(aq) + H_2O(l) \rightarrow H_3O^+(aq) + NO_3^-(aq)$$
$$\text{acid} \qquad \text{base}$$

When bases, such as metal oxides (e.g. $Na_2O$), react with water they accept a proton from water, for example:

$$(Na^+)_2O^{2-}(s) + H_2O(l) \rightarrow 2Na^+(aq) + 2OH^-(aq)$$

The $O^{2-}$ ion accepts a proton from the water. In this reaction the water acts as an acid because it donates a proton.

When sulfuric acid reacts with copper(II) oxide (CuO) we have:

$$Cu^{2+}O^{2-}(s) + H_2SO_4(aq) \rightarrow Cu^{2+}(aq) + SO_4^{2-}(aq) + H_2O(l)$$

$H_2SO_4$ is an acid, it donates protons to CuO, the base.

Ammonia is also a base:

$$NH_3(aq) + H_2O(l) \rightleftharpoons NH_4^+(aq) + OH^-(aq)$$
$$\text{base} \qquad \text{acid}$$

The ammonia accepts a proton ($H^+$) from the water.

For reactions such as:

$$HCl(aq) + NaOH(aq) \rightarrow NaCl(aq) + H_2O(l)$$

What actually happens is that the $H_3O^+$ ion formed when the acid reacts with water donates a proton to the base, $OH^-$.

$$H_3O^+(aq) + OH^-(aq) \rightarrow 2H_2O(l)$$
$$\text{acid} \qquad \text{base}$$

But for International GCSE purposes, we almost always use the simplified version:

$$H^+(aq) + OH^-(aq) \rightarrow H_2O(l)$$

### HINT

Water can act as proton acceptor (base) or proton donor (acid) in reactions on this page. Do not get confused with what is happening here; remember, water is still neutral! We are simply understanding the reactions that occur in terms of proton transfer.

## CHAPTER QUESTIONS

SKILLS | CRITICAL THINKING

SKILLS | PROBLEM SOLVING

1 State which of the following will react with dilute sulfuric acid:

   a copper, copper(II) oxide, copper(II) hydroxide, copper(II) carbonate

   b In the case of each of the substances which does react, write the equation (including state symbols) for the reaction. All of these substances are insoluble solids.

2 Read this description of the chemistry of metal **A** and some of its compounds, and then answer the questions.

Metal **A** has no reaction with dilute hydrochloric acid or dilute sulfuric acid. It forms a black oxide, **B**, which reacts with hot dilute sulfuric acid to give a blue solution, **C**. Metal **A** also forms a green compound, **D**, which reacts with dilute nitric acid to give a colourless gas, **E**, and another blue solution, **F**. The colourless gas, **E**, turned limewater milky.

a State the names of **A**, **B**, **C**, **D**, **E** and **F**.

b Write equations for the reactions between:

  i **B** and dilute sulfuric acid

  ii **D** and dilute nitric acid.

3 a Nickel, Ni, is a silvery metal just above hydrogen in the reactivity series. Nickel(II) compounds in solution are green.

  i Describe what you would see if you warmed some nickel with dilute sulfuric acid in a test-tube.

  ii State the name of any gas formed.

  iii Write the equation for the reaction between nickel and dilute sulfuric acid.

b Nickel(II) carbonate is a green, insoluble powder. A spatula measure of nickel(II) carbonate is added to some dilute hydrochloric acid in a test-tube.

  i Describe what you would observe in this experiment.

  ii Write an equation for the reaction that occurs.

  iii Write an ionic equation for the reaction that occurs.

4 Organise the following compounds into two lists: those that are soluble in water and those that are insoluble.

sodium chloride, lead(II) sulfate, zinc nitrate, calcium carbonate, iron(III) sulfate, lead(II) chloride, potassium sulfate, copper(II) carbonate, silver chloride, aluminium nitrate, barium sulfate, ammonium chloride, magnesium nitrate, calcium sulfate, sodium phosphate, nickel(II) carbonate, chromium(III) hydroxide, potassium dichromate(VI)

5 a Describe in detail the preparation of a pure, dry sample of copper(II) sulfate crystals, $CuSO_4 \cdot 5H_2O$, starting from copper(II) oxide.

b Write equations for i the reaction producing copper(II) sulfate solution and ii the crystallisation reaction.

## CHEMISTRY ONLY

6 a Read the following description of a method for making sodium sulfate crystals, $Na_2SO_4 \cdot 10H_2O$, and then explain the reasons for each of the underlined phrases or sentences.

25 $cm^3$ of sodium carbonate solution was transferred to a conical flask <u>using a pipette</u>, and a <u>few drops of methyl orange were added</u>. Dilute sulfuric acid was run in from a burette <u>until the solution became orange</u>. The volume of acid added was noted. That same volume of dilute sulfuric acid was added to a fresh 25 $cm^3$ sample of sodium carbonate solution in a clean flask, but <u>without the methyl orange</u>. The mixture was <u>evaporated until a sample taken on the end of a glass rod crystallised on cooling in the air</u>. <u>The solution was left to cool</u>. The crystals formed were separated from the remaining solution and dried.

**SKILLS**   PROBLEM SOLVING

**SKILLS**   REASONING, PROBLEM SOLVING

b   Write equations for **i** the reaction producing sodium sulfate solution and **ii** the crystallisation reaction.

7   Suggest solutions that could be mixed together to make each of the following insoluble salts. In each case write the equation for the reaction you choose.

   a   Silver chloride

   b   Calcium carbonate

   c   Lead(II) sulfate

   d   Lead(II) chloride

**SKILLS**   EXECUTIVE FUNCTION, PROBLEM SOLVING

8   Describe in detail the preparation of a pure, dry sample of barium carbonate. Write the equation for the reaction you use.

9   There are three main methods of making salts:

   A   Reacting an acid with an excess of a suitable solid.

   B   Using a titration.

   C   Using a precipitation reaction.

For each of the following salts, write down the letter of the appropriate method, and state the names of the substances you would react together. You should state whether they are used as solids or solutions. Write an equation for each reaction.

   a   Zinc sulfate

   b   Barium sulfate

   c   Potassium nitrate

   d   Copper(II) nitrate

   e   Lead(II) chromate(VI) (a bright yellow insoluble solid; chromate(VI) ions have the formula $CrO_4^{2-}$).

**END OF CHEMISTRY ONLY**

**SKILLS**   REASONING

10   All of the following equations represent acid–base reactions. For each of the equations, state which substance is the acid and which the base.

   a   $CO_3^{2-}(s) + H_3O^+(aq) \rightarrow HCO_3^-(aq) + H_2O(l)$

   b   $MgO(s) + H_2SO_4(aq) \rightarrow MgSO_4(aq) + H_2O(l)$

   c   $HNO_3(aq) + NH_3(aq) \rightarrow NH_4^+(aq) + NO_3^-(aq)$

   d   $H_2SO_4(aq) + H_2O(l) \rightarrow HSO_4^-(aq) + H_3O^+(aq)$

   e   $NH_3(g) + HCl(g) \rightarrow NH_4Cl(s)$

   f   $CH_3COOH(aq) + NH_3(aq) \rightarrow CH_3COO^-(aq) + NH_4^+(aq)$

# 18 CHEMICAL TESTS

In this chapter we will look at how to identify substances using chemical tests.

▲ Figure 18.1 Although the reactions we learn here are useful for identifying substances in the lab and we are also introduced to some interesting chemistry, analysis of substances is now routinely carried out using sophisticated machines.

## LEARNING OBJECTIVES

- Describe tests for these gases:
    - hydrogen
    - oxygen
    - carbon dioxide
    - ammonia
    - chlorine

- Describe how to carry out a flame test
- Know the colours formed in flame tests for these cations:
    - $Li^+$ (red)
    - $Na^+$ (yellow)
    - $K^+$ (lilac)
    - $Ca^{2+}$ (orange-red)
    - $Cu^{2+}$ (blue-green)

- Describe tests for these cations:
    - $NH_4^+$ using sodium hydroxide solution and identifying the gas evolved
    - $Cu^{2+}$, $Fe^{2+}$ and $Fe^{3+}$ using sodium hydroxide solution

- Describe tests for these anions:
    - $Cl^-$, $Br^-$ and $I^-$ using acidified silver nitrate solution
    - $SO_4^{2-}$ using acidified barium chloride solution
    - $CO_3^{2-}$ using hydrochloric acid and identifying the gas evolved

- Describe a test for the presence of water using anhydrous copper(II) sulfate
- Describe a physical test to show whether or not a sample of water is pure

## TESTING FOR GASES

### HYDROGEN, H$_2$

#### THE TEST FOR HYDROGEN GAS

A lighted splint is held to the mouth of the tube. The hydrogen explodes with a squeaky pop.

The hydrogen combines explosively with oxygen in the air to make water.

$$2H_2(g) + O_2(g) \rightarrow 2H_2O(l)$$

### OXYGEN, O$_2$

#### THE TEST FOR OXYGEN GAS

A glowing splint is put into the tube containing the gas. Oxygen relights a glowing splint.

### CARBON DIOXIDE, CO$_2$

#### KEY POINT

If you keep bubbling the carbon dioxide through the limewater, the white precipitate will eventually disappear again. This is because the $CO_2$ reacts with the calcium carbonate to form calcium hydrogencarbonate, which is soluble in water:

$$CaCO_3(s) + H_2O(l) + CO_2(g) \rightarrow$$
$$Ca(HCO_3)_2(aq)$$

#### THE TEST FOR CARBON DIOXIDE GAS

The carbon dioxide is bubbled through limewater. Carbon dioxide turns limewater milky/chalky/cloudy.

Limewater is calcium hydroxide solution. Carbon dioxide reacts with it to form a white precipitate of calcium carbonate.

$$Ca(OH)_2(aq) + CO_2(g) \rightarrow CaCO_3(s) + H_2O(l)$$

### CHLORINE, Cl$_2$

#### THE TEST FOR CHLORINE GAS

A piece of damp litmus paper or universal indicator paper is put into the test-tube or held over its mouth. Chlorine is a green gas that bleaches (turns white) the damp litmus paper or universal indicator paper.

If blue litmus paper or universal indicator paper are used for this test, they go red first (the chlorine dissolves in the water to form an acidic solution) and then white.

 Safety Note: Wear eye protection. Avoid inhaling or 'sniffing' chlorine or ammonia, especially if you have a breathing problem such as asthma.

▶ Figure 18.2 Chlorine bleaches damp litmus paper.

**AMMONIA, NH₃**

THE TEST FOR AMMONIA GAS

Hold a piece of damp universal indicator paper or red litmus paper at the mouth of the test-tube. Ammonia turns the universal indicator paper/litmus paper blue.

Ammonia is the only alkaline gas that you will meet at International GCSE.

## TESTING FOR WATER

**USING ANHYDROUS COPPER(II) SULFATE**

Water turns white anhydrous copper(II) sulfate blue.

Anhydrous copper(II) sulfate lacks water of crystallisation and is white. Dropping water onto it replaces the water of crystallisation and turns it blue.

$$\text{anhydrous copper sulfate} + \text{water} \rightarrow \text{hydrated copper sulfate}$$
$$CuSO_4(s) + 5H_2O(l) \rightarrow CuSO_4 \cdot 5H_2O(s)$$
$$\text{white} \qquad\qquad\qquad\qquad \text{blue}$$

If you aren't sure about water of crystallisation, see Chapter 17.

This test works for anything that contains water, so would work with sodium chloride solution or sulfuric acid. It does *not* show that the water is pure.

**!** Safety Note: Avid skin contact with the copper salts.

▲ Figure 18.3 Testing for water with anhydrous copper(II) sulfate.

**PHYSICAL TEST TO SHOW THAT WATER IS PURE**

You can find out whether a sample of water is pure by measuring the freezing (melting) point or boiling point. Pure water freezes at exactly 0 °C and boils at exactly 100 °C at 1 atmosphere pressure. If the water is impure, it will usually freeze at a lower temperature and boil at a higher temperature.

## TESTING FOR IONS

All salts contain at least one cation (positive ion) and anion (negative ion). In this section we will learn how to test for these anions and cations. We do the tests for these anions and cations separately.

**FLAME TESTS**

**KEY POINT**

If a new piece of wire is not being used, it is usually cleaned by dipping it into concentrated hydrochloric acid and then holding it in a non-luminous Bunsen flame. This is repeated until the wire doesn't give any colour to the flame.

A flame test is used to show the presence of certain metal ions (cations) in a compound. A platinum or nichrome wire is dipped into concentrated hydrochloric acid and then into the salt you want to test, so that some salt sticks on the end. The wire and the salt are then held just within a non-luminous (roaring) Bunsen burner flame and the colour observed.

**HINT**

The colour for sodium is described as yellow in the specification, but you will also see it described as orange. Both are usually acceptable.

**KEY POINT**

Lithium isn't the only ion to give a red flame colour. Strontium gives a very similar colour. The only way to be sure is to compare the flame colour next to the flame colour of a known compound of lithium or strontium.

Safety Note: Wear eye protection and take great care with the concentrated acid. Do not get it on your skin and watch out for 'spitting' when heating it on the wire.

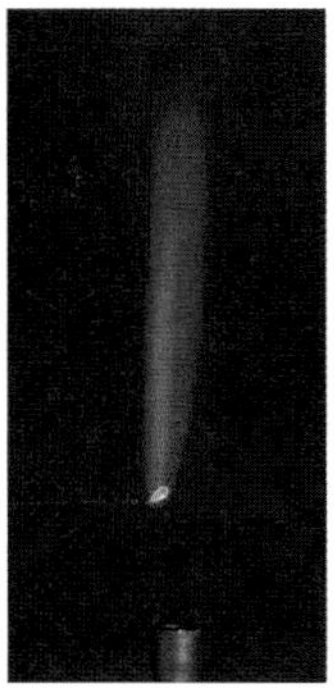

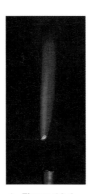

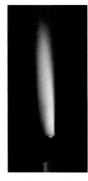

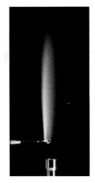

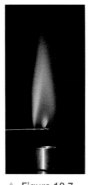

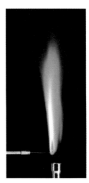

▲ Figure 18.4 Red shows lithium ions.

▲ Figure 18.5 Yellow shows the presence of sodium ions.

▲ Figure 18.6 Lilac shows potassium ions.

▲ Figure 18.7 Orange-red shows calcium ions.

▲ Figure 18.8 Blue-green shows the presence of copper(II) ions.

## TESTING FOR CATIONS (POSITIVE IONS) USING SODIUM HYDROXIDE SOLUTION

Of the common hydroxides, only sodium, potassium and ammonium hydroxides dissolve in water to any extent. Most metal hydroxides are insoluble. That means that if you add sodium hydroxide solution to a solution containing the metal ions, you will get a precipitate of the metal hydroxide.

To carry out this test, you dissolve the salt in distilled (pure) water and put about 1 $cm^3$ in a test-tube. Add about the same volume of dilute sodium hydroxide solution. What you are looking for in most of these tests is the formation of a precipitate; the solution will go cloudy as a solid forms.

### THE TEST FOR COPPER(II) IONS

**HINT**

Ionic equations, which just show the things that actually react, can be a useful way of showing what is happening in a precipitation reaction.

Formation of a blue precipitate shows the presence of copper(II) ions. The precipitate is copper(II) hydroxide.

$$Cu^{2+}(aq) + 2OH^-(aq) \rightarrow Cu(OH)_2(s)$$

Any copper(II) salt in solution will react with sodium hydroxide solution in this way. For example, with copper(II) sulfate solution, the full equation is:

$$CuSO_4(aq) + 2NaOH(aq) \rightarrow Cu(OH)_2(s) + Na_2SO_4(aq)$$

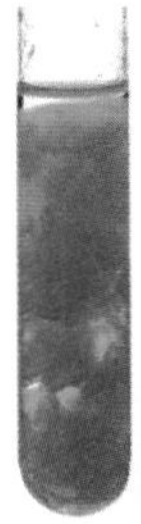

▲ Figure 18.9 The blue precipitate of copper(II) hydroxide.

### THE TEST FOR IRON(III) IONS

Safety Note: Wear eye protection for all these tests and avoid skin contact with the reactants and products.

Formation of an orange-brown precipitate shows the presence of iron(III) ions ($Fe^{3+}$). The precipitate is iron(III) hydroxide.

$$Fe^{3+}(aq) + 3OH^-(aq) \rightarrow Fe(OH)_3(s)$$

Any iron(III) compound in solution will give this precipitate. An example of a full equation might be:

$$FeCl_3(aq) + 3NaOH(aq) \rightarrow Fe(OH)_3(s) + 3NaCl(aq)$$

Notice how much more complicated the full equations for these reactions are. They also hide what is going on. Use ionic equations for precipitation reactions wherever possible.

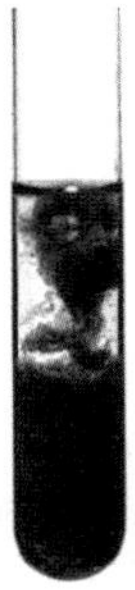

▲ Figure 18.10 The orange-brown precipitate of iron(III) hydroxide.

## THE TEST FOR IRON(II) IONS

### KEY POINT

The green precipitate darkens on standing and turns orange-brown around the top of the tube. This is due to the iron(II) hydroxide being oxidised to iron(III) hydroxide by the air.

### KEY POINT

You must carry out these three tests on a solution of the salt you are testing. If you tested the solid you would not be able to see whether a precipitate had been formed because there is already solid present.

Formation of a green precipitate shows the presence of iron(II) ions ($Fe^{2+}$). The precipitate is iron(II) hydroxide.

$$Fe^{2+}(aq) + 2OH^-(aq) \rightarrow Fe(OH)_2(s)$$

This could be the result of reacting, say, iron(II) sulfate solution with sodium hydroxide solution.

$$FeSO_4(aq) + 2NaOH(aq) \rightarrow Fe(OH)_2(s) + Na_2SO_4(aq)$$

Safety Note: Wear eye protection for all these tests and avoid skin contact with the reactants and products.

▲ Figure 18.11 The green precipitate of iron(II) hydroxide.

## THE TEST FOR AMMONIUM IONS

### KEY POINT

Recognising gases by smelling them has to be done with great care. In this case, there is usually so little ammonia gas present in the cold that it is safe to smell if you take the normal precautions. You shouldn't, however, make any attempt to smell the mixture when it is warm.
Safety Note: Do not attempt to smell the ammonia gas if you have a breathing problem such as asthma.

This test shows the presence of an ammonium salt. Sodium hydroxide solution reacts with ammonium salts (either solid or in solution) to produce ammonia gas. In the cold, there is just enough ammonia gas produced for you to be able to smell it. If you warm it, you can test the gas coming off with a piece of damp red litmus paper or universal indicator paper. Ammonia is alkaline and turns the litmus paper/universal indicator blue. The ionic equation is:

$$NH_4^+(aq) + OH^-(aq) \rightarrow NH_3(g) + H_2O(l)$$

This test works on a sample of a solid or a solution as you are not looking for a precipitate.

A full equation for adding sodium hydroxide solution to solid ammonium chloride is:

$$NH_4Cl(s) + NaOH(aq) \rightarrow NaCl(aq) + NH_3(g) + H_2O(l)$$

Table 18.1 A summary of the reactions of positive ions with sodium hydroxide solution

| Ion | Result of adding NaOH(aq) | Name of precipitate formed | Formula of precipitate |
|---|---|---|---|
| $Cu^{2+}$ | blue precipitate | copper(II) hydroxide | $Cu(OH)_2$ |
| $Fe^{2+}$ | green precipitate | iron(II) hydroxide | $Fe(OH)_2$ |
| $Fe^{3+}$ | orange-brown precipitate | iron(III) hydroxide | $Fe(OH)_3$ |
| $NH_4^+$ | ammonia gas produced | no precipitate | |

## TESTING FOR CARBONATES ($CO_3{}^{2-}$)

### EXTENSION WORK

The ionic equation shows any carbonate reacting with any acid.
$$CO_3{}^{2-}(s) + 2H^+(aq) \rightarrow CO_2(g) + H_2O(l)$$

Add a little dilute hydrochloric acid to your sample of salt and look for fizzing/bubbles of gas. Fizzing indicates that a gas is given off. You can test the gas by bubbling it through limewater to show that it is carbon dioxide.

Most carbonates are insoluble in water and so you will usually be doing this test on a sample of a solid. Soluble carbonates are sodium carbonate, potassium carbonate and ammonium carbonate, and the test also works on solutions of these.

You can use any acid to do this test, but you have to be careful, especially with sulfuric acid, as some acid–carbonate combinations can produce an insoluble salt that coats the solid carbonate and stops the reaction. The test works well with dilute hydrochloric acid or dilute nitric acid.

Safety Note: Wear eye protection for all these tests and avoid skin contact with the reactants and products.

For example, using zinc carbonate and dilute nitric acid:

$$ZnCO_3(s) + 2HNO_3(aq) \rightarrow Zn(NO_3)_2(aq) + CO_2(g) + H_2O(l)$$

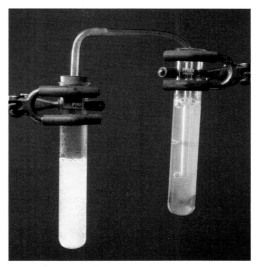

▲ Figure 18.12 We can test to see if carbon dioxide is produced by bubbling the gas through limewater (in the tube on the right) and looking to see whether the limewater goes cloudy/milky (white precipitate of calcium carbonate formed).

## TESTING FOR SULFATES (SO₄²⁻)

You could also use nitric acid and barium nitrate solution in the same way.

Make a solution of your suspected sulfate in distilled (pure) water, add some *dilute hydrochloric acid* and then add some *barium chloride solution*. A sulfate will produce a white precipitate of barium sulfate.

$$Ba^{2+}(aq) + SO_4^{2-}(aq) \rightarrow BaSO_4(s)$$

You add acid to the solution to react with and remove other anions which might also produce white precipitates when you add the barium chloride solution. For example, if you didn't add acid, you would also obtain a white precipitate if there were carbonate ions present. This is because barium carbonate is also white and insoluble. The acid reacts with the carbonate ions and removes them. It is important to remember that the acid must be added *before* the barium chloride solution.

You must never acidify the solution with sulfuric acid because sulfuric acid contains sulfate ions. If you add those, you are bound to get a precipitate of barium sulfate because you have added sulfate ions.

## TESTING FOR CHLORIDES (Cl⁻), BROMIDES (Br⁻) AND IODIDES (I⁻)

Safety Note: Wear eye protection for all these tests and avoid skin contact with the reactants and products.

All of these precipitates tend to discolour to greys and pale purples on exposure to light.

Make a solution of your suspected chloride, bromide or iodide and add enough *dilute nitric acid* to make it acidic. Then add some *silver nitrate solution*.

A *white* precipitate (of silver chloride) shows the presence of chloride ions:

$$Ag^+(aq) + Cl^-(aq) \rightarrow AgCl(s)$$

A *cream* precipitate (of silver bromide) shows the presence of bromide ions:

$$Ag^+(aq) + Br^-(aq) \rightarrow AgBr(s)$$

A *yellow* precipitate (of silver iodide) shows the presence of iodide ions:

$$Ag^+(aq) + I^-(aq) \rightarrow AgI(s)$$

▲ Figure 18.13 Precipitates of silver chloride, silver bromide and silver iodide

The acid is added to react with and remove other anions which might also produce precipitates with silver nitrate solution, for example carbonate and hydroxide ions. You must not use hydrochloric acid (or hydrobromic acid or hydroiodic acid) as you will be adding halide ions and you are certain to obtain a precipitate.

Table 18.2 A summary of the tests for negative ions (anions)

| Ion | Test | Positive result | Product of reaction |
|---|---|---|---|
| $CO_3^{2-}$ | add dilute hydrochloric acid then bubble any gas produced through limewater | fizzing, limewater goes cloudy (white precipitate) | $CO_2$ $CaCO_3$ (with limewater) |
| $SO_4^{2-}$ | add dilute hydrochloric acid followed by barium chloride solution | white precipitate | $BaSO_4$ |
| $Cl^-$ | add dilute nitric acid followed by silver nitrate solution | white precipitate | $AgCl$ |
| $Br^-$ | add dilute nitric acid followed by silver nitrate solution | cream precipitate | $AgBr$ |
| $I^-$ | add dilute nitric acid followed by silver nitrate solution | yellow precipitate | $AgI$ |

## CHAPTER QUESTIONS

**SKILLS** CRITICAL THINKING

1 Name the gas being described in each of the following cases.

  a A green gas that bleaches damp litmus paper.

  b A gas that dissolves readily in water to produce a solution with a pH of about 11.

  c A gas that produces a white precipitate with limewater.

  d A gas that pops when a lighted splint is placed in it.

  e A gas that relights a glowing splint.

**SKILLS** EXECUTIVE FUNCTION

2 Describe in detail how you would carry out the following tests. In each case state whether you would carry out the test on the solid or a solution, and describe what you would expect to see happen.

  a A flame test for lithium ions in lithium chloride.

  b A test for ammonium ions in ammonium sulfate.

  c A test for sulfate ions in ammonium sulfate.

  d A test for carbonate ions in calcium carbonate.

  e A test for iodide ions in potassium iodide.

3 **A** is an orange solid, which dissolves in water to give an orange solution. When sodium hydroxide solution is added to a solution of **A**, an orange-brown precipitate, **B**, is formed. Adding dilute nitric acid and silver nitrate solution to a solution of **A** gives a white precipitate, **C**.

**SKILLS** CRITICAL THINKING

  a State the names of **A**, **B** and **C**.

**SKILLS** PROBLEM SOLVING

  b Write equations (full or ionic) for the reactions producing **B** and **C**.

SKILLS PROBLEM SOLVING, CRITICAL THINKING

4 **D** is a green crystalline solid that dissolves in water to give a very pale green solution. Addition of sodium hydroxide solution to a solution of **D** produces a green precipitate, **E**, which turns orange-brown around the top after standing in air. Addition of dilute hydrochloric acid and barium chloride solution to a solution of **D** gives a white precipitate, **F**.

    a State the names of **D**, **E** and **F**.

SKILLS PROBLEM SOLVING

    b Write equations (full or ionic) for the reactions producing **E** and **F**.

5 **G** is a colourless crystalline solid which reacts with dilute nitric acid to give a colourless solution, **H**, and a colourless, odourless gas, **I**, which turns limewater milky. **G** has a lilac flame colour.

    a State the names of **G**, **H** and **I**.

SKILLS CRITICAL THINKING

SKILLS PROBLEM SOLVING

    b Write an equation (full or ionic) for the reaction between **G** and dilute nitric acid.

6 **J** is a colourless solid. When it is warmed with sodium hydroxide solution it produces a gas, **K**, which turns damp universal indicator paper blue. When nitric acid, followed by silver nitrate solution, is added to a solution of **J**, a cream precipitate of **L** is formed.

    a State the names of **J**, **K** and **L**.

SKILLS CRITICAL THINKING

SKILLS PROBLEM SOLVING

    b Write an ionic equation for the formation of the cream precipitate.

SKILLS EXECUTVE FUNCTION

7 A student has found that her sample of potassium nitrate is contaminated with small amounts of a green solid. She picks out a small piece of the green solid and finds that it is insoluble in water.

SKILLS CRITICAL THINKING

    a Describe how you would make a pure sample of potassium nitrate from the impure mixture.

    b The student believes that the green solid could be copper(II) carbonate. Describe a series of tests that the student could use to confirm this.

8 A student finds a bottle containing a colourless liquid in the laboratory.

    a Describe a test to show that the bottle contains water.

    b Describe a test to show whether it is pure water.

SKILLS EXECUTIVE FUNCTION, REASONING

    c The student carried out the test in b and it showed that the water was not pure. He then carried out some more tests to try to identify the solution. The following results were obtained:

| Test | Result |
| --- | --- |
| add sodium hydroxide solution | no visible reaction but the test-tube became warm |
| add dilute hydrochloric acid | no reaction |
| add dilute hydrochloric acid followed by barium chloride solution | no reaction |
| add nitric acid followed by silver nitrate solution | white precipitate |

Explain what could be present in the bottle and how the student could confirm this.

# UNIT QUESTIONS

You may need to refer to the Periodic Table on page 320.

SKILLS REASONING 8

**1** Lithium, sodium and potassium are three elements in Group 1 of the Periodic Table. All these elements are stored under oil so that they cannot react with air and water.

a Explain in terms of the structure of the atoms why these three elements are all placed in Group 1 of the Periodic Table. **(1)**

SKILLS CRITICAL THINKING 5

b State the name of the product obtained when potassium reacts with air. **(1)**

c Only three of the statements about the reaction of lithium with water given below are true. Put a cross next to each of the correct statements. **(3)**

| | |
|---|---|
| The water turns blue | |
| The piece of lithium floats | |
| The lithium burns with a lilac flame | |
| Hydrogen gas is formed | |
| The lithium explodes | |
| The final solution is alkaline | |
| Lithium reacts more violently than sodium | |

d Some data about the physical properties of lithium, sodium and potassium are shown in the table.

| | Melting point/°C | Boiling point/°C | Density/g/cm$^3$ |
|---|---|---|---|
| Li | 181 | 1342 | 0.53 |
| Na | 98 | 883 | 0.97 |
| K | 63 | 760 | 0.86 |

Caesium is another element in Group 1 of the Periodic Table.

7

i Which of the following are likely to be properties of caesium? Put crosses in two boxes. **(2)**

| | |
|---|---|
| It forms a 1+ ion in compounds | |
| It has a melting point of 60 °C | |
| It is a non-metal | |
| It reacts with water and air | |
| It is an alkali | |

SKILLS REASONING 8

ii Explain why the data in the table cannot be used to predict whether caesium has a higher or lower density than potassium. **(2)**

**(Total 9 marks)**

SKILLS ▶ REASONING

**2** A student carried out some reactions between solutions of chlorine, bromine and iodine, and solutions of potassium halides. Some of their results are shown in the table:

| | Potassium chloride solution | Potassium bromide solution | Potassium iodide solution |
|---|---|---|---|
| Chlorine solution | | orange solution formed | |
| Bromine solution | | | |
| Iodine solution | no reaction | no reaction | |

a Explain why the student did not carry out the experiments shaded in dark grey in the table. **(2)**

SKILLS ▶ INTERPRETATION

b Complete the table. **(3)**

SKILLS ▶ PROBLEM SOLVING

c Write a word equation for the reaction between chlorine and potassium bromide. **(1)**

d Complete and balance the ionic equation for the reaction between chlorine and potassium bromide solutions: **(3)**

$$Cl_2(aq) + Br^-(aq) \rightarrow$$

SKILLS ▶ REASONING

e Explain in terms of electrons why the reaction between chlorine and potassium bromide solution is described as a redox reaction. **(3)**

f Chlorine can be prepared in the laboratory by the reaction between concentrated hydrochloric acid and potassium manganate(VII). The equation for the reaction is:

$$2KMnO_4 + 16HCl \rightarrow 2KCl + 2MnCl_2 + 8H_2O + ....Cl_2$$

SKILLS ▶ PROBLEM SOLVING

i The equation is not balanced. Give the number that should be put in front of $Cl_2$ in order to balance it. **(1)**

SKILLS ▶ EXECUTIVE FUNCTION

ii Describe a test for chlorine gas. **(2)**

Test:

Positive result:

iii When chlorine dissolves in water some chloride ions are formed. Describe a chemical test to show that a solution contains chloride ions. **(3)**

Test:

Positive result: **(Total 18 marks)**

SKILLS ▶ CRITICAL THINKING

**3** a State the name of the most common gas in the air. **(1)**

b A student carries out an experiment to measure the percentage of oxygen in the air. He used the following apparatus:

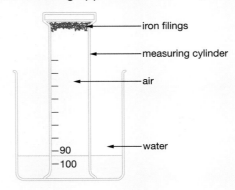

He carried out the following procedure:

■ Put some damp iron filings at the bottom of a 100 cm³ measuring cylinder.

■ Invert the measuring cylinder in a beaker of water.

■ Take the initial reading of the level of water in the measuring cylinder.

■ Take a reading of the level of water in the measuring cylinder every day for 1 week.

The student's results were:

| | |
|---|---|
| initial reading on measuring cylinder/cm³ | 94 |
| reading on measuring cylinder after 1 day/cm³ | 85 |
| reading on measuring cylinder after 2 days/cm³ | 80 |
| reading on measuring cylinder after 3 days/cm³ | 77 |
| reading on measuring cylinder after 4 days/cm³ | 76 |
| reading on measuring cylinder after 5 days/cm³ | 75 |
| reading on measuring cylinder after 6 days/cm³ | 75 |
| reading on measuring cylinder after 7 days/cm³ | 75 |

**SKILLS** REASONING   (5)

i   Explain why the water level rises.    **(2)**

  (6)

ii   Explain why the reading on the measuring cylinder eventually remains constant.    **(1)**

**SKILLS** CRITICAL THINKING

iii   Give the name of the iron compound that is formed in this experiment.    **(1)**

**SKILLS** REASONING

iv   The student trapped the gas in the tube and put a lighted splint into it. Explain why the splint goes out.    **(2)**

**SKILLS** PROBLEM SOLVING

v   Using the results in the table, calculate the percentage oxygen present in the air in the measuring cylinder after 1 day.    **(4)**

**SKILLS** CRITICAL THINKING   (7)

vi   The student decided that 1 week is too long to wait for results so suggested using a more reactive metal such as calcium or lithium in this experiment. Explain one reason, other than that it might be too dangerous, why the experiment will not work with these metals.    **(2)**

**SKILLS** EXECUTIVE FUNCTION   (5)

vii   Suggest one change that the student could make to the experiment so he does not have to wait so long for the results.    **(1)**

c   Sulfur reacts with oxygen. A teacher burns a piece of sulfur in a gas jar of oxygen.

**SKILLS** CRITICAL THINKING

i   State the colour of the flame.    **(1)**

**SKILLS** PROBLEM SOLVING   (9)

ii   Write a balanced equation for the reaction that occurs.    **(1)**

**SKILLS** REASONING   (6)

iii   A few drops of litmus solution are put into the gas jar at the end of the experiment. State and explain what colour will be seen.    **(2)**

d   Sodium reacts with oxygen to form sodium oxide.

**SKILLS** PROBLEM SOLVING   (9)

i   Write a chemical equation for this reaction.    **(2)**

**SKILLS** CRITICAL THINKING   (5)

ii   When sodium oxide reacts with water an alkaline solution is formed. State the name of the ion responsible for making the solution alkaline.    **(1)**

**(Total 21 marks)**

**4** This question is about copper(II) sulfate solution.

a Describe a procedure to make a solution of copper(II) sulfate starting from copper(II) oxide. **(3)**

b Describe a chemical test that can be used to show that the solution contains sulfate ions. **(2)**

c A student added some sodium hydroxide solution to the copper(II) sulfate solution. State the name of the blue precipitate formed. **(1)**

d A student carried out an experiment to investigate displacement reactions of three different metals with copper(II) sulfate solution. She followed this method:

■ Put some copper(II) sulfate solution into a polystyrene cup.

■ Measure the initial temperature.

■ Add some metal.

■ Stir the solution.

■ Record the maximum temperature.

i The student uses the same amount in moles of each metal in the experiments. State two other variables the student should keep the same in this experiment to allow her to be able to compare the different metals. **(2)**

ii The student recorded the following data.

| Metal | Temperature change/°C |
|-------|----------------------|
| zinc | 15 |
| silver | 0 |
| nickel | 7 |

Arrange the four metals, zinc, silver, copper and nickel, in order of reactivity, starting with the most reactive. **(1)**

iii The student did not carry out the experiment with copper and copper(II) sulfate. Predict, giving a reason, what the temperature rise would be for this experiment. **(2)**

iv The student carried out the experiment using 3.25 g of zinc (0.05 mol). She then used the same amount in moles of nickel and silver. Another student decided to use the same mass (3.25 g) of each metal. The results of the two students were very similar. Suggest reasons for this. **(3)**

v The ionic equation for the reaction between copper(II) sulfate solution and nickel is:

$$Ni(s) + Cu^{2+}(aq) \rightarrow Ni^{2+}(aq) + Cu(s)$$

Use the ionic equation to explain whether nickel is oxidised or reduced in this reaction. **(2)**

e The student wants to make some anhydrous copper(II) sulfate. Describe how she could do this starting with copper(II) sulfate solution. **(2)**

f The student carried out a flame test on the anhydrous copper(II) sulfate.

i Describe how to carry out a flame test. **(2)**

ii State the colour of the flame produced when the student carried out a flame test on copper(II) sulfate. **(1)**

**(Total 21 marks)**

## CHEMISTRY ONLY

**5** Molybdenite is an ore of molybdenum. The mineral it contains is $MoS_2$.

The following set of reactions are used to produce pure molybdenum(VI) oxide ($MoO_3$), from which molybdenum metal can be obtained.

The ore is first roasted in air to form $MoO_3$. The equation for this reaction is:

$$2MoS_2(s) + 7O_2(g) \rightarrow 2MoO_3(s) + 4SO_2(g) \ldots 1$$

The crude compound is now heated with concentrated ammonia solution:

$$MoO_3(s) + 2NH_3(aq) + H_2O(l) \rightarrow (NH_4)_2MoO_4(aq) \ldots 2$$

Ammonium molybdate is crystallised from the solution and when these crystals are heated pure $MoO_3$ is formed:

$$(NH_4)_2MoO_4(s) \rightarrow MoO_3(s) + H_2O(l) + 2NH_3(g) \ldots 3$$

The molybdenum can be extracted from $MoO_3$ by heating with carbon.

a Explain the meaning of the word *ore*. (1)

b Put ticks in the appropriate boxes in the table to classify each of the reactions 1 and 3 above. Each reaction type may occur once, twice or not at all. (2)

|  | Thermal decomposition | Neutralisation | Redox |
|---|---|---|---|
| Reaction 1 |  |  |  |
| Reaction 3 |  |  |  |

c Molybdenum(VI) oxide can be heated with carbon to form molybdenum metal. The other product of the reaction is carbon monoxide. Write a chemical equation for this reaction. (2)

d Explain which substance is the *reducing agent* when molybdenum(VI) oxide is heated with carbon. (2)

e Explain what the reaction in c suggests about the position of molybdenum in the reactivity series. (2)

f Molybdenum can also be extracted from $MoO_3$ by heating with aluminium. Use your knowledge of the reactivity series to give a reason why this method is more expensive than heating with carbon. (2)

g Describe a chemical test you could use to show that the product of reaction 2 contains ammonium ions. (2)

**(Total 13 marks)**

**6** This question is about lead(II) iodide. This is a bright yellow substance that is insoluble in water. In the past it was used as a pigment, but due to its toxicity this no longer happens. It is now used in the development of new types of solar cell.

a Describe a method a student could use to produce a pure, dry sample of lead(II) iodide. (5)

b The student adds nitric acid followed by silver nitrate solution to their sample of lead(II) iodide. The student stated that the formation of a yellow precipitate showed that it contained iodide ions. Explain why the student's conclusion is incorrect. **(2)**

**(Total 7 marks)**

## END OF CHEMISTRY ONLY

 **7**

This question is about hydrogen chloride and hydrochloric acid.

a Hydrochloric acid is made when hydrogen chloride dissolves in water. Draw a dot-and-cross diagram for hydrogen chloride. **(2)**

b When hydrogen chloride dissolves in water an acid–base reaction occurs. The equation for this reaction is:

$$HCl(aq) + H_2O(l) \rightarrow Cl^-(aq) + H_3O^+(aq)$$

Explain why this is classified as an acid–base reaction and identify the acid and the base. **(3)**

## CHEMISTRY ONLY

c Hydrochloric acid can be used to make chloride salts. Complete the table by stating what you would add to dilute hydrochloric acid to make each of the salts. In each case, state whether you would add it as a solid or in solution and state any other products formed. The first one has been done for you. **(4)**

| Reagent | Solid or solution? | | Salt | Other product |
|---|---|---|---|---|
| magnesium | solid | hydrochloric acid | magnesium chloride | hydrogen |
| copper(II) oxide | | hydrochloric acid | copper(II) chloride | |
| | solution | hydrochloric acid | sodium chloride | water |
| | solution | hydrochloric acid | silver chloride | nitric acid |

## END OF CHEMISTRY ONLY

d A student wants to show that an unknown solution she has been given contains chloride ions. She puts a small amount of the solution in a test-tube and adds dilute nitric acid followed by silver nitrate solution.

i Describe what would be seen if the solution contains chloride ions. **(1)**

ii Explain why hydrochloric acid cannot be added instead of nitric acid. **(2)**

e Write an equation, including state symbols, for the reaction that occurs when hydrochloric acid is added to solid copper(II) carbonate. **(3)**

**(Total 15 marks)**

**⑤** **8** This question is about chemical tests.

Samples of a very pale green solution, **X**, were tested as follows:

| Test | Observation |
|------|-------------|
| A sample of **X** was acidified with dilute nitric acid and silver nitrate solution was added | A white precipitate (**A**) was formed |
| A small amount of sodium hydroxide solution was added to a sample of solution **X** | A dark green precipitate (**B**) was formed |
| Chlorine was bubbled through a sample of **X** | The pale green solution turned yellow (solution **C**) |
| A small amount of sodium hydroxide solution was added to solution **C** | An orange-brown precipitate (**D**) was formed |

a  Use the results from the first two tests to identify **X**.  (2)

**⑥** b  State the name and formula of **B**.  (2)

c  State the name and formula of **D**.  (2)

**⑦** d  Explain what type of reaction occurred when chlorine was bubbled through **X**.  (2)

**(Total 8 marks)**

## CHEMISTRY ONLY

**⑥** **9**

sodium sulfate solution

platinum electrodes

⊕  Ⓐ  ⊖

powerpack

This question is about sodium sulfate.

a  State the formula of sodium sulfate.  (1)

b  Describe a practical method by which a student could make a solution of sodium sulfate starting from an acid and an alkali. You should state the names of any pieces of apparatus used and of all chemicals required.  (5)

c  A student used the apparatus shown in the diagram to investigate the effect of changing the time of electrolysis on the volume of oxygen produced when the solution of sodium sulfate was electrolysed.

The student followed this method:

■ Set the current to be 0.50 A.

■ Record the initial volume of gas in the left-hand arm of the apparatus.

■ Leave the current running for 2 minutes and record the final volume of gas in the left-hand arm of the apparatus.

■ Repeat for times of 4 minutes, 6 minutes, 8 minutes and 10 minutes.

His results are shown in the table.

| Time/min | 2 | 4 | 6 | 8 | 10 |
|---|---|---|---|---|---|
| Volume of gas/cm$^3$ | 3.6 | 7.2 | 9.0 | 14.4 | 18.0 |

i   Plot these results on a piece of graph paper and draw a straight line
    of best fit.    (3)

ii  The student suggested that one of the points is anomalous.
    Identify the anomalous point and suggest the volume of gas that
    should have been obtained for that time.    (2)

iii State the relationship between the volume of gas produced and the time of
    electrolysis.    (2)

iv  Describe a chemical test for the gas produced at the cathode    (2)
    Test:
    Positive result:

**(Total 15 marks)**

**END OF CHEMISTRY ONLY**

# UNIT 3
# PHYSICAL CHEMISTRY

In a world where the population has more than trebled in the last 100 years, fighting against hunger is an extremely important challenge. At the beginning of the 20th century scientists developed industrial processes to make artificial fertilisers that can be used to increase crop yields. In order to develop these processes, chemists and chemical engineers from around the world have had to understand the principles of energy changes, rates of reaction and equilibrium. The world's resources are limited and it is vital that these processes are as efficient and environmentally friendly as possible. This is also the work of scientists.

▲ Figure 19.1 Burning fuel produces enough energy to launch a rocket.

# 19 ENERGETICS

Some chemical reactions produce heat. Others need to be heated constantly to make them occur at all. This chapter explores some examples of both kinds of reaction and examines how energy changes during reactions can be measured by experiments or calculated using bond energies.

## LEARNING OBJECTIVES

- Know that chemical reactions in which heat energy is given out are described as exothermic, and those in which heat energy is taken in are described as endothermic

- Describe simple calorimetry experiments for reactions such as combustion, displacement, dissolving and neutralisation

- Calculate the heat energy change from a measured temperature change using the expression $Q = mc\triangle T$

- Calculate the molar enthalpy change ($\triangle H$) from the heat energy change, $Q$

- Practical: Investigate the temperature changes accompanying some of the following types of change:
  - salts dissolving in water
  - neutralisation reactions
  - displacement reactions
  - combustion reactions

### CHEMISTRY ONLY

- Draw and explain energy level diagrams to represent exothermic and endothermic reactions.

- Know that bond breaking is an endothermic process and that bond making is an exothermic process.

- Use bond energies to calculate the enthalpy change during a chemical reaction.

## EXOTHERMIC REACTIONS

Some chemical reactions give out energy in the form of heat. A reaction that *gives out heat to the surroundings* is said to be exothermic. If you are holding a test-tube in which an exothermic reaction is occurring, you will notice that the test-tube *gets warmer*.

An example of an exothermic reaction is adding water to calcium oxide. If you add water to solid calcium oxide, the heat produced is enough to boil the water and produce steam. Calcium hydroxide is produced.

$$CaO(s) + H_2O(l) \rightarrow Ca(OH)_2(s)$$

In an *exothermic* reaction, *the products of the reaction have less (chemical) energy than the reactants*. In the reaction, *chemical energy* (stored in the bonds of chemicals) *is converted to heat energy*, which is released to the surroundings. The temperature of the reaction mixture and its surroundings *goes up*.

> **DID YOU KNOW?**
> Calcium oxide is known as quicklime. Adding water to it is described as *slaking* it. The calcium hydroxide produced is known as slaked lime.

 Safety Note: It is less hazardous to use a lump of calcium oxide, rather than powered calcium oxide

▲ Figure 19.2 Calcium oxide reacting with water.

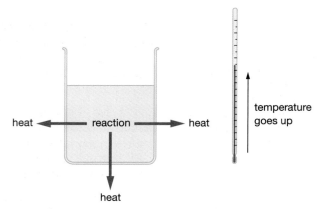

▲ Figure 19.3 In an exothermic reaction, chemical energy is converted to heat energy. Heat is released so the temperature goes up.

You will come across lots of exothermic reactions in this course. Some examples are given below.

## COMBUSTION REACTIONS

Any reaction that produces a flame is exothermic. Burning things produces heat energy.

For instance, hydrogen burns in oxygen, producing water and lots of heat:

$$2H_2(g) + O_2(g) \rightarrow 2H_2O(l)$$

Apart from burning, other exothermic changes include:

- the reactions of metals with acids
- neutralisation reactions
- displacement reactions.

▲ Figure 19.4 The burning of hydrogen is used in oxy-hydrogen cutting equipment underwater.

## THE REACTIONS OF METALS WITH ACIDS

### REMINDER

This reaction is described in detail in Chapter 14 (pages 152–153).

When magnesium reacts with dilute sulfuric acid, for example, the mixture gets very warm:

$$Mg(s) + H_2SO_4(aq) \rightarrow MgSO_4(aq) + H_2(g)$$

## NEUTRALISATION REACTIONS

### REMINDER

You can read about this reaction in Chapter 16 (page 170). We will investigate how much heat energy is released in a typical neutralisation reaction later in this chapter (pages 217–219).

About the only interesting thing that you can observe happening when sodium hydroxide solution reacts with dilute hydrochloric acid is that the temperature rises:

$$NaOH(aq) + HCl(aq) \rightarrow NaCl(aq) + H_2O(l)$$

## DISPLACEMENT REACTIONS

The thermite reaction between powdered aluminium and iron(III) oxide is a displacement (competition) reaction. This reaction releases a large amount of heat, which can be used in railway welding:

$$2Al(s) + Fe_2O_3(s) \rightarrow 2Fe(l) + Al_2O_3(s)$$

## ENTHALPY CHANGE OF A REACTION

**KEY POINT**

$\Delta H$ is pronounced 'delta *H*'. The Greek letter $\Delta$ is used to mean 'change in'. $\Delta H$ means 'change in heat'. Note: it is not possible to measure how much enthalpy (*H*) something has – you can only measure the change in enthalpy ($\Delta H$) when it reacts.

**REMINDER**

The mole is a unit for the amount of a substance. You can read more about it in Chapter 5.

**KEY POINT**

The units for $\Delta H$ can be written as kJ/mol or kJ $mol^{-1}$.

You can measure the amount of heat energy taken in or released in a chemical reaction. It is called the **enthalpy change** of the reaction and is given the symbol $\Delta H$. The enthalpy change is *the amount of heat energy taken in or given out in a chemical reaction*. It is the difference between the energy of the products and the energy of the reactants.

$\Delta H$ is given a minus or a plus sign to show whether heat is being given out or absorbed by the reaction. You always look at it from the point of view of the reactants. For an exothermic reaction, $\Delta H$ is given a *negative* number because the reactants are *losing* energy as heat. That heat is transferred to the surroundings, which then get warmer. $\Delta H$ is measured in units of kJ/mol (kilojoules per mole).

In an equation, the amount of heat given out or taken in can be shown as, for example:

$$Mg(s) + H_2SO_4(aq) \rightarrow MgSO_4(aq) + H_2(g) \quad \Delta H = -466.9\,kJ/mol$$

The $\Delta H$ written next to an equation represents the enthalpy change of the reaction, i.e. 466.9 kJ of heat is given out when *one mole* of magnesium reacts in this way. You know heat has been given out because $\Delta H$ has a negative sign.

## CHEMISTRY ONLY

## SHOWING AN EXOTHERMIC CHANGE ON AN ENERGY LEVEL DIAGRAM

**KEY POINT**

The term **stability** is usually used to describe the relative energies of the reactants and the products in a chemical reaction. The more energy a chemical has, the less stable it is.

**REMINDER**

Remember that in a chemical reaction, the reactants are the chemicals you start with.

In an exothermic reaction, the reactants have more (chemical) energy than the products; we say that the products are *more stable* than the reactants. As the reaction happens, energy is given out in the form of heat. That energy warms up both the reaction itself and its surroundings.

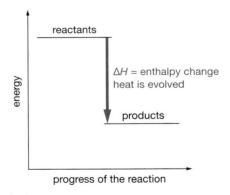

▲ Figure 19.5 An exothermic change

## END OF CHEMISTRY ONLY

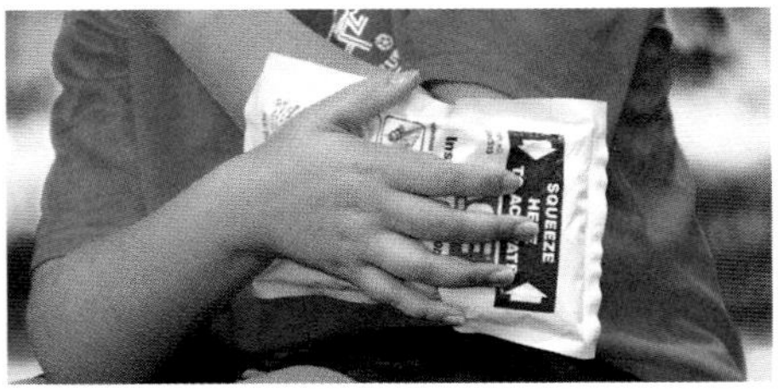

▲ Figure 19.6 A boy using a cold pack to relieve pain in his elbow. Endothermic reactions occur in the chemical cold pack to absorb heat from the surroundings and reduce the temperature of his arm.

## ENDOTHERMIC REACTIONS

*A reaction that absorbs heat from the surroundings* is said to be **endothermic**. If you hold a test-tube in which an endothermic reaction is occurring you will notice that it *gets colder*.

In an endothermic reaction, *the products have more (chemical) energy than the reactants*. In order to supply the extra energy that is needed to convert the reactants (lower energy) into the products (higher energy), heat energy

**REMINDER**

Not all reactions that have to be heated are endothermic reactions. Sometimes we just heat something to make it go faster.

needs to be absorbed from the surroundings. This *heat energy is converted to chemical energy* (energy stored in the bonds of chemicals). The temperature of the reaction mixture and the surroundings *goes down* because heat energy has been converted into a different form of energy.

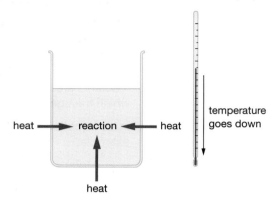

▲ Figure 19.7 In an endothermic reaction, heat energy is converted to chemical energy. Heat is absorbed so the temperature goes down.

You have seen the *thermal decomposition* of metal carbonates before in Chapter 13. These are examples of endothermic reactions. You have to heat a carbonate constantly to make it decompose.

For example, copper(II) carbonate (green) decomposes on heating to produce copper(II) oxide (black).

$CuCO_3(s) \rightarrow CuO(s) + CO_2(g)$

Similarly, zinc carbonate decomposes to form zinc oxide when heated.

$ZnCO_3(s) \rightarrow ZnO(s) + CO_2(g)$

**CHEMISTRY ONLY**

## SHOWING AN ENDOTHERMIC CHANGE ON AN ENERGY LEVEL DIAGRAM

▲ Figure 19.8 An endothermic change

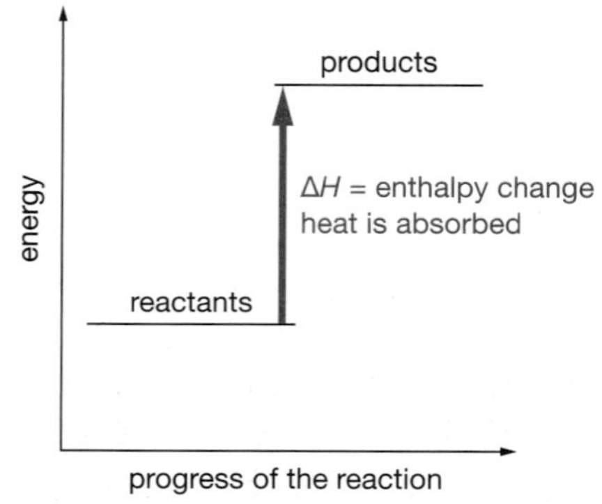

In an endothermic change, the products have more energy than the reactants so we say that the products are *less stable* than the reactants. That extra energy has to come from somewhere, and it is taken from the surroundings. In the case of the thermal decomposition of carbonates in the laboratory, it comes from the Bunsen burner.

Because the reactants are *gaining* energy, the enthalpy change of the reaction $\Delta H$ is given a *positive* sign.

For example:

$CaCO_3(s) \rightarrow CaO(s) + CO_2(g) \qquad \Delta H = +178$ kJ/mol

This means that 178 kJ of heat energy must be absorbed to convert 1 mole of calcium carbonate into calcium oxide and carbon dioxide.

**END OF CHEMISTRY ONLY**

## MEASURING ENTHALPY CHANGES OF REACTIONS

Here we discuss how we can measure how much heat is taken in or given out by a chemical reaction, in other words the enthalpy change of a reaction.

## SPECIFIC HEAT CAPACITY

When we heat something up, it gets hotter. The specific heat capacity tells us about *how much* energy has to be put in to increase the temperature of something. The specific heat capacity of a substance is defined as *the amount of heat needed to raise the temperature of 1 gram of a substance by 1 °C.*

For water, the value is 4.18 J/g/°C (joules per gram per degree Celsius). This means that 4.18 J of heat energy is needed if we want to increase the temperature of 1 g of water by 1 °C. If you want the temperature of 1 g of water to go up by 2 °C, then 4.18 × 2 = 8.36 J of heat energy must be supplied. If now you have 2 g of water, then 2 × 8.36 J of energy would be needed to raise the temperature by 2 °C.

The amount of heat energy required is *directly proportional* to the mass ($m$) and the temperature change ($\Delta T$) of the substance. The following equation can be used to calculate how much heat energy needs to be supplied to raise the temperature of mass $m$ by $\Delta T$°C:

heat energy change = mass × specific heat capacity × temperature change

$$Q = m \times c \times \Delta T$$

## CALORIMETRY EXPERIMENTS FOR DETERMINING THE ENTHALPY CHANGES OF REACTIONS

It is fairly uncomplicated to measure the amount of heat absorbed or given out in several kinds of chemical reactions and physical changes. The technique used to do this is called calorimetry and it is based on the idea that if we use the heat from a reaction to heat another substance, such as water, we can then use the equation introduced above ($Q = mc\Delta T$) to calculate the amount of heat released. Here the mass, the specific heat capacity and the temperature change are all referring to *the substance heated*. If we know how many moles of reactants are used in the reaction, we can then work out the molar enthalpy change, $\Delta H$, of the reaction in the unit kJ/mol.

The following activity illustrates how we can use calorimetry to determine the molar enthalpy change of combustion of an alcohol, i.e. how much heat energy is released when 1 mole of alcohol burns.

### ACTIVITY 1

### ▼ PRACTICAL: MEASURING ENTHALPY CHANGES IN COMBUSTION REACTIONS

One of the most common calorimetry experiments at International GCSE is to measure the amount of heat given off when a number of small alcohols are burned. You could use methanol, ethanol, propan-1-ol and butan-1-ol (see Chapter 22 for an introduction to the naming of alcohols).

The alcohols are burned in a small spirit burner, and the heat produced is used to heat some water in a copper can (the calorimeter).

The following procedure could be used:
- Measure 100 cm$^3$ of cold water using a measuring cylinder and transfer the water to a copper can.
- Take the initial temperature of the water.
- Weigh a spirit-burner containing ethanol with its lid *on*. The lid should be kept on when the wick is not lit to prevent the alcohol from evaporating.

Safety Note: Wear eye protection. Do not carry a lit spirit burner and do not fill or re-fill one when there is a naked flame nearby.

- Arrange the apparatus as shown in Figure 19.9 so that the spirit-burner can be used to heat the water in the copper can. The apparatus is shielded as far as possible to prevent draughts.
- Light the wick to heat the water. Stop heating when you have a reasonable temperature rise of water (say, about 40.0 °C). The flame can be extinguished by putting the lid back on the wick.
- Stir the water thoroughly and measure the maximum temperature of the water.
- Weigh the spirit-burner again with its lid *on*.
- The experiment can be repeated with the same alcohol to check for reliability, and then carried out again with whatever other alcohols are available.

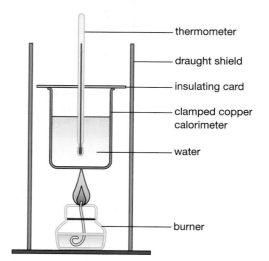

thermometer

draught shield

insulating card

clamped copper calorimeter

water

burner

▲ Figure 19.9 A calorimetry experiment to measure the enthalpy change of combustion of alcohols.

**SAMPLE DATA**

| volume of water/cm³ | 100 |
|---|---|
| mass of burner + ethanol before experiment/g | 137.36 |
| mass of burner + ethanol after experiment/g | 136.58 |
| original temperature of water/°C | 21.5 |
| final temperature of water/°C | 62.8 |

Combustion is an exothermic reaction so the temperature of the water goes up. As ethanol is burned, the total mass of the burner and ethanol goes down.

Using the above data we can determine how much heat is released when one mole of an alcohol is burned, that is, the molar enthalpy change of combustion.

**CALCULATIONS FOR ACTIVITY 1**

We are going to use the equation $Q = mc\Delta T$ so we need to find out what each quantity is.

Temperature change of water = $\Delta T$ = 62.8 − 21.5 = 41.3 °C.

Mass of water being heated = $m$ = 100 g, the density of water is approximately 1 g/cm³ at room temperature, so 100 cm³ of water has a mass of 100 g.

$c$ is the specific heat capacity of the water (it is the water that is being heated):
$c = 4.18\,J/g/°C$

Heat gained by water $= Q = mc\Delta T = 100 \times 4.18 \times 41.3 = 17\,260\,J$.

Divide $Q$ by 1000 to give energy in kJ $= 17.26\,kJ$.

This means $17.26\,kJ$ of heat energy is released by the combustion of the ethanol in this experiment. To calculate the amount of heat produced when 1 mole of ethanol, $CH_3CH_2OH$, burns, you need to find out how many moles of ethanol are burned in your experiment.

Mass of ethanol burned = decrease in the mass of the burner
$$= 137.36 - 136.58 = 0.78\,g$$

Ethanol has the formula $C_2H_5OH$. We can calculate the relative molecular mass of ethanol by adding up the relative atomic masses:
$(2 \times 12) + (5 \times 1) + 16 + 1 = 46$

Number of moles ($n$) of ethanol, $CH_3CH_2OH$, burned

$$n = \frac{mass\ (m)}{relative\ molecular\ mass\ (M_r)} = \frac{0.78}{46} = 0.01696\,mol$$

Molar enthalpy change of combustion of ethanol ($\Delta H$)

$$\Delta H = \frac{heat\ energy\ change\ (Q)}{number\ of\ moles\ of\ ethanol\ burned\ (n)}$$

$$= \frac{17.26}{0.01696} = 1020\,kJ/mol$$

The amount of heat released in the complete combustion of 1 mole of ethanol is therefore:

$$CH_3CH_2OH + 3O_2 \rightarrow 2CO_2 + 3H_2O \qquad \Delta H = -1020\ kJ/mol$$

*The negative sign shows that heat is released and the combustion reaction is exothermic.*

## EVALUATION OF THE EXPERIMENTAL RESULTS

How accurate is the figure of $-1020\,kJ/mol$ for the enthalpy change of the combustion of ethanol? The accepted value found in data booklets for ethanol is $-1370\,kJ/mol$, which means that $1370\,kJ$ of heat should be given out when 1 mole of ethanol burns. You can see that the value we obtained is less exothermic than expected; our reaction seemed to give out less heat than expected.

There are many sources of error in this experiment, in particular *large amounts of heat loss*.

The warm water gives out heat to the air, heat is lost from the flame, which goes straight into the air rather than into the water, and heat is used to raise the temperature of the copper can and the thermometer.

Another major source of error is the *incomplete combustion* of alcohol. We will talk about incomplete combustion in more detail in Chapter 23. Incomplete combustion of an alcohol occurs when there is not enough oxygen present. Incomplete combustion releases less heat than complete combustion. We can see that the combustion in Activity 1 is incomplete because the flame of the wick is often yellow orange rather than blue, and soot (carbon) is produced at the bottom of the copper can. If the combustion is complete, the flame should be blue and carbon dioxide should be produced instead of carbon.

A bomb calorimeter can be used to obtain more accurate values of the enthalpy changes of combustion. The substance to be combusted is ignited electrically and burns in the presence of pure oxygen. The heat transferred to both water and the calorimeter itself is determined to give a more accurate value of the enthalpy change.

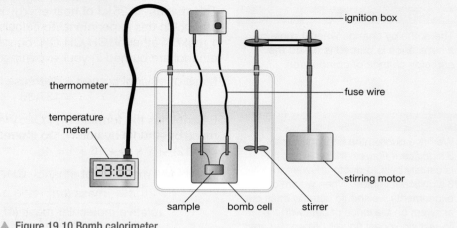

▲ Figure 19.10 Bomb calorimeter

That doesn't mean you can't use this experiment to make useful comparisons. If you repeat it with other alcohols, under conditions that are as similar as possible, you can find how the heat evolved (given out) changes as the alcohol gets bigger.

In Figure 19.11 we can see that the combustion reaction gets more exothermic as the alcohol chain becomes longer. In other words, longer alcohols give out more heat energy per mole when they burn than shorter ones.

The increase in the heat evolved as alcohol chain lengthens is regular. The difference between one alcohol and the next is always an extra $CH_2$ (see Chapter 26, page 288, for more detail on the structures of alcohols), and so the number of extra bonds broken and made (see below) increases in a regular way. That means that the heat evolved will also increase in a regular way.

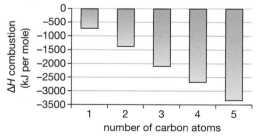

▲ Figure 19.11 Molar enthalpy change of the combustion of alcohols.

# WORKING OUT ENTHALPY CHANGES FOR REACTIONS INVOLVING SOLUTIONS USING CALORIMETRY EXPERIMENTS

You can use very similar methods for measuring molar enthalpy changes in displacement reactions (e.g. zinc and copper(II) sulfate solution), dissolving (e.g. dissolving ammonium chloride in water to form a solution) and neutralisation reactions (e.g. between potassium hydroxide solution and dilute hydrochloric acid). These experiments also involve heating water, but this time the water is part of the solutions we are using. We will look at how to do this in the next few practicals.

Safety Note: Avoid skin contact with the chemicals.

## ACTIVITY 2

### ▼ PRACTICAL: MEASURING ENTHALPY CHANGES FOR DISPLACEMENT REACTIONS

In order to determine the enthalpy change of the reaction of zinc and copper(II) sulfate, the following procedure could be used:

- Place a polystyrene cup in a 250 cm³ glass beaker.
- Transfer 50 cm³ of 0.200 mol/dm³ copper(II) sulfate solution into the polystyrene cup using a measuring cylinder.
- Weigh 1.20 g of zinc using a weighing boat on a balance.
- Record the initial temperature of the copper(II) sulfate solution.
- Add the zinc.
- Stir the solution as quickly as possible.
- Record the maximum temperature reached.

▲ 19.12 A calorimetry experiment to measure the enthalpy change of a displacement reaction.

**SAMPLE DATA**

| | |
|---|---|
| initial temperature of copper(II) sulfate solution/°C | 17.0 |
| maximum temperature of the reaction mixture/°C | 27.3 |

We can use this data to calculate the enthalpy change for this displacement reaction, when 1 mole of copper(II) sulfate reacts with zinc.

**CALCULATIONS FOR ACTIVITY 2**

Heat given out in this reaction: $Q = mc\Delta T = 50 \times 4.18 \times (27.3 - 17.0)$

$$= 2152.7 \, J$$

$$= 2.1527 \, kJ$$

Here we assume the following:

1. The density of the copper sulfate solution is the same as that of water, so 1 cm³ of solution has a mass of 1 g.

2. The specific heat capacity of the mixture is the same as that of water. This is a fairly reasonable assumption because the reaction mixture is mostly water.

1 mol of Zn reacts with 1 mol of $CuSO_4$, therefore 0.0100 mol of $CuSO_4$ reacts with 0.0100 mol of Zn. We added 0.0185 mol of Zn, so Zn is in excess.

Remember that we have to divide the volume by 1000 to convert it to $dm^3$ because the concentration is given in $mol/dm^3$.

You could repeat this experiment with metals of different reactivities. The more reactive a metal is, the more heat should be released in the displacement reaction. Make sure you keep everything else the same: the number of moles of the metals, the size of the solid particles, and the volume and concentration of the copper(II) sulfate solution. Do not use metals that are more reactive than magnesium, otherwise you are measuring the heat released when the metal reacts with water instead!

Safety Note: Wear eye protection and avoid skin contact with the salts and their solutions.

In this experiment we have used excess zinc. *Excess* means more than enough zinc is present to ensure all the copper(II) sulfate reacts. If you calculate the number of moles of copper(II) sulfate and the number of moles of zinc used in this procedure, you should spot that the number of moles of zinc used is more than that of copper(II) sulfate:

$$\text{number of moles } (n) \text{ of zinc added} = \frac{\text{mass } (m)}{\text{relative atomic mass } (A_r)}$$

$$= \frac{1.20}{65}$$

$$= 0.0185 \, mol$$

number of moles $(n)$ of copper(II) sulfate added = volume $(V)$ × concentration $(C)$

$$= 0.050 \times 0.200$$

$$= 0.0100 \, mol$$

Now we need to calculate how much heat is released when 1 mole of copper sulfate reacts with excess zinc:

Molar enthalpy change of reaction $(\Delta H)$

$$\Delta H = \frac{\text{heat energy change } (Q)}{\text{number of moles of copper sulfate reacted } (n)}$$

$$= \frac{2.1527}{0.0100} = 215 \, kJ/mol$$

The amount of heat released in the displacement reaction when 1 mole of $CuSO_4$ reacts with excess Zn is therefore:

$$Zn(s) + CuSO_4(aq) \rightarrow ZnSO_4(aq) + Cu(s) \qquad \Delta H = -215 \, kJ/mol$$

*We have added the negative sign because we know that the temperature of the reaction mixture went up. The negative sign shows that this is an exothermic reaction; heat is released.*

## ACTIVITY 3

### ▼ PRACTICAL: MEASURING ENTHALPY CHANGES WHEN SALTS DISSOLVE IN WATER

We can also use calorimetry experiments to work out the amount of heat given out/taken in when salts dissolve in water. The following procedure could be used:

- Place a polystyrene cup in a $250 \, cm^3$ glass beaker.
- Transfer $100 \, cm^3$ of water into the polystyrene cup using a measuring cylinder.
- Record the initial temperature of the water.
- Weigh 5.20 g of ammonium chloride using a weighing boat on a balance.
- Add the ammonium chloride to water and stir the solution vigorously until all the ammonium chloride has dissolved.
- Record the minimum temperature.

The set-up is very similar to the one used in Activity 2, see Figure 19.12.

**SAMPLE DATA**

| | |
|---|---|
| initial temperature of water/°C | 18.3 |
| minimum temperature of salt solution/°C | 15.1 |

Note: the temperature of the water decreases, so the reaction is endothermic and heat is absorbed from the surroundings for the dissolving process to occur.

We can use this data to calculate the molar enthalpy change for dissolving ammonium chloride.

**CALCULATIONS FOR ACTIVITY 3**

Heat absorbed: $Q = mc\Delta T = 100 \times 4.18 \times (18.3 - 15.1) = 1337.6\,J = 1.3376\,kJ$

Here we assume the following:

1. The specific heat capacity of the diluted solution of ammonium chloride is the same as that of water. This is quite a reasonable assumption because the mixture is mostly water.

2. The mass of the solution is 100 g. The mass of the ammonium chloride is relatively small and it is ignored in the calculation. There are other major sources of error in the experiment, for example heat absorbed from the surrounding air, which makes much more difference to the results.

Now we need to calculate how much heat is absorbed when 1 mole of ammonium chloride dissolves in excess water.

The relative formula mass of ammonium chloride ($NH_4Cl$) is $14 + 4 \times 1 + 35.5 = 53.5$.

Number of moles ($n$) of ammonium chloride dissolved

$$n = \frac{\text{mass } (m)}{\text{relative formula mass } (M_r)} = \frac{5.20}{53.5} = 0.0972\,mol$$

Molar enthalpy change of solution ($\Delta H$)

$$\Delta H = \frac{\text{heat energy change } (Q)}{\text{number of moles of ammonium chloride dissolved } (n)}$$

$$= \frac{1.3376}{0.0972}$$

$$= 13.8\,kJ/mol$$

The amount of heat absorbed when dissolving 1 mole of ammonium chloride in water is therefore:

$$NH_4Cl(s) \rightarrow NH_4Cl(aq) \qquad \Delta H = +13.8\,kJ/mol$$

*The positive sign shows that heat is absorbed and the dissolving of ammonium chloride in water is an endothermic process (the temperature goes down).*

**KEY POINT**

Dissolving is a physical process rather than a chemical reaction. When some salts dissolve in water heat is given out (exothermic), but when others dissolve the process is endothermic.

Safety Note: Wear eye protection and avoid skin contact with the acid and alkali.

**ACTIVITY 4**

## ▼ PRACTICAL: MEASURING ENTHALPY CHANGES OF NEUTRALISATION BETWEEN AN ALKALI AND AN ACID

The reaction between an alkali and an acid is essentially between $OH^-$ and $H^+$ ions to form water (see Chapter 16):

$$OH^-(aq) + H^+(aq) \rightarrow H_2O(l)$$

Let's look at the reaction between potassium hydroxide (an alkali) and hydrochloric acid. Suppose you know the concentration of the potassium hydroxide but are not so sure of the concentration of the dilute hydrochloric acid solution (the label has gone missing!). The following method could be used to find out the concentration of the acid and how much heat is released during the neutralisation reaction.

■ Place a polystyrene cup in a 250 cm³ glass beaker.
■ Transfer 25 cm³ of 2.00 mol/dm³ potassium hydroxide into the polystyrene cup using a measuring cylinder.
■ Record the initial temperature.
■ Fill a burette with 50.00 cm³ of dilute hydrochloric acid.
■ Use the burette to add 5.00 cm³ of dilute hydrochloric acid to the potassium hydroxide.
■ Stir vigorously and record the maximum temperature reached.
■ Continue adding further 5.00 cm³ portions of dilute hydrochloric acid to the cup, stirring and recording the maximum temperature each time, until a total volume of 50.00 cm³ has been added.

We can plot a graph of the temperature of the mixture versus the volume of acid added.

### SAMPLE DATA

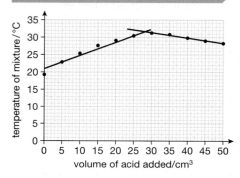

▲ Figure 19.13 Temperature change in the neutralisation reaction between KOH and HCl.

In this reaction the temperature increases at first but then decreases. The reaction between the acid and the alkali is exothermic. At the beginning the temperature goes up because the acid reacts with the alkali, giving out heat. But when all the alkali has been used up, we are just adding cold acid to our warm solution (there is no reaction because there is nothing for the acid to react with) and the temperature goes down.

Two lines of best fit can be drawn on our graph. The point where the lines cross represents complete neutralisation. From the graph, we can identify the maximum temperature reached during the experiment as 31.8 °C. At this point, 28.00 cm³ of acid has been used.

### CALCULATIONS FOR ACTIVITY 4

To work out how much heat is released in the neutralisation reaction, we can use the same calculation as before.

The total volume of the solution at the neutralisation point is 25 + 28 = 53 cm³.

Heat given out is shown by $Q = mc\Delta T = 53 \times 4.18 \times (31.8 - 19.3) = 2769.3$ J.

If we divide by 1000 we get 2.7693 kJ.

Here we assume the following:

1. The density of the reaction mixture is the same as that of water, so 1 cm³ of solution has a mass of 1 g.
2. The specific heat capacity of the mixture is the same as that of water. This is a fairly reasonable assumption because the neutralised solution is mostly water.

Now we need to calculate how much heat is released when 1 mole of potassium hydroxide reacts with 1 mole of hydrochloric acid:

Number of moles of KOH $(n)$ = concentration $(C)$ × volume $(V)$

$$= 2.00 \times \frac{25}{1000} = 0.0500 \text{ mol}$$

The number of moles of HCl will be exactly the same because we have taken the temperature at the neutralisation point.

Molar enthalpy change of neutralisation ($\Delta H$)

$$\Delta H = \frac{\text{heat energy change } (Q)}{\text{number of moles of KOH or HCl reacted } (n)}$$

$$= \frac{2.7693}{0.0500} = 55.4 \, \text{kJ/mol}$$

The amount of heat released in the neutralisation reaction when 1 mole of KOH reacts with 1 mole of HCl is therefore:

$$KOH(aq) + HCl(aq) \rightarrow KCl(aq) + H_2O(l) \qquad \Delta H = -55.4 \, \text{kJ/mol}$$

*The negative sign shows that heat is released in this exothermic neutralisation reaction.*

## CHEMISTRY ONLY

We can also work out the unknown HCl concentration.

We know that there was 0.0500 mol KOH originally present.

28.00 cm³ of HCl was required for neutralisation therefore this contained 0.0500 mol of HCl:

$$\text{Concentration of HCl } (C) = \frac{\text{number of moles of HCl } (n)}{\text{volume of HCl used for neutralisation } (V)}$$

$$= \frac{0.0500}{0.02800} = 1.79 \, \text{mol/dm}^3$$

## WHY DO REACTIONS EITHER GIVE OUT OR ABSORB HEAT?

During chemical reactions bonds in the reactants have to be broken and new ones formed to make the products. *Breaking bonds needs energy (endothermic) and making bonds releases energy (exothermic).*

Think about what happens when hydrogen burns in oxygen to make water:

$$2H_2(g) + O_2(g) \rightarrow 2H_2O(l)$$

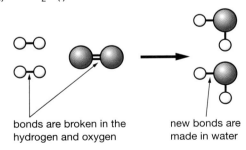

bonds are broken in the hydrogen and oxygen

new bonds are made in water

▲ Figure 19.14 Hydrogen burning in oxygen to make water

### KEY POINT

If you want to be really accurate about this, the water would originally be formed as steam. Heat is also given out when the steam condenses to form liquid water. This is due to formation of *intermolecular forces* of attraction between the molecules in liquid water. Formation of attractive forces between molecules is also exothermic.

Energy has to be *supplied* to *break the bonds* in the hydrogen molecules and in the oxygen molecule. Energy is *released* when *new bonds are formed* between the hydrogen and oxygen atoms in the water molecules.

In this particular reaction, the products are more stable than the reactants. This means the total amount of energy released when covalent bonds are formed in the product (water molecules) is more than is required to break the covalent bonds in the reactants (hydrogen and oxygen molecules). The reaction is exothermic.

When you heat calcium carbonate, breaking up the original bonds in the compound uses more energy than you obtain when the new ones are made.

The reactants are more stable than the products. That means that when the reaction is finished, more energy has been absorbed than is released, an endothermic change.

## CALCULATION OF ENTHALPY CHANGES OF REACTION USING BOND ENERGIES

### BOND ENERGIES

Breaking chemical bonds requires energy. The stronger the bond, the more energy is needed to break it. Bond energies are measured in kJ/mol (kilojoules per mole). For example, the Cl–Cl bond energy is 243 kJ/mol. This means that it will take 243 kJ to break all the Cl–Cl bonds in *1 mole* of chlorine *gas*. The bond energy also represents the amount of energy released when 1 mole of the bonds form. Figure 19.15 shows that it needs an input of 243 kJ of energy to break 1 mole of chlorine molecules into atoms. If the atoms recombine into their original molecules, then exactly the same amount of energy will be released again.

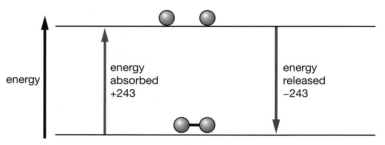

▲ Figure 19.15 Breaking and making bonds between chlorine atoms

■ *Breaking bonds needs energy: endothermic.*

■ *Making bonds releases energy: exothermic.*

The H–Cl bond energy is 432 kJ/mol. If you see '2HCl(g)' in an equation, it will take 2 × 432 kJ to break all the H–Cl bonds in the two moles of hydrogen chloride gas. Or, 2 × 432 kJ will be released when 2HCl(g) is formed from H and Cl atoms.

### KEY POINT

**Bond energy** is defined as *the amount of energy needed to break 1 mole of covalent bonds in gaseous molecules.*

Table 19.1 The energy needed to break bonds

| Bond | C–H | C–Cl | C–I | Cl–Cl | I–I | H–Cl | H–I |
|---|---|---|---|---|---|---|---|
| Bond energy/kJ/mol | 413 | 346 | 234 | 243 | 151 | 432 | 298 |

You can see that some bonds are much stronger than others, for example the bond between iodine and hydrogen is about twice as strong as the bond between two iodine atoms.

Bond energies can be used to calculate how much heat will be absorbed or released during reactions of *covalent* molecules. If a reaction involves *ionic* compounds, different energy terms have to be used, which are beyond International GCSE.

### CALCULATION OF THE HEAT RELEASED OR ABSORBED DURING A REACTION

You can estimate the heat energy released or absorbed in a reaction by calculating how much energy would be needed to break the substances up into individual atoms, and then how much would be given out when those atoms recombine into new arrangements.

Bond energy calculations only ever give *estimates* of the amount of heat evolved or absorbed. They never give exact values. The strength of a bond varies slightly depending on what is around it in the molecule. Quoted bond energies are usually *average* values, which give fairly good, but not perfect, answers for enthalpy changes of reactions. The enthalpy changes of reactions measured using calorimetry experiments are more specific for particular compounds.

For example:

| | | | |
|---|---|---|---|
| if: | energy needed to break all the bonds in reactants | = | +1000 kJ (positive sign as the process if endothermic) |
| and: | energy released when new bonds are made in products | = | −1200 kJ (negative sign as the process is exothermic) |
| then: | overall change | = | −200 kJ (the overall reaction is exothermic) |

This reaction would release 200 kJ of energy.

## THE REACTION BETWEEN METHANE AND CHLORINE

$$CH_4(g) + Cl_2(g) \rightarrow CH_3Cl(g) + HCl(g)$$

Methane reacts with chlorine in the presence of ultraviolet light to produce chloromethane and hydrogen chloride (see Chapter 24, page 280, for more details on this reaction). You can imagine all the bonds being broken in the methane and chlorine, and then being reformed in new ways in the products:

The real reaction doesn't actually happen by all the bonds being broken in this way. That isn't important. The overall amount of heat released or absorbed is the same (as long as the reactants and the products are exactly the same) however you do the reaction, even if one of the routes you use to calculate it is entirely imaginary! This is summarised in an important law called *Hess's law*, which you will meet if you do chemistry at a higher level.

You can calculate the energy needed to break all the bonds and the energy given out as new ones are made.

Bonds that need to be broken (endothermic):

| | | | |
|---|---|---|---|
| 4 C–H bonds | = 4 × (+413) | = | + 1652 kJ |
| 1 Cl–Cl bond | = 1 × (+243) | = | + 243 kJ |
| total | | = | + 1895 kJ |

New bonds made (exothermic):

| | | | |
|---|---|---|---|
| 3 C–H bonds | = 3 × (−413) | = | −1239 kJ |
| 1 C–Cl bond | = 1 × (−346) | = | −346 kJ |
| 1 H–Cl bond | = 1 × (−432) | = | −432 kJ |
| total | | = | −2017 kJ |

Remember that when energy is given out, you show this by putting a negative sign in front of the value.

We can see that more energy is released when bonds are formed than is required to break them, so the reaction releases energy overall; it is exothermic.

The overall energy change is +1895 + (−2017) kJ = −122 kJ.

The negative sign of the answer shows that, overall, heat is given out as the bonds rearrange.

You can show all this happening on an energy level diagram, as in Figure 19.16.

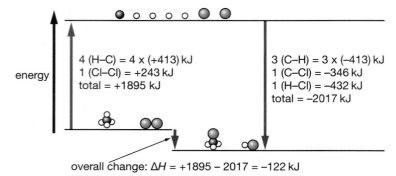

4 (H–C) = 4 x (+413) kJ
1 (Cl–Cl) = +243 kJ
total = +1895 kJ

3 (C–H) = 3 x (–413) kJ
1 (C–Cl) = –346 kJ
1 (H–Cl) = –432 kJ
total = –2017 kJ

energy

overall change: $\Delta H$ = +1895 – 2017 = –122 kJ

▲ Figure 19.16 Methane reacts with chlorine to produce chloromethane and hydrogen chloride.

## AN EXAMPLE INVOLVING DOUBLE BONDS

The bond energy for the O=O double bond is 498 kJ/mol. This is the energy required to break the whole of the double bond.

The equation for the combustion of methane is:

$$CH_4(g) \quad + \quad 2O_2(g) \quad \rightarrow \quad CO_2(g) \quad + \quad 2H_2O(g)$$

Calculate the enthalpy change for this reaction given the bond energies in the table and draw an energy level diagram for the reaction.

| Bond | C–H | O=O | C=O | H–O |
|---|---|---|---|---|
| Bond energy/kJ/mol | 413 | 498 | 743 | 464 |

We start by drawing out the structures showing all the bonds so that you can make sure that we get the number of bonds correct:

Bonds that need to be broken (endothermic):
4 C–H bonds = 4 × (+413) = +1652 kJ
2 O=O bonds = 2 × (+498) = +996 kJ
total = +2648 kJ

New bonds made (exothermic):
2 C=O bonds = 2 × (–743) = –1486 kJ
4 O–H bonds = 4 × (–464) = –1856 kJ
total = –3342 kJ

Enthalpy change = +2648 + (–3342) = –694 kJ/mol.

More energy is released when bonds are formed than is required to break them; the reaction is exothermic.

Energy is given out when the reactants are converted to the products. Therefore we know that the products must have less energy than the reactants; the products are more stable. We can show this on an energy level diagram (Figure 19.17).

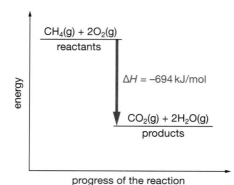

▲ Figure 19.17 An energy level diagram for the combustion of methane

**END OF CHEMISTRY ONLY**

## LOOKING AHEAD – ENTROPY

Many reactions we meet at International GSCE level are exothermic, for example combustion and neutralisation. Despite requiring a small amount of energy to get the reactions started (*activation energy*), these reactions release heat energy overall because the products are more stable than the reactants. This shows that enthalpy is certainly a driving force for a reaction to occur. However, endothermic processes do occur. How do they happen? Dissolving ammonium nitrate into water causes the temperature of the solution to decrease, but it dissolves anyway. This is possible because of a different factor called entropy. *Entropy is a measure of the number of ways for the energy in a system to be distributed*. It is greater in a more disordered system because the energy can be dispersed more flexibly. It follows that solids have lower entropy than liquids, which have lower entropy than gases, as shown in Figure 19.18.

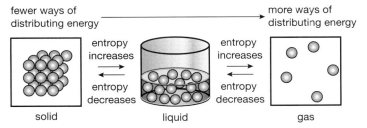

▲ Figure 19.18 Entropy is greater in a more disordered system.

If the entropy of a system increases during a reaction, it is more likely for that reaction to occur. When dissolving ammonium nitrate in water, although the process is endothermic, the $NH_4^+$ and $NO_3^-$ ion particles in the giant ionic lattice become separated from each other. The ions have more freedom of movement in water and the entropy of the system increases, which outweighs the effect of enthalpy.

The reaction between baking soda and vinegar is another example of an endothermic reaction that occurs spontaneously. The reaction produces a gas which leads to an increase in the entropy of the system.

▲ Figure 19.19 The reaction between baking soda and vinegar is used to make papier mâché volcanos.

## CHAPTER QUESTIONS

SKILLS ▸ CRITICAL THINKING, PROBLEM SOLVING

1 a Explain what is meant by an *exothermic reaction* and write balanced chemical equations for any two exothermic changes (apart from the combustion of heptane given in part b).

## CHEMISTRY ONLY

SKILLS ▸ INTERPRETATION

b Heptane, $C_7H_{16}$, is a hydrocarbon found in petrol. The equation for the combustion of heptane is:

$$C_7H_{16}(l) + 11O_2(g) \rightarrow 7CO_2(g) + 8H_2O(l) \qquad \Delta H = -4817 \, kJ/mol$$

Draw an energy level diagram to show the combustion of heptane. Label the axes and show clearly the reactants, the products and the enthalpy change of reaction on your diagram.

SKILLS ▸ REASONING

c Explain in terms of breaking and making bonds why the reaction in b is exothermic.

## END OF CHEMISTRY ONLY

2 a Explain what is meant by an *endothermic reaction*.

## CHEMISTRY ONLY

SKILLS ▸ INTERPRETATION

b Nitrogen and oxygen gases can combine to form nitrogen monoxide as follows:

$$N_2(g) + O_2(g) \rightarrow 2NO(g) \qquad \Delta H = +180 \, kJ/mol$$

Draw an energy level diagram for this process. Label the axes and show clearly the reactants, the products and the enthalpy change of reaction on your diagram.

## END OF CHEMISTRY ONLY

SKILLS ▸ CRITICAL THINKING

3 Identify each of the following changes as exothermic or endothermic. In some cases you will have to rely on your previous knowledge of chemistry. Several reactions are likely to be completely new to you.

a The reaction between sodium and water.

b Burning methane (major constituent of natural gas).

c The reaction between sodium carbonate and ethanoic acid. A thermometer placed in the reaction mixture shows a temperature drop.

d $S(s) + O_2(g) \rightarrow SO_2(g)$            $\Delta H = -297 \, kJ/mol$

e $CuSO_4(s) \rightarrow CuO(s) + SO_3(g)$      $\Delta H = +220 \, kJ/mol$

f If you dissolve solid sodium hydroxide in water, the solution becomes very hot.

SKILLS ▸ INTRAPERSONAL

4 Self-heating cans are used to provide warm food in situations where it is inconvenient to use a more conventional form of heat. By doing an internet (or other) search, find out how self-heating cans work. Write a short explanation of your findings (not exceeding 200 words). You should include equation(s) for any reaction(s) involved, and a diagram or picture if it is useful.

## CHEMISTRY ONLY

**SKILLS** REASONING, PROBLEM SOLVING

5 Use the bond energies in the table to calculate the amount of heat released or absorbed when the following reactions take place. In each case, say whether the change is exothermic or endothermic.

| Bond | Bond energy/kJ/mol |
|------|-------------------|
| C–H | 413 |
| C–Br | 290 |
| Br–Br | 193 |
| H–Br | 366 |
| H–H | 436 |
| Cl–Cl | 243 |
| H–Cl | 432 |
| O=O | 498 |
| O–H | 464 |
| N≡N | 944 |
| N–H | 388 |

a $CH_4(g) + Br_2(g) \rightarrow CH_3Br(g) + HBr(g)$

(The structure of $CH_3Br$ is the same as that of $CH_3Cl$)

b $H_2(g) + Cl_2(g) \rightarrow 2HCl(g)$

c $2H_2(g) + O_2(g) \rightarrow 2H_2O(g)$

d $N_2(g) + 3H_2(g) \rightarrow 2NH_3(g)$

## END OF CHEMISTRY ONLY

**SKILLS** EXECUTIVE FUNCTION

6 A student investigated the amount of heat given out when hexane, $C_6H_{14}$, burns using the apparatus in Figure 19.9. Hexane is a highly flammable liquid which is one of the components of petrol (gasoline).

In each case, the student calculated the amount of heat evolved per mole of hexane. Her first two experiments produced answers of –3200 and –3900 kJ/mol of heat evolved. She then decided to do a third experiment. Her results were as follows:

| | |
|---|---|
| volume of water in copper calorimeter/cm³ | 100 |
| mass of burner + hexane before experiment/g | 35.62 |
| mass of burner + hexane after experiment/g | 35.23 |
| original temperature of water/°C | 19.0 |
| final temperature of water/°C | 55.0 |

a Suggest a reason why the student decided to do a third experiment.

b Apart from wearing eye protection, suggest two other safety precautions the student should take during the experiment.

c Use the results table to calculate the amount of heat in **kJ** evolved by the burning hexane during the experiment.

(Specific heat capacity of water = 4.18 J/g/°C; density of water = 1 g/cm³.)

d Calculate the amount of heat energy released per gram of hexane.

e Calculate the molar enthalpy change of combustion of hexane.
($A_r$: H = 1, C = 12)

f To calculate the average value of the heat evolved per mole when hexane burns, the student took an average of her results. She decided not to use the figure of −3900 in calculating the average because it was so different from the other two. Suggest two reasons why this reaction might have been more exothermic than the others.

g A data book gave a figure of −4194 kJ of heat evolved when 1 mole of hexane burns. Suggest two reasons why all the results in the student's experiment were less negative than this.

7 When 5.15 g of lithium chloride (LiCl) is dissolved in 50 cm³ of water the temperature of the solution goes from 17.0 °C to 33.5 °C.

a Calculate the heat energy released in this experiment. (Specific heat capacity of water = 4.18 J/g/°C; mass of 1 cm³ of solution = 1 g.)

b Calculate the amount, in moles, of lithium chloride dissolved in the solution.

c Using your answers from a and b, calculate the enthalpy change when 1 mol of lithium chloride dissolves in water in kJ/mol.

d A data book gives a figure of −37.2 kJ/mol for heat released when 1 mole of lithium chloride is dissolved in water. Suggest two reasons why the value you calculated in c might be different from the value given in the data book.

## CHEMISTRY ONLY

8 To completely neutralise 200 cm³ of 0.500 mol/dm³ sodium hydroxide (NaOH), a student adds 100 cm³ of 0.500 mol/dm³ sulfuric acid (H₂SO₄). The temperature of the solution goes up by 4.50 °C.

The equation for the reaction is

$$2NaOH(aq) + H_2SO_4(aq) \rightarrow Na_2SO_4(aq) + 2H_2O(l)$$

a Calculate the amount, in moles, of NaOH in the sodium hydroxide solution.

b Calculate the heat released in this reaction. (Specific heat capacity of solution = 4.18 J/g/°C; mass of 1 cm³ of solution = 1 g.)

c Using your answers from a and b, calculate the enthalpy change when 1 mol of sodium hydroxide is neutralised by sulfuric acid in kJ/mol.

d Draw an energy level diagram for this reaction. Label the axes and show clearly the reactants, the products and the enthalpy change of reaction on your diagram.

e Predict the temperature rise if the experiment were repeated using 200 cm³ of 1.00 mol/dm³ sodium hydroxide (NaOH) with 100 cm³ of 1.00 mol/dm³ sulfuric acid.

**END OF CHEMISTRY ONLY**

# 20 RATES OF REACTION

Reactions can vary in speed between those that happen within fractions of a second – explosions, for example – and those that never happen at all. Gold can be exposed to the air for thousands of years and not react in any way.

This chapter looks at the factors controlling the speeds of chemical reactions.

▲ Figure 20.1 Some reactions are very fast.

▲ Figure 20.2 Some reactions happen over several minutes.

▲ Figure 20.3 Rusting takes days or weeks.

▲ Figure 20.4 The weathering of limestone and the formation of stalagmites and stalactites takes a very long time.

## LEARNING OBJECTIVES

- Describe experiments to investigate the effects of changes in the surface area of a solid, the concentration of a solution, the temperature and the use of a catalyst on the rate of a reaction

- Describe the effects of changes in the surface area of a solid, the concentration of a solution, the pressure of a gas, the temperature and the use of a catalyst on the rate of a reaction

- Explain the effects of changes in the surface area of a solid, the concentration of a solution, the pressure of a gas and the temperature on the rate of a reaction in terms of particle collision theory

- Know that a catalyst is a substance that increases the rate of a reaction, but is chemically unchanged at the end of the reaction

- Know that a catalyst works by providing an alternative pathway with lower activation energy

### CHEMISTRY ONLY

- Draw and explain reaction profile diagrams showing $\triangle H$ and activation energy.

- Practical: Investigate the effect of changing the surface area of marble chips and of changing the concentration of hydrochloric acid on the rate of reaction between marble chips and dilute hydrochloric acid

- Practical: Investigate the effect of different solids on the catalytic decomposition of hydrogen peroxide solution

## EXPERIMENTS TO MEASURE THE RATE OF REACTION

The **rate** of a reaction is the speed at which the amount of reactants decreases or the amount of products increases. It is measured as *a change in the concentration (or amount) of reactants or products per unit time* (per second, per minute etc.).

$$\text{rate of reaction} = \frac{\text{change in concentration, volume or mass}}{\text{time}}$$

Safety Note: Wear eye protection and avoid skin contact with the acid.

**ACTIVITY 5**

▼ **PRACTICAL: AN INVESTIGATION OF THE RATE OF REACTION BETWEEN MARBLE CHIPS AND DILUTE HYDROCHLORIC ACID**

Marble chips are made of calcium carbonate and react with hydrochloric acid to produce carbon dioxide gas. Calcium chloride solution is also formed.

$$CaCO_3(s) + 2HCl(aq) \rightarrow CaCl_2(aq) + H_2O(l) + CO_2(g)$$

Figure 20.5 shows some apparatus that can be used to measure how the mass of carbon dioxide produced changes with time. Part (a) is drawn as the apparatus would look before the reaction starts.

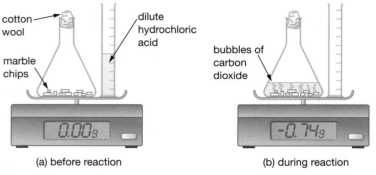

(a) before reaction                    (b) during reaction

▲ Figure 20.5 Investigating the reaction between calcium carbonate and hydrochloric acid.

The following procedure could be used:

■ Use a measuring cylinder to measure 25 cm³ of 2.00 mol/dm³ dilute hydrochloric acid.

■ Add 5.00 g of large marble chips to a conical flask and place a piece of cotton wool at the opening of the flask. The cotton wool is there to allow the carbon dioxide to escape during the reaction, but to stop any acid spitting out. The marble is in *excess* – some of it will be left over when the acid is all used up.

■ Place everything on a balance and reset it to zero.

■ Add the acid to the marble chips and record the reading on the balance every 30 seconds.

Part (b) shows what happens during the reaction. The acid has been poured into the flask and everything has been replaced on the balance. Once the reaction starts, the balance shows a negative mass. The mass goes down because the carbon dioxide escapes through the cotton wool.

When we plot a graph of mass of carbon dioxide lost against time, we obtain something similar to the one in Figure 20.6.

*The steeper the slope (gradient) of the line, the faster the reaction.* We can see from Figure 20.6 that about 0.47 g of carbon dioxide is produced in the first minute. Only about 0.20 g of extra carbon dioxide is produced in the second minute, the reaction is slowing down.

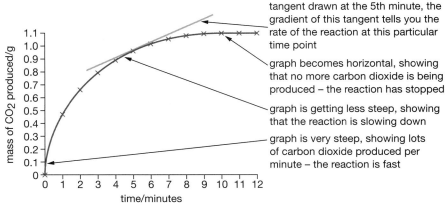

Figure 20.6 The mass of carbon dioxide lost when calcium carbonate reacts with hydrochloric acid.

We can calculate the *average rate* of the reaction during any time interval by using

$$\text{rate} = \frac{\text{mass of } CO_2 \text{ lost}}{\text{time}}$$

For example, the average rate of the reaction in the *first* minute

$$= \frac{0.47}{1} = 0.47 \, \text{g/min}$$

The average rate of the reaction in the *second* minute

$$= \frac{0.20}{1} = 0.20 \, \text{g/min}$$

The average rate of the reaction over the *first two* minutes

$$= \frac{0.67}{2} = 0.34 \, \text{g/min}$$

We can see the reaction is fastest at the beginning. It then slows down, until it eventually stops because *all the hydrochloric acid has been used up*.

We can measure how fast the reaction is going *at any time point* by finding the slope (gradient) of the line at that point. This is the rate of the reaction *at that point* (rather than the average). This is done by drawing a tangent to the line at the time you are interested in and finding its slope (gradient). For example, at 5 minutes the carbon dioxide is being lost at the rate of about 0.05 g per minute (see Figure 20.6).

We can also follow the rate of this reaction by measuring the *volume* of carbon dioxide given off. The apparatus shown in Figure 20.7 can be used.

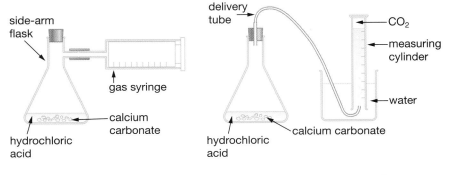

Figure 20.7 Either method can be used to measure the volume of $CO_2$ given off.

**EXPLAINING WHAT'S HAPPENING**

We can explain the shape of the curve by thinking about the particles present and how they interact. This is called the **collision theory**.

Reactions can happen only when particles collide. Not all collisions end up in a reaction. Many particles just bounce off each other. In order for a reaction to actually happen, *the particles have to collide with a minimum amount of energy*, called the **activation energy**. *The collisions with energy greater than or equal to the activation energy* are usually called **successful collisions**.

In the reaction between calcium carbonate and hydrochloric acid, particles in the acid have to collide with the particles at the surface of the marble chips. As the acid particles are getting used up, the collision rate decreases, and so the reaction slows down. The marble is in a large excess so that its shape doesn't change very much during the reaction.

**KEY POINT**

If you have studied Chapter 16, you may realise that the particles we are talking about in the acid are $H^+$ ions. A reaction occurs when the $H^+$ ions in the acid collide successfully with the $CO_3^{2-}$ ions in the calcium carbonate. You should not talk about HCl molecules colliding with $CaCO_3$ molecules. If in doubt about what particles are actually colliding, it is probably safer to just use the word *particle*.

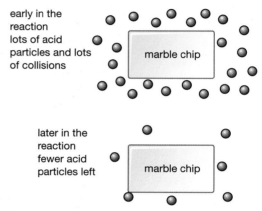

▲ Figure 20.8 As the reaction proceeds, the collision rate of acid particles with the marble chips decreases.

**A DIFFERENT FORM OF GRAPH**

At International GCSE you normally plot graphs showing the mass or volume of product *formed* during a reaction. It is possible, however, that you will see graphs showing the fall in the concentration of one of the *reactants* – in this case, the concentration of the dilute hydrochloric acid.

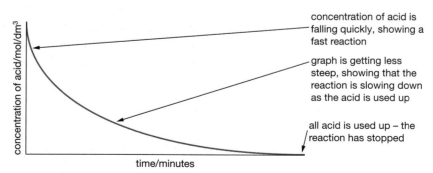

▲ Figure 20.9 The fall in concentration of hydrochloric acid over time

Where the graph is falling most quickly (is steepest), it shows that the reaction is fastest.

Eventually, the graph becomes horizontal because the reaction has stopped when all the acid has been consumed.

## CHANGING THE SURFACE AREA OF THE REACTANTS

### ACTIVITY 6

**HINT**

If we are going to investigate the effect of changing the size of the marble chips, it is important that everything else stays exactly the same to make this a valid (fair) test.

▼ **PRACTICAL: INVESTIGATING THE EFFECT OF CHANGING THE SURFACE AREA OF MARBLE CHIPS ON THE RATE OF REACTION BETWEEN MARBLE CHIPS AND DILUTE HYDROCHLORIC ACID**

We can repeat the experiment in Activity 5 using exactly the same quantities of everything, but using much smaller marble chips. The reaction with the smaller chips happens faster.

We can plot both sets of results (Activities 5 and 6) on the same graph (Figure 20.10). Notice that the same mass of carbon dioxide is produced because we are using the same quantities of everything in both experiments. However, the reaction with the smaller chips starts off much faster and finishes sooner.

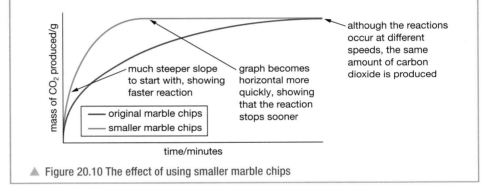

▲ Figure 20.10 The effect of using smaller marble chips

Reactions between solids and liquids (or solids and gases) are faster if the solids are present as a lot of small pieces rather than a few big ones. The more finely divided the solid, the faster the reaction. This is because the *surface area* in contact with the gas or liquid is *much greater* and there are more particles of the solid exposed on the surface. Only the particles on the surface are available for collisions. The *frequency of successful collisions increases* as the surface area of the solid increases.

one big lump

same lump split into smaller pieces

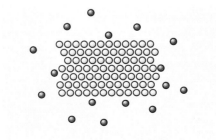
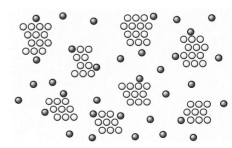

liquid (or gas) particles can't get at the particles hidden in the middle of the solid

far fewer particles are now hidden away

▲ Figure 20.11 The more divided the solid, the faster the reaction

▲ Figure 20.12 A catalytic converter has a honeycomb structure to give a very large surface area for the exhaust gases to flow through.

Large surface areas are frequently used to speed up reactions outside the lab. For example, a **catalytic converter** for a car uses expensive metals such as platinum, palladium and rhodium coated onto a honeycomb structure in a very thin layer to give the maximum possible surface area.

In the presence of these metals, harmful substances such as carbon monoxide and nitrogen oxides are converted into relatively harmless carbon dioxide and nitrogen. The large surface area means the reaction is very rapid. This is important because the gases in the exhaust system are in contact with the catalytic converter for only a very short time.

## CHANGING THE CONCENTRATION OF THE REACTANTS

### THE EFFECT OF CHANGING THE CONCENTRATION

We can repeat our original experiment with large marble chips and hydrochloric acid. Everything is kept the same except we use hydrochloric acid of half the concentration.

We find that reducing the concentration of the acid makes the reaction slower. We can see this on our graph because the graph (red line) is less steep than for our original experiment (blue line).

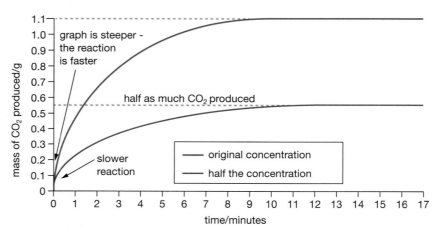

▲ Figure 20.13 The effect of changing the concentration of the acid

In our original experiment we used $25\,cm^3$ of $2.00\,mol/dm^3$ hydrochloric acid. In this experiment we used $25\,cm^3$ of $1.00\,mol/dm^3$ hydrochloric acid. In the second experiment we have started with half the number of hydrochloric acid particles (half as many moles) and so we will produce half as much carbon dioxide.

In general, if you increase the concentration of the reactants, the reaction becomes faster. *Increasing the concentration increases the number of acid particles within a fixed volume*, therefore the particles are closer together and collide more frequently. There are *more successful collisions* between the acid particles and the marble chips *every second*.

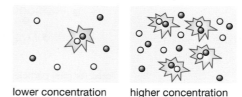

lower concentration　　higher concentration

▲ Figure 20.14 At a higher concentration, more particles collide per second

**Safety Note:** Wear eye protection and avoid skin contact with the acid.

To calculate the concentration of hydrochloric acid you need to look at the proportions of acid and water. The equation to calculate the concentration of hydrochloric acid is:

concentration ($mol/dm^3$)

$$= \frac{\text{volume of hydrochloric acid } (cm^3)}{\text{total volume of solution } (25\,cm^3)}$$

$\times$ original concentration ($2.00\,mol/dm^3$)

**DID YOU KNOW?**

The words *proportional* and *directly proportional* mean the same thing. If the graph we plotted was a straight line that did not go through the origin, this is not a proportional relationship. In that case we would describe the relationship as *linear*.

**EXTENSION WORK**

The relationship between the rate of a reaction and the concentration of the reactants is not always as uncomplicated as it seems. In some reactions, the rate of the reaction is proportional to the concentration of the reactants, as shown by the example in Figure 20.15. We say the reaction is first order with respect to a particular reactant.

If you go on to do chemistry at a higher level, you will come across reactions in which increasing the concentration of one of the reactants has no effect on how fast the reaction happens. These reactions are of zero order with respect to the reactants. In some other cases the rate of a reaction might be proportional to the square of the concentration of a reactant. In this case we say the reaction is of second order with respect to the reactant.

## ACTIVITY 7

### ▼ PRACTICAL: INVESTIGATING THE EFFECT OF CHANGING THE CONCENTRATION OF THE ACID ON THE RATE OF REACTION BETWEEN MARBLE CHIPS AND DILUTE HYDROCHLORIC ACID

We can repeat our previous experiment in Activity 5 using the original large marble chips, but using hydrochloric acid of different concentrations. Everything else would be the same, that is, the mass of the marble chips (5.00 g) and the total volume of the acid (25 cm$^3$). We can dilute the acid by adding distilled water to the original 2.00 mol/dm$^3$ solution but making sure the total volume of water and acid remains at 25 cm$^3$.

For example, if 12.5 cm$^3$ of the original HCl is mixed with 12.5 cm$^3$ of distilled water, we get an acid solution of half of the original concentration (1.00 mol/dm$^3$). If only 5 cm$^3$ of the original HCl is mixed with 20 cm$^3$ of distilled water, the acid solution is now one-fifth (5 out of 25) of the original concentration (0.40 mol/dm$^3$).

In this experiment we are going to calculate the average rate during the first 30 seconds, so we record the mass of carbon dioxide lost in 30 seconds.

We can calculate the average rate of reaction within the first 30 seconds by dividing the mass loss by 30. For example, if we obtain a loss of 0.32 g:

$$\text{average rate} = \frac{0.32}{30} = 0.011\,g/s$$

To identify the effect of changing concentration on rate, we can plot the results for the different concentrations on a piece of graph paper with rate on the *y*-axis and concentration of acid on the *x*-axis (Figure 20.15). The line of best fit should go through the origin (0, 0). If there is no acid, there should be no rate. A straight line going through the origin shows that the rate of the reaction is *directly proportional* to the concentration of the hydrochloric acid. If the concentration doubles, the rate should double. If the concentration triples, the rate should triple.

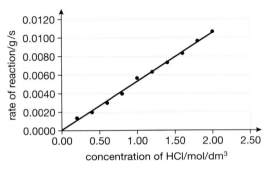

▲ Figure 20.15 The more concentrated the acid, the faster the reaction

## CHANGING THE TEMPERATURE OF THE REACTION

We can do the original experiment again, but this time at a higher temperature. We keep everything else exactly the same as before.

Reactions get faster as the temperature is increased; the graph is *steeper* and *finishes sooner*. The same mass of gas is given off because we have used the same quantities of everything in the mixture.

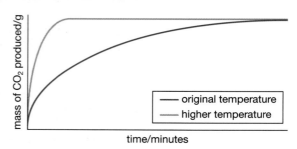

▲ Figure 20.16 The effect of changing the temperature of the reaction

There are two factors that need to be considered when explaining why increasing temperature increases the rate of reaction:

- Increasing the temperature means that the particles are moving faster, and so collide more frequently. That will make the reaction go faster, but it only accounts for a small part of the increase in rate.

- We learned above that in order for a collision to cause a reaction (*a successful collision*), the particles have to collide with a minimum amount of energy, called the *activation energy*. A relatively small increase in temperature produces *a very large increase in the number of particles with energy greater than or equal to the activation energy*. This means that it is going to be much more probable that two particles which have sufficient energy to react collide with each other. So *the frequency of successful collisions increases*.

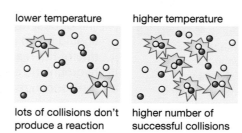

▲ Figure 20.17 A small increase in temperature produces a large increase in the number of collisions with energy greater than the activation energy.

## CHANGING THE PRESSURE ON THE REACTION

Changing the pressure of a reaction in which the reactants are only solids or liquids makes almost no difference to the rate of reaction. But increasing the pressure of a reaction where the reactants are *gases* does speed up the reaction.

If we have a fixed mass of a gas, we increase the pressure by squeezing it into a smaller volume.

This forces the particles *closer together*, so they *collide more frequently*. This is exactly the same as increasing the concentration of the gas.

same number of
particles squeezed
into smaller volume

lower pressure    higher pressure

▲ Figure 20.18 Increased pressure means gas particles collide more frequently.

# CATALYSTS

**WHAT ARE CATALYSTS?**

**Catalysts** are *substances that speed up chemical reactions*, but aren't used up in the process. They are still there, *chemically unchanged, at the end of the reaction*. Because they aren't consumed, small amounts of catalyst can be used to process lots and lots of reactant particles. Different reactions need different catalysts.

**THE CATALYTIC DECOMPOSITION OF HYDROGEN PEROXIDE**

▲ Figure 20.19 Bombardier beetles use hydrogen peroxide as part of their defence mechanism.

Bombardier beetles defend themselves by spraying a hot, unpleasant liquid at their attackers. Part of the reaction involves splitting hydrogen peroxide into water and oxygen, using the enzyme catalase. **Enzymes** are *biological catalysts*. This reaction happens almost explosively, and produces a lot of heat. The same enzyme can be found in potatoes, or even liver tissues.

There are a lot of other things that also catalyse the decomposition of hydrogen peroxide. Some examples are manganese(IV) oxide (also called manganese dioxide), $MnO_2$, and lead(IV) oxide, $PbO_2$. Manganese(IV) oxide is what is normally used in the lab to speed up the decomposition of hydrogen peroxide.

The reaction happening with the hydrogen peroxide is:

hydrogen peroxide → water + oxygen

$$2H_2O_2(aq) \rightarrow 2H_2O(l) + O_2(g)$$

Notice that we don't write catalysts into the equation because they are chemically unchanged at the end of the reaction. If you like, you can write their name or formula over the top of the arrow.

**SHOWING THAT A SUBSTANCE IS A CATALYST**

▶ Figure 20.20 The beakers both contain hydrogen peroxide solution, the right-hand one has $MnO_2$ added to speed up oxygen production.

It isn't difficult to show that manganese(IV) oxide speeds up the decomposition of hydrogen peroxide to produce oxygen. Figure 20.20 shows two beakers, both of which contain hydrogen peroxide solution. Without the catalyst, there is only a trace of bubbles in the solution. With it, oxygen is given off quickly.

How can you show that the manganese(IV) oxide is chemically unchanged by the reaction? It still looks the same, but has any been used up? You can only find out by weighing it before you add it to the hydrogen peroxide solution and then reweighing it at the end.

You can separate the manganese(IV) oxide from the liquid by filtering it through a weighed filter paper, allowing the paper and residue to dry, and then reweighing to calculate the mass of the remaining manganese(IV) oxide. You should find that the mass hasn't changed.

Safety Note: Wear eye protection and avoid skin contact with the solution.

### HINT

The experiment in Figure 20.21 is very simple and easy to set up. However, using this setup has the disadvantage that some oxygen will escape at the beginning of the reaction when $MnO_2$ is added to $H_2O_2$ before the bung can be put back on the conical flask. Using a weighing bottle like the one in Figure 20.22 is a simple way of mixing the chemicals together without losing any oxygen before you can get the bung in. When you are ready to start the reaction, shake the flask so that the weighing bottle falls over and the manganese(IV) oxide comes into contact with the hydrogen peroxide. You need to keep shaking so that an even mixture is formed.

### DID YOU KNOW?

'*vol*', meaning volume, is a measurement for concentration for hydrogen peroxide. $1\,cm^3$ of 2 vol hydrogen peroxide solution decomposes completely to give $2\,cm^3$ of oxygen.

## ACTIVITY 8

### ▼ PRACTICAL: INVESTIGATE THE EFFECT OF DIFFERENT SOLIDS ON THE CATALYTIC DECOMPOSITION OF HYDROGEN PEROXIDE SOLUTION

Figure 20.21 shows apparatus that can be used to measure how the volume of oxygen produced changes with time.

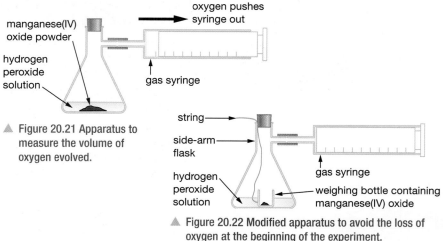

▲ Figure 20.21 Apparatus to measure the volume of oxygen evolved.

▲ Figure 20.22 Modified apparatus to avoid the loss of oxygen at the beginning of the experiment.

The following procedure could be used:

- Measure $100\,cm^3$ of 2 *vol* hydrogen peroxide and transfer to a $250\,cm^3$ conical flask.
- Weigh out 0.20 g of manganese(IV) oxide on a balance.
- Add the manganese(IV) oxide to the hydrogen peroxide and quickly replace the bung with the gas syringe already attached. Swirl the reaction mixture at a constant speed.
- Record the amount of oxygen produced every 20 seconds for 3 minutes and plot a graph of volume of oxygen versus time.
- Repeat the reaction with 0.20 g of lead(IV) oxide and copper(II) oxide but keep everything else the same.

The sample data below show manganese(IV) oxide is a very effective catalyst for the decomposition of hydrogen peroxide. In comparison, lead(IV) oxide is less effective as the rate of the reaction is much slower and copper(II) oxide does not act as a catalyst for this reaction at all.

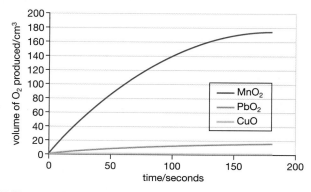

▲ Figure 20.23 The catalytic effect of different solids on the decomposition of $H_2O_2$.

▲ Figure 20.24 Traffic passes easily through a road tunnel under a mountain.

## HOW DOES A CATALYST WORK?

Not all collisions result in a reaction happening. Collisions have to involve at least a certain minimum amount of energy, called the *activation energy* (page 230).

If a reaction is slow, it means that very few collisions have this amount of energy; when most collisions happen, the particles simply bounce off each other.

*Catalysts work by providing an alternative route for the reaction, involving a lower activation energy.*

If the activation energy is lower, many more collisions are likely to be successful. The reaction happens faster because many more particles *have energy greater than or equal to the activation energy for the alternative route*. The energy of the particles has not changed but it has been made easier for them to react.

You can illustrate this with a simple everyday example. Imagine you have a mountain between two valleys. Only a few very energetic people will climb over the mountain from one valley to the next. Now imagine building a road tunnel through the mountain. Lots of people will be able to travel easily from one valley to the next.

Be careful how you phrase the statement explaining how a catalyst works. You should say 'Catalysts provide an alternative route with a lower activation energy'.

They *do not* 'lower the activation energy' any more than building a tunnel lowers the mountain. The original route is still there, and if particles collide with enough energy they will still use it, just as very energetic people will still choose to climb over the top of the mountain.

## CHEMISTRY ONLY

You can show the effect of a catalyst on a reaction profile diagram (Figure 20.25 and Figure 20.26). Adding a catalyst gives the reaction an alternative way to happen that involves a lower activation energy.

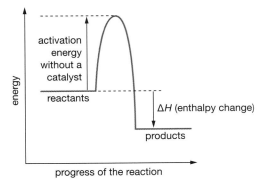

▲ Figure 20.25 A reaction profile diagram

**KEY POINT**

The reaction here is an exothermic reaction: the products have lower energy than the reactants.

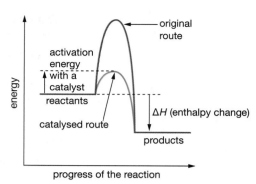

▲ Figure 20.26 Activation energy with a catalyst

**END OF CHEMISTRY ONLY**

**CATALYSTS IN INDUSTRY**

Catalysts are especially important in industrial reactions because they help substances to react quickly at lower temperatures and pressures than would otherwise be needed. This saves money.

**CHAPTER QUESTIONS**

**SKILLS** ❯ CREATIVITY

1 A student carried out an experiment to investigate the rate of a reaction between an excess of dolomite (magnesium carbonate) and $50\,cm^3$ of dilute hydrochloric acid. The dolomite was in small pieces. The reaction is:

$$MgCO_3(s) + 2HCl(aq) \rightarrow MgCl_2(aq) + H_2O(l) + CO_2(g)$$

He measured the volume of carbon dioxide given off at regular intervals, with the results shown in the table below.

| Time/s | 0 | 30 | 60 | 90 | 120 | 150 | 180 | 210 | 240 | 270 | 300 | 330 | 360 |
|---|---|---|---|---|---|---|---|---|---|---|---|---|---|
| Volume/cm³ | 0 | 27 | 45 | 59 | 70 | 78 | 85 | 90 | 94 | 97 | 99 | 100 | 100 |

**SKILLS** ❯ INTERPRETATION

**SKILLS** ❯ ANALYSIS, REASONING

**SKILLS** ❯ PROBLEM SOLVING

**SKILLS** ❯ REASONING

a Draw a diagram of the apparatus you would use for this experiment and label it.

b Plot these results on a piece of graph paper, with time on the *x*-axis and volume of gas on the *y*-axis.

c At what time is the gas being given off most quickly? Explain why the reaction is fastest at that time.

d Use your graph to find out how long it took to produce $50\,cm^3$ of gas.

e Use your graph to calculate the average rate of this reaction in the first 80 seconds.

f In each of the following questions, state and explain what would happen to the initial rate of the reaction and to the total volume of gas given off if various changes were made to the experiment.

   i The mass of dolomite and the volume and concentration of acid were kept constant, but the dolomite was in one big lump instead of small pieces.

   ii The mass of dolomite was unchanged and it was still in small pieces. $50\,cm^3$ of hydrochloric acid was used, which had half the original concentration.

iii  The dolomite was unchanged again. This time 25 cm³ of the original acid was used instead of 50 cm³.

iv  The acid was heated to 40 °C before the dolomite was added to it.

2  The effect of concentration and temperature on the rate of a reaction can be explored using the reaction between magnesium ribbon and dilute sulfuric acid:

$$Mg(s) + H_2SO_4(aq) \rightarrow MgSO_4(aq) + H_2(g)$$

A student dropped a 2 cm length of magnesium ribbon into 25 cm³ of dilute sulfuric acid (a large excess of acid) in a boiling tube. She stirred the contents of the tube continuously and timed how long it took for the magnesium to disappear.

a  What would you expect to happen to the time taken for the reaction if the student repeated the experiment using the same length of magnesium with a mixture of 20 cm³ of acid and 5 cm³ of water? Explain your answer in relation to the collision theory.

b  What would you expect to happen to the time taken for the reaction if she repeated the experiment using the original quantities of magnesium and acid, but first heated the acid to 50 °C? Explain your answer in relation to the collision theory.

c  Give a reason why is it important to keep the reaction mixture stirred continuously.

## CHEMISTRY ONLY

d  The reaction between magnesium ribbon and dilute sulfuric acid is exothermic. Draw a reaction profile diagram for this reaction, label the axes clearly and show both the activation energy and enthalpy change on your diagram.

## END OF CHEMISTRY ONLY

3  Catalysts speed up reactions, but can be recovered chemically unchanged at the end of the reaction.

a  Explain how a catalyst works to increase the rate of reactions.

b  Describe how you would find out whether iron(III) oxide was a catalyst for the decomposition of hydrogen peroxide solution. Note: you need to show not only that it speeds the reaction up, but that it is chemically unchanged at the end.

## CHEMISTRY ONLY

c  Draw a clearly labelled reaction profile diagram showing the effect of using a catalyst in a reaction.

## END OF CHEMISTRY ONLY

# 21 REVERSIBLE REACTIONS AND EQUILIBRIA

This chapter explores the idea of 'reversibility' in a reaction, and how you can control the conditions of such reactions in order to obtain as much as possible of what you want.

## LEARNING OBJECTIVES

- Know that some reactions are reversible and that this is indicated by the symbol $\rightleftharpoons$ in equations
- Describe reversible reactions such as the dehydration of hydrated copper(II) sulfate and the effect of heat on ammonium chloride

### CHEMISTRY ONLY

- Know that a reversible reaction can reach dynamic equilibrium in a sealed container.
- Know that the characteristics of a reaction at dynamic equilibrium are:
  - the forward and reverse reactions occur at the same rate
  - the concentrations of reactants and products remain constant.

- Understand why a catalyst does not affect the position of equilibrium in a reversible reaction.
- Know the effect of changing either temperature or pressure on the position of equilibrium in a reversible reaction:
  - an increase (or decrease) in temperature shifts the position of equilibrium in the direction of the endothermic (or exothermic) reaction
  - an increase (or decrease) in pressure shifts the position of equilibrium in the direction that produces fewer (or more) moles of gas.

## REVERSIBILITY AND DYNAMIC EQUILIBRIA

**TWO REVERSIBLE REACTIONS**

▲ Figure 21.1 Copper(II) sulfate crystals are split into anhydrous copper(II) sulfate and water on gentle heating.

### DEHYDRATION OF COPPER(II) SULFATE CRYSTALS

If you heat blue copper(II) sulfate crystals gently, the blue crystals turn to a white powder and water is driven off. Heating causes the crystals to lose their *water of crystallisation* and white *anhydrous* copper(II) sulfate is formed. 'Anhydrous' simply means 'without water'.

$$CuSO_4 \cdot 5H_2O(s) \rightarrow CuSO_4(s) + 5H_2O(l)$$
    blue              white

Heat is needed for this reaction to occur

If you add water to the white solid, it turns blue again; it also becomes very warm. See Figure 18.3 (page 192).

$$CuSO_4(s) + 5H_2O(l) \rightarrow CuSO_4 \cdot 5H_2O(s)$$
    white                  blue

Heat is released from this reaction

The original change has been exactly reversed. Even the heat that you put in originally has been given out again. This is called a **reversible** reaction and is indicated by a special arrow:

$$CuSO_4 \cdot 5H_2O(s) \rightleftharpoons CuSO_4(s) + 5H_2O(l)$$

### HEATING AMMONIUM CHLORIDE

If you heat ammonium chloride, the white crystals disappear from the bottom of the tube and reappear further up. Heating ammonium chloride splits it into the colourless gases ammonia and hydrogen chloride.

$$NH_4Cl(s) \rightarrow NH_3(g) + HCl(g)$$
white solid    colourless gases      Heat is needed for this reaction to occur

These gases recombine further up the tube, where it is cooler, to form a white solid:

$$NH_3(g) + HCl(g) \rightarrow NH_4Cl(s)$$
colourless gases    white solid      Heat is released from this reaction

The reaction reverses when the conditions are changed from hot to cool.

◀ Figure 21.2 Heating ammonium chloride

This can, again, be shown with the reversible arrow:

$$NH_3(g) + HCl(g) \rightleftharpoons NH_4Cl(s)$$

## CHEMISTRY ONLY

## REVERSIBLE REACTIONS IN A SEALED CONTAINER

A *sealed container* means that no substances are added to the reaction mixture and no substances escape from it. On the other hand, heat may be either given off or absorbed.

Imagine a substance that can exist in two forms, one of which we'll represent by a blue square and the other by a yellow square. Suppose you start off with a sample which is entirely blue.

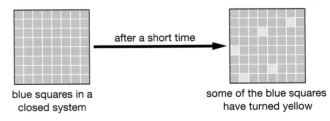

blue squares in a closed system      some of the blue squares have turned yellow

▲ Figure 21.3 Blue squares converting to yellow ones

Because you are starting with a high concentration of blue squares, at the beginning of the reaction the rate at which they turn yellow will be relatively high in terms of the number of squares changing colour per second. The number of blue squares changing colour per second (the rate of change) will fall as the blue gradually gets used up.

But the yellow squares can also change back to blue ones again. This is because it is a *reversible* reaction, *in which the products can react with each other and go back to form reactants*. At the start, there aren't any yellow squares, so the rate of change from yellow into blue is zero. As their number increases, the rate at which yellow change to blue also increases.

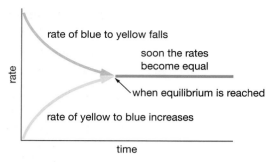

▲ Figure 21.4 The rates of the forward reaction and the reverse reaction become equal when equilibrium is reached.

Soon the rates of both reactions become equal. At that point, blue squares are changing into yellow ones at exactly the same rate that yellow ones are turning blue.

What would you see in the reaction mixture when that happens? The total numbers of blue squares and of yellow squares would remain constant, but the reaction would still be happening. If you followed the fate of any one particular square, sometimes it would be blue and sometimes yellow.

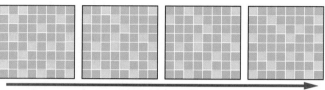

▲ Figure 21.5 The reaction continues, but total numbers of blue and yellow squares remain constant.

This is an example of a **dynamic equilibrium**. It is dynamic in the sense that the reactions are still continuing, but *the rate of the forward reaction is equal to the rate of the reverse reaction*. It is an equilibrium in the sense that the total amounts or *concentrations of the various things present (reactants and products) are now constant*.

Notice that you can set up a dynamic equilibrium only if the system is closed *in a sealed container*. If, for example, you removed the yellow squares as soon as they were formed, they would never get the chance to turn blue again. What was a reversible reaction would now go entirely in one direction as blue squares turn yellow without being replaced.

Consider the reversible reaction that occurs when calcium carbonate decomposes:

$$CaCO_3(s) \rightleftharpoons CaO(s) + CO_2(g)$$

If this were not done in a closed container then all the $CO_2$ would escape and the reverse reaction would not occur.

Reversible reactions will eventually reach a state of equilibrium if they are left in a sealed container.

## THE POSITION OF EQUILIBRIUM

Taking a general case where A reacts reversibly to give B:

$$A \rightleftharpoons 2B$$

The reaction from A to B (the left-to-right reaction) is described as the **forward reaction**. The reaction between from B to A (the right-to-left reaction) is called the **reverse reaction**.

> **HINT**
>
> Be careful with the wording of your answer in the exam - the concentrations of the reactants and products at equilibrium are *constant*, they are not *equal*.

If we let this reaction come to equilibrium then measure the amount of each substance present we might find that we have

A ⇌ 2B
90% 10%

Because there is more A than B present at equilibrium, we say that the position of equilibrium lies to the left. It is important to realise that equilibrium does not mean 50% of reactants and products; the key thing is that the concentrations of A and B stay constant at equilibrium. These concentrations will remain constant until we change the conditions, for instance the temperature or the pressure. If we increase the temperature and allow the reaction time to reach equilibrium again, we might find that we have:

A ⇌ 2B
80% 20%

Because the change we made decreased the concentration of what is on the left (A) and increased the concentration of what is on the right (B), we say that *the position of equilibrium has shifted to the right*. It does not matter that we still have more A than B, what we are looking at is how things change when we change the conditions.

## HOW TO PREDICT THE EFFECT OF CHANGING CONDITIONS ON THE POSITION OF EQUILIBRIUM

This section looks at what can be done to change the position of equilibrium to produce as much as possible of what you want in the equilibrium mixture.

When the conditions in a system in dynamic equilibrium are changed, for example the temperature or the pressure, the position of equilibrium might shift left or right to give less or more of the products. A useful way to help you to decide what happens if various conditions are changed is to remember that *the reaction always sets about counteracting any changes you make*. This is similar to the idea that if a room is too warm you might take off your jacket so that you can cool down, or if the room is too cold you might put some more clothes on!

Things we might try to do to influence the reaction include:

■ changing the pressure

■ changing the temperature

■ adding a catalyst (in fact, this turns out to have no effect on the position of equilibrium).

### CHANGING THE PRESSURE

**KEY POINT**

The particles in solids and liquids are very close together and they are affected very little by changes in pressure. Their concentration does not really change as the pressure changes.

This only really applies to reactions in which at least one of the reactants or products is a *gas*, and where *the total numbers of gaseous molecules on both sides of the equation are different*. In the example here, there are three gaseous molecules on the left, but only two on the right.

A(g) + 2B(g) ⇌ C(g) + D(g)

Pressure is caused by molecules hitting the walls of their container. If you have fewer molecules in the same volume at the same temperature, you will have a lower pressure.

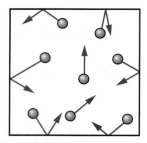

▲ Figure 21.6 Pressure is caused by molecules hitting the walls of their container – the fewer molecules, the lower the pressure.

If you increase the pressure, the reaction will respond by reducing it again. It can reduce the pressure by producing fewer gaseous molecules to hit the walls of the container, in this case by converting A + 2B (3 molecules) into C + D (2 molecules). Increasing the pressure will always cause the position of equilibrium to shift in the direction which produces the smaller number of gaseous molecules.

Consider the reaction between nitrogen gas and hydrogen gas to make ammonia:

$$N_2(g) + 3H_2(g) \rightleftharpoons 2NH_3(g)$$

If we want to make as much ammonia as possible, we need to make the pressure as high as possible. There are four molecules of gas on the left-hand side but only two on the right-hand side. Increasing the pressure will therefore shift the position of equilibrium to the right (the ammonia side), the side with fewer gas molecules.

*If there is the same number of gaseous molecules on both sides of the equation, changing the pressure will make no difference to the position of equilibrium.* For example, in this reaction:

$$H_2(g) + I_2(g) \rightleftharpoons 2HI(g)$$

increasing or decreasing the pressure has no effect on the position of equilibrium – it does not affect how much hydrogen, iodine and hydrogen iodide we have at equilibrium.

To summarise:

*Increasing pressure*: the position of equilibrium shifts to the side which has *fewer gas molecules.*

*Decreasing pressure*: the position of equilibrium shifts to the side which has *more gas molecules.*

## CHANGING THE TEMPERATURE

### REMINDER

Exothermic and endothermic reactions and $\Delta H$ values are explained in Chapter 19.

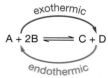

▲ Figure 21.7 If a reaction is exothermic in the forward reaction it will be endothermic in the reverse direction.

### KEY POINT

At equilibrium, the rate of the forward reaction is equal to the rate of the reverse reaction so as much heat is being given out as is taken in – nothing appears to change.

### KEY POINT

In an endothermic reaction the heat is converted to a different form (chemical energy), so it is being removed.

When we write a reversible reaction showing an enthalpy change, the $\Delta H$ always shows the enthalpy change for the *forward* reaction. The value of $\Delta H$ is given as if the reaction was a one-way process.

$$A + 2B \rightleftharpoons C + D \quad \Delta H = -100\,kJ/mol$$

So, in this case $\Delta H$ being negative tells us that the *forward* reaction is *exothermic*.

The reverse reaction will be *endothermic* by exactly the same amount.

Suppose you changed the conditions by decreasing the temperature of the equilibrium, for example if the reaction was originally in equilibrium at 500 °C you lower the temperature to 100 °C. The reaction will respond in a way that increases the temperature again. How can it do that?

If some A and B are converted into C and D, more heat is given out because this is an exothermic change. The extra heat that is produced will warm the reaction mixture up again. In other words, decreasing the temperature will cause the position of equilibrium to move to the right-hand side and more C and D to be formed.

| | A + 2B $\rightleftharpoons$ C + D | |
|---|---|---|
| Equilibrium at 500 °C | 80% | 20% |
| Equilibrium at 100 °C | 60% | 40% |

Increasing the temperature will have exactly the opposite effect. The reaction equilibrium will change to remove the extra heat by absorbing it in an endothermic change. This time the reverse reaction is favoured and the position of equilibrium moves to the left.

To summarise:

*Increasing temperature*: the position of equilibrium shifts in the *endothermic direction*.

*Decreasing temperature*: the position of equilibrium shifts in the *exothermic direction*.

## ADDING A CATALYST

Adding a catalyst speeds up reactions. In a reversible change, it speeds up the forward and reverse reactions by *the same proportion*. For example, if it speeds up the forward reaction ten times, it speeds up the reverse reaction ten times as well.

The net effect of this is that there is *no change* in the position of equilibrium if you add a catalyst. The catalyst is added to increase the rate at which equilibrium is reached.

## AN EXAMPLE TO ILLUSTRATE HOW CHANGING REACTION CONDITIONS CAN AFFECT THE POSITION OF EQUILIBRIUM IN A REVERSIBLE REACTION

Nitrogen dioxide, $NO_2$, is a dark brown, poisonous gas. It can join together in pairs (**dimerise**) to make molecules of dinitrogen tetroxide, $N_2O_4$, which is colourless. There is a dynamic equilibrium between the two forms:

$$2NO_2(g) \rightleftharpoons N_2O_4(g) \qquad \Delta H = -57 \, kJ/mol$$
$$\text{brown} \qquad \text{colourless}$$

## THE EFFECT OF PRESSURE

### HINT

**Warning!** You have to be very careful here about predicting exactly what colour changes you would see. If you increase the pressure, for example, you squeeze the same number of molecules into a smaller space and so the colour will darken initially. Then it fades a bit as the equilibrium re-establishes, but not to its original colour. The gases are still compressed.

If you increase the pressure, the position of equilibrium will shift to reduce it again by producing fewer gaseous molecules. In other words, the position of equilibrium will move to the right and the reaction will produce more dinitrogen tetroxide.

If you lower the pressure, the position of equilibrium will shift to increase it again by producing more gaseous molecules. Therefore the position of equilibrium shifts to the left, and you will obtain a higher proportion of the brown nitrogen dioxide in the equilibrium mixture.

## THE EFFECT OF TEMPERATURE

Notice from the equation that the change from nitrogen dioxide to dinitrogen tetroxide is exothermic. The negative sign for $\Delta H$ shows that heat is given out by the forward reaction.

If you decrease the temperature, the position of equilibrium will shift to produce more heat to counteract the change you have made. In other words, lowering the temperature causes the position of equilibrium to shift in the exothermic direction and there will be more dinitrogen tetroxide in the equilibrium mixture. The colour of the reaction mixture will fade.

If you increase the temperature, the position of equilibrium will shift to lower it again – the position of equilibrium shifts in the reverse, endothermic direction. In other words, more nitrogen dioxide will be formed and the colour of the gas will darken.

You can see this happening in Figure 21.8. Of the three tubes containing this equilibrium mixture, one is at lab temperature, one is in ice and one is in hot water. Notice that the hot one is very dark brown and therefore contains a high proportion of nitrogen dioxide.

The one in the ice is slightly paler than the one in the air, showing that it must have a slightly greater proportion of the colourless dinitrogen tetroxide.

▲ Figure 21.8 Tubes containing the $NO_2/N_2O_4$ equilibrium at different temperatures.

## THE EFFECT OF CONCENTRATION

**EXTENSION WORK**

Changing the concentrations of reactants or products in a reversible reaction also has an effect on the position of equilibrium.

Take the reaction in Figure 21.9 as an example. If you add more A, the system responds by removing it again. That produces more C and D, which is what you probably want. You might choose to increase the amount of A if it was essential to convert as much B as possible into product because it was expensive, for example.

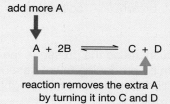

add more A

$A + 2B \rightleftharpoons C + D$

reaction removes the extra A
by turning it into C and D

▲ Figure 21.9 Adding more of substance A

more A and B react to replace
the C you have removed

$A + 2B \rightleftharpoons C + D$

remove C

▲ Figure 21.10 Removing substance C as soon as it is formed

Alternatively, if you remove C as soon as it is formed, the reaction will respond by replacing it again by reacting more A and B (Figure 21.10). Removing a substance as soon as it is formed is a useful way of moving the position of equilibrium to generate more product.

## END OF CHEMISTRY ONLY

## CHAPTER QUESTIONS

**SKILLS** REASONING

1 The thermal decomposition of ammonium chloride is a reversible reaction.

$NH_4Cl(s) \rightleftharpoons NH_3(g) + HCl(g)$

a Explain how the equation shows that the reaction is reversible.

**SKILLS** CRITICAL THINKING

b Give one observation associated with the forward reaction.

c Name all the substances present in the reaction mixture if an equilibrium state is reached.

**SKILLS**   CRITICAL THINKING    (8)

**SKILLS**   REASONING    (7) (8)

## CHEMISTRY ONLY

2 For the following reactions at equilibrium, predict in which direction (**Left** or **Right** or **No change**) the position of equilibrium might shift if the conditions were changed:

|  |  | *Change in conditions* |
|---|---|---|
| $N_2O_4(g) \rightleftharpoons 2NO_2(g)$ | Endothermic | Increase in temperature |
| $2NH_3(g) \rightleftharpoons N_2(g) + 3H_2(g)$ | Endothermic | Increase in pressure |
| $N_2(g) + O_2(g) \rightleftharpoons 2NO(g)$ | Endothermic | Decrease in temperature |
| $H_2(g) + I_2(g) \rightleftharpoons 2HI(g)$ | Endothermic | Decrease in pressure |
| $CO(g) + 2H_2(g) \rightleftharpoons CH_3OH(g)$ | Exothermic | Adding a catalyst |

3 An important stage in the Contact process for the production of sulfuric acid involves a reversible reaction:

$$2SO_2(g) + O_2(g) \rightleftharpoons 2SO_3(g) \quad \Delta H = -196\,kJ/mol$$

At 450 °C the reaction mixture consists of a *dynamic equilibrium* involving sulfur dioxide, oxygen and sulfur trioxde.

a Explain what is meant by the term *dynamic equilibrium*. Be sure that you have explained what both of the words mean.

b Predict and explain the effect of an increase in pressure on the proportion of the sulfur trioxide present in the equilibrium mixture.

c Predict and explain the effect of lowering the temperature of the mixture on the position of equilibrium of the reaction.

4 Hydrogen can be made by the reaction between methane (natural gas) and steam. The reaction can be carried out by passing a mixture of methane and steam over a nickel catalyst at pressures between 2 and 30 atmospheres and a temperature of about 1000 °C.

$$CH_4(g) + H_2O(g) \rightleftharpoons CO(g) + 3H_2(g) \quad \Delta H = +210\,kJ/mol$$

a The pressure used is relatively low. What would be the effect on the conversion of the methane into carbon monoxide and hydrogen if the pressure was higher?

b Explain why a high temperature is used to get a good conversion of methane into hydrogen.

c Explain the use of the nickel catalyst in the production of hydrogen, referring to both the rate of the reaction and the position of equilibrium.

5 Ammonia, $NH_3$, is manufactured by passing a mixture of nitrogen and hydrogen over an iron catalyst at a pressure of 200 atmospheres or more, and a temperature of 450 °C.

$$N_2(g) + 3H_2(g) \rightleftharpoons 2NH_3(g) \quad \Delta H = -92\,kJ/mol$$

a Explain why this reaction will produce a higher percentage conversion into ammonia if the pressure is very high.

b 200 atmospheres is a high pressure, but not very high. Give a reason why most ammonia manufacturers don't use a pressure of, for example, 1000 atmospheres. (Hint: think about the risk of using high pressure.)

c Predict and explain whether you would get the best yield of ammonia in the equilibrium mixture at a low or a high temperature.

d The temperature used, 450 °C, is neither very high nor very low. Suggest a reason why a manufacturer might choose a temperature which gave less than an ideal percentage conversion. (Hint: think about rates of reaction.)

**END OF CHEMISTRY ONLY**

# UNIT QUESTIONS

SKILLS ▶ INTERPRETATION  **1**

Hydrogen, $H_2$, and bromine, $Br_2$, react vigorously to form hydrogen bromide, HBr, according to the following equation:

$$H_2(g) + Br_2(g) \rightarrow 2HBr(g)$$

a Draw a dot-and-cross diagram to show the arrangement of the electrons in a hydrogen bromide molecule, showing electrons in the outer shell only. **(2)**

## CHEMISTRY ONLY

SKILLS ▶ PROBLEM SOLVING

b Use the bond energies given below to calculate the enthalpy change for the reaction shown in the above equation. **(3)**

| Bond | Bond energy/kJ/mol |
|------|--------------------|
| H–H | 436 |
| Br–Br | 196 |
| H–Br | 368 |

SKILLS ▶ INTERPRETATION

c i Complete the reaction profile diagram by showing the products of the reaction. **(1)**

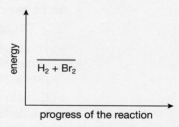

ii Label the diagram to show the enthalpy change, $\Delta H$, and the activation energy of the reaction. **(2)**

**(Total 8 marks)**

## END OF CHEMISTRY ONLY

**2**

A group of students wanted to investigate the energy changes when salts dissolve in water. The teacher suggested that they should measure the temperature changes that occur when the salts were dissolved.

This is the method they followed:

■ Add 100 cm³ of water to a beaker.

■ Record the temperature of the water.

■ Weigh 5.00 g of salt and add it to the water in the beaker.

■ Stir the mixture with a glass rod vigorously until all the solid has dissolved.

■ Record the maximum (or minimum) temperature of the solution.

a The diagram below shows the readings on the thermometer before and after the student dissolved a salt, potassium chloride, in water.

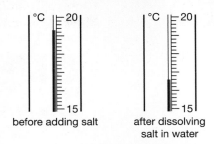

before adding salt       after dissolving salt in water

i Write down the thermometer readings and calculate the temperature change. **(3)**

Temperature before    …….. °C

Temperature after    …….. °C

Temperature change    …….. °C

ii Which of the following statements is true about the dissolving of potassium chloride in water (choose one answer)? **(1)**

  **A** The process is endothermic and $\Delta H$ is positive.

  **B** The process is exothermic and $\Delta H$ is positive.

  **C** The process is endothermic and $\Delta H$ is negative.

  **D** The process is exothermic and $\Delta H$ is negative.

b Another student repeated the experiment with a different salt, calcium chloride, using the same method and recorded these results.

Volume of water = 100 cm$^3$

Starting temperature of water = 15.9 °C

Maximum temperature of solution = 23.2 °C

Mass of salt = 5.00 g

i Calculate the heat energy released in this experiment. The specific heat capacity of the solution $c$ = 4.2 J/g/°C and the mass of 1 cm$^3$ of mixture = 1 g. **(3)**

ii How many moles of calcium chloride dissolved in the solution? **(2)**

iii Work out the molar enthalpy change, in kJ/mol, for dissolving calcium chloride in water. **(2)**

c Another student dissolved magnesium chloride in water. She compared her result with a data book value.

Student's value = −105 kJ/mol

Data book value = −141 kJ/mol

There are no errors in the calculation of her result.

Suggest two reasons why the student's value differs from the data book value. **(2)**

**(Total 13 marks)**

**3**

In an experiment to investigate the rate of decomposition of hydrogen peroxide solution in the presence of manganese(IV) oxide, a student mixed 10 cm³ of hydrogen peroxide solution with 30 cm³ of water and added 0.20 g of manganese(IV) oxide. She measured the volume of oxygen evolved at 60 s intervals. The results of her experiment are recorded in the table below.

| Time/s | 0 | 60 | 120 | 180 | 240 | 300 |
|---|---|---|---|---|---|---|
| Volume/cm³ | 0 | 30 | 48 | 57 | 60 | 60 |

a Write a balanced equation for the decomposition of hydrogen peroxide. **(2)**

b Explain why the manganese(IV) oxide was added in a weighing bottle rather than directly into the hydrogen peroxide solution. **(1)**

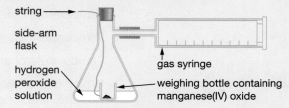

c Plot a graph of her results and draw a line of best fit. Make sure that you label the axes. **(4)**

d Use your graph to find out the following:

i How long it took to produce 50 cm³ of oxygen. **(1)**

ii The volume of gas produced after 100 seconds. **(1)**

iii The average rate (with unit) in the first 150 seconds. **(2)**

e Explain why the graph becomes horizontal after 240 seconds. **(2)**

f Manganese(IV) oxide acts as a catalyst for this reaction.

i A catalyst speeds up a chemical reaction by (choose one answer): **(1)**

**A** increasing the number of collisions between the reactant particles.

**B** providing an alternative reaction pathway with lower activation energy.

**C** increasing the energy of the reactant particles.

**D** changing the enthalpy of the reaction.

ii Describe a method you could use to show that the manganese(IV) oxide is acting as a catalyst in this reaction. **(4)**

g Suppose the experiment had been repeated using the same quantities of everything, but with the reaction flask immersed in ice.
Sketch the graph you would expect to get. Use the same grid as in c.
Label the new graph **G**. **(2)**

h On the same grid as in c and g sketch the graph you would expect to get if you repeated the experiment at the original temperature using 5 cm³ of hydrogen peroxide solution, 35 cm³ of water and 0.20 g of manganese(IV) oxide. Label this graph **H**. **(2)**

**(Total 22 marks)**

**4** A group of students investigated the effect of changing the concentration of dilute hydrochloric acid on the rate of its reaction with marble chips (calcium carbonate).

The equation for this reaction is

$$CaCO_3(s) + 2HCl(aq) \rightarrow CaCl_2(aq) + H_2O(l) + CO_2(g)$$

They used the following method:

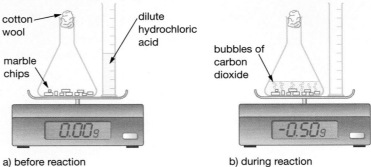

a) before reaction         b) during reaction

The students recorded the time taken for the mass of the flask and contents to decrease by 0.50 g.

The experiment was then repeated using different concentrations of hydrochloric acid.

a To ensure it was a valid (fair) test, the students kept the number of marble chips constant in each experiment. Suggest two other properties of the marble chips that should be kept the same in each experiment. **(2)**

b What is the purpose of the cotton wool? **(1)**

c The teacher gave the students some hydrochloric acid that was labelled 100%. The table below shows the results.

| Students | Mass of CaCO₃/g | Volume of 2.00 mol/ dm³ HCl/cm³ | Volume of water/cm³ | Concentration of HCl/% | Time to lose 0.50 g /s |
|----------|----------|----------|----------|----------|----------|
| 1 | 5.00 | 25 | 0 | 100 | 105 |
| 2 | 5.00 | 20 | 5 | 80 | 150 |
| 3 | 5.00 | 15 | | 60 | 175 |
| 4 | 5.00 | 10 | 15 | 40 | 272 |
| 5 | 5.00 | 5 | 20 | 20 | 520 |

i The results of Student 3 are incomplete.
Calculate the volume of water the student should have used for the result to be comparable with the other four (choose one answer). **(1)**

    **A** 5

    **B** 10

    **C** 15

    **D** 20

ii Plot a graph of the results on a separate piece of graph paper, with concentration of acid on the x-axis and time on the y-axis.
Draw a line of best fit. **(4)**

iii Use your graph in ii to find the time taken for the loss of 0.50 g of mass from the flask when the concentration of acid is 70%. **(1)**

iv One of the points on the graph is anomalous. Identify a reason for this (choose one answer). **(1)**

   **A** The student started the stopwatch too late.

   **B** The student stopped the stopwatch before mass loss reached 0.50 g.

   **C** The student added too much water at the beginning of the experiment.

   **D** The student spilt some water before adding it into the reaction mixture.

d Another group of students repeated the experiment, but this time they measured the mass loss after 1 minute.

The table below shows the results obtained by the students.

| Mass of carbon dioxide given off/g | 0.36 | 0.72 | 0.88 | 1.28 | 1.44 | 1.65 |
|---|---|---|---|---|---|---|
| Concentration of acid /mol/dm³ | 0.20 | 0.40 | 0.50 | 0.70 | 0.80 | 0.90 |

i Describe the relationship between the mass of carbon dioxide given off in 1 minute and the concentration of the acid. **(2)**

ii Explain this relationship in terms of particles. **(3)**

**(Total 15 marks)**

**5** During the manufacture of nitric acid from ammonia, the ammonia is oxidised to nitrogen monoxide, NO, by oxygen in the air.

$$4NH_3(g) + 5O_2(g) \rightarrow 4NO(g) + 6H_2O(g)$$

The ammonia is mixed with air and passed through a stack of large circular gauzes made of platinum–rhodium alloy at red heat (about 900 °C). The platinum–rhodium gauzes act as a catalyst for the reaction.

a Gas particles have to collide before they can react. Use the collision theory to help you to answer the following questions.

   i Because the gases are in contact with the catalyst for only a very short time, it is important that the reaction happens as quickly as possible. Explain why increasing the temperature to 900 °C makes the reaction very fast. **(3)**

   ii Explain what will happen to the reaction rate if the pressure is increased. **(2)**

   iii Explain why the platinum–rhodium alloy is used as gauzes rather than as pellets. **(2)**

b Platinum and rhodium are extremely expensive metals. Explain why the manufacturer can justify their initial cost. **(2)**

**(Total 9 marks)**

**6** At temperatures above 150 °C, brown nitrogen dioxide gas dissociates (splits up reversibly) into colourless nitrogen monoxide and oxygen:

$$2NO_2(g) \rightleftharpoons 2NO(g) + O_2(g) \qquad \Delta H = +114 \text{ kJ/mol}$$

a Write down the meaning of the symbols $\rightleftharpoons$ and $\Delta H$. **(2)**

b What does the positive sign of $\Delta H$ indicate about the reaction? **(1)**

**SKILLS** ▶ INTERPRETATION  (9)

**CHEMISTRY ONLY**

c Complete the reaction profile diagram for this reaction.

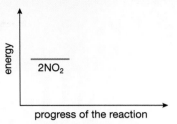

Label the diagram clearly with the activation energy and the $\Delta H$ of the reaction. **(3)**

**SKILLS** ▶ REASONING  (8)

d Predict what would happen to the position of equilibrium and the colour of the mixture if you:
  i   increase the temperature **(3)**
  ii  increase the pressure **(3)**
e Reactions can be accelerated using a catalyst. Write down and explain the effect of a catalyst on the position of equilibrium in this reaction. **(2)**

**(Total 14 marks)**

(6) **7**

Nitrogen and hydrogen are used in the manufacture of ammonia ($NH_3$).

The reaction is reversible and can reach a state of dynamic equilibrium:

$$N_2(g) + 3H_2(g) \rightleftharpoons 2NH_3(g)$$

**SKILLS** ▶ CRITICAL THINKING

a What are the characteristics of a reaction at dynamic equilibrium (choose two answers)? **(2)**
  **A** The rate of the forward reaction is the same as the rate of the reverse reaction.
  **B** The concentrations of the reactants and the products are always equal.
  **C** The position of equilibrium can be shifted by adding a catalyst.
  **D** The concentrations of the reactants and the products are constant.

**SKILLS** ▶ REASONING

b The graph below shows how the percentage of $NH_3$ in the equilibrium mixture varies with temperature and pressure (in atmospheres (atm)).

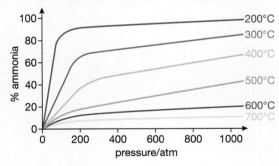

(8)
  i   Increasing pressure increases the percentage of $NH_3$ at equilibrium. Explain why this is the case. **(3)**
(5)
  ii  Describe the relationship between temperature and the percentage of $NH_3$ at equilibrium. **(1)**
(8)
  iii Using your answer from ii, comment on the sign of enthalpy change ($\Delta H$) for the forward reaction. **(2)**
  iv  Predict and explain what happens to the rate of the reaction when the temperature is increased. **(3)**

**(Total 11 marks)**

**END OF CHEMISTRY ONLY**

# UNIT 4
# ORGANIC CHEMISTRY

Organic chemistry is the study of the compounds of carbon (there are some inorganic carbon compounds as well). There are more organic compounds than all the other compounds put together. These compounds include naturally occurring ones that are found in our bodies, for example proteins, DNA and fats. There are also artificially made compounds such as plastics, dyes and drugs. The artificial compounds are derived from chemicals obtained from crude oil. When all the crude oil has been used we will have to find new sources for more than just fuel for cars and making electricity. Organic chemists are involved in the synthesis of a huge variety of new compounds, including drugs and medicines used to treat diseases such as cancer, AIDS, influenza and asthma. The synthesis of these drugs can involve a large number of steps and requires the knowledge and understanding of a great number of organic chemistry reactions. It can take many years and millions of dollars to develop a new drug.

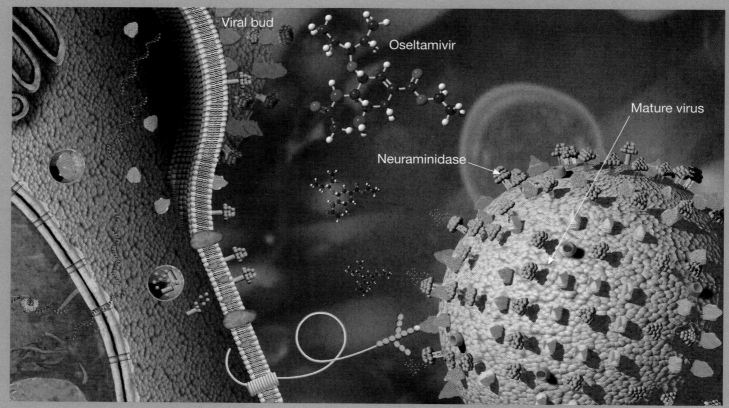

▲ Figure 22.1 Oseltamivir (Tamiflu®) is a drug that targets the influenza virus.

# 22 INTRODUCTION TO ORGANIC CHEMISTRY

There are millions of different organic compounds. They all contain carbon and hydrogen, and often other elements such as oxygen, nitrogen and chlorine. Carbon atoms can join together to form chains and rings, which is why there are so many carbon compounds. This chapter introduces you to some of the important ideas that you need to know before you can start to understand organic chemistry.

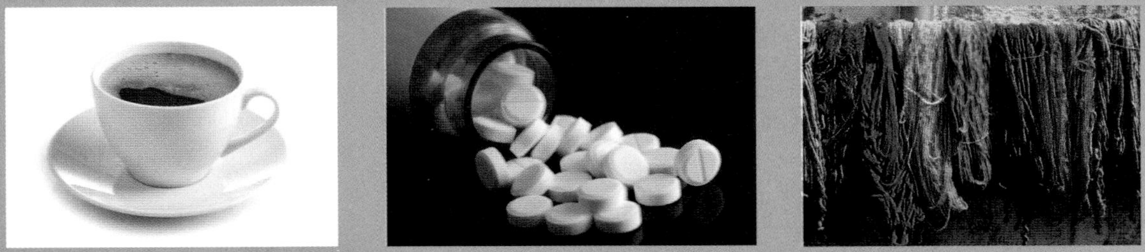

▲ Figure 22.2 What do caffeine, aspirin and dyes have in common? They are all organic compounds.

## LEARNING OBJECTIVES

- Know that a hydrocarbon is a compound made up of hydrogen and carbon only

- Understand how to represent organic molecules using empirical formulae, molecular formulae, general formulae, structural formulae and displayed formulae

- Know what is meant by the terms homologous series, functional group and isomerism

- Understand how to name compounds (containing up to six carbon atoms) using the rules of International Union of Pure and Applied Chemistry (IUPAC) nomenclature

- Understand how to write the possible structural and displayed formulae of an organic molecule given its molecular formula

- Understand how to classify reactions of organic compounds as substitution, addition and combustion

In the next few chapters we will be looking at organic compounds. The term 'organic' was originally used because it was believed that organic compounds could only come from living things. Now it is used for any carbon compound except for the very simplest (carbon dioxide, carbon monoxide, the carbonates and the hydrogencarbonates).

Organic compounds can exist as chains, branched chains or rings of carbon atoms with hydrogens attached.

When you start doing organic chemistry, you are suddenly faced with a whole lot of new compounds with strange names and unfamiliar ways of drawing them. It can be quite scary!

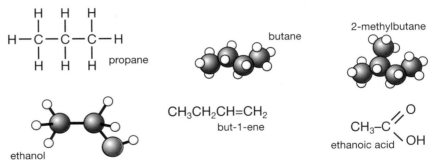

▲ Figure 22.3 Organic chemistry involves a lot of new compounds.

**HINT**

Your school may have models you can use. Otherwise, you can make your own out of modelling clay and matchsticks or small nails to use as bonds.

The secret at the beginning is to spend a lot of time exploring the subject – understanding names, drawing structures and making models.

## HYDROCARBONS

The simplest organic compounds are **hydrocarbons**. *These are molecules that contain carbon and hydrogen only.*

**HINT**

The word *only* is important here. Virtually all organic compounds contain carbon and hydrogen but hydrocarbons contain *only* carbon and hydrogen.

The carbon atoms are joined together with single, double or triple bonds. Carbon atoms are joined to hydrogen atoms by single bonds.

**KEY POINT**

Organic compounds can be described as **saturated** (containing only single C–C bonds) or **unsaturated** (containing double or triple C–C bonds). This is discussed in Chapters 24 and 25.

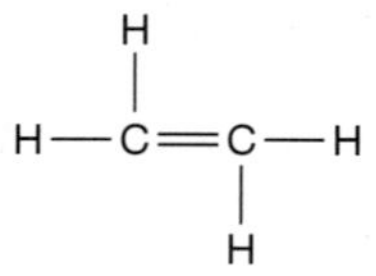

▲ Figure 22.4 Examples of hydrocarbons

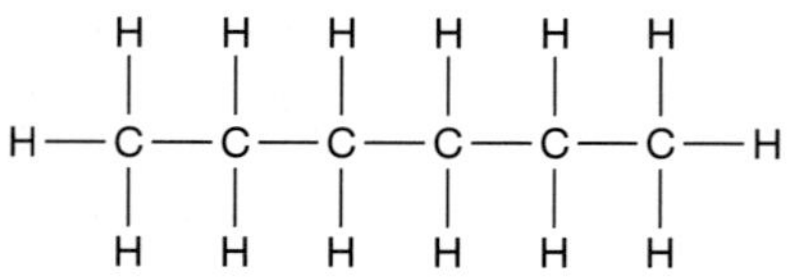

## TYPES OF FORMULA FOR ORGANIC MOLECULES

**EMPIRICAL FORMULAE**

**REMINDER**

Empirical and molecular formula are introduced in Chapter 5.

An **empirical formula** tells you *the simplest whole number ratio of the atoms in a compound.* It can be calculated from experimental data. Without more information, it is not possible to identify the 'true' or 'molecular' formula. Sometimes the empirical formula is the same as the molecular formula. For example, the empirical formula of methane is $CH_4$, which is the same as its molecular formula. However, the empirical formula of ethane is $CH_3$ whereas its molecular formula is $C_2H_6$.

**MOLECULAR FORMULAE**

A **molecular formula** *counts the actual number of each type of atom present in a molecule.* For example, the molecular formula of butane is $C_4H_{10}$ and the molecular formula of ethene is $C_2H_4$.

*The molecular formula is a multiple of the empirical formula.* For example, if a compound has empirical formula $CH_2$, the molecular formula could be $C_2H_4$, $C_3H_6$, $C_4H_8$, etc. The ratio of C:H must always be the same as in the empirical formula.

The molecular formula tells you nothing about the way the atoms are joined together. Both the compounds in Figure 22.5 have the same molecular formula.

▲ Figure 22.5 These two compounds have the same molecular formula but different structures.

Molecular formulae are used very rarely in organic chemistry because they don't give any useful information about the bonding in the molecule. You might use them in equations for the combustion of simple hydrocarbons, where the structure of the molecule doesn't matter. For example:

$$2C_4H_{10}(g) + 13O_2(g) \rightarrow 8CO_2(g) + 10H_2O(l)$$

In almost all other cases, you use a structural or a displayed formula.

## GENERAL FORMULAE

There are many different families of organic compounds, known as **homologous series**. Examples of homologous series are alkanes, alkenes, alcohols, carboxylic acids and esters. These homologous series will be discussed in more detail in later chapters.

Table 22.1 The first few members of the alkane, alkene and alcohol homologous series

| Alkanes | Alkenes | Alcohols |
|---------|---------|----------|
| $CH_4$ | | $CH_4O$ |
| $C_2H_6$ | $C_2H_4$ | $C_2H_6O$ |
| $C_3H_8$ | $C_3H_6$ | $C_3H_8O$ |
| $C_4H_{10}$ | $C_4H_8$ | $C_4H_{10}O$ |

A homologous series is a *series of compounds with similar chemical properties because they have the same functional group. Each member differs from the next by one $-CH_2-$.*

Members of a homologous series can be represented using a **general formula**. The first few members of the alkane series are shown in Table 22.1.

In the case of alkanes, if there are *n* carbons in a molecule, there are always $2n + 2$ hydrogens. The general formula of alkanes is $C_nH_{2n+2}$. For methane, $CH_4$, there is 1 carbon atom in the molecule and $2 \times 1 + 2 = 4$ hydrogen atoms. For propane $C_3H_8$, there are 3 carbon atoms in the molecule and $2 \times 3 + 2 = 8$ hydrogen atoms. We can therefore work out that dodecane, which has 12 carbon atoms, will have 26 hydrogen atoms and the molecular formula $C_{12}H_{26}$.

Different homologous series usually have different general formulae. For example, the general formula for alkenes is $C_nH_{2n}$ and that for alcohols is $C_nH_{2n+2}O$.

> **HINT**
>
> More features of a homologous series are discussed on page 278.

> **HINT**
>
> A way of thinking about this is: the functional group is what makes the behaviour of a particular compound different from that of an alkane.

A **functional group** is *an atom or a group of atoms that determine the chemical properties of a compound*. All compounds in the same homologous series have the same functional group. For example, the functional group for alcohols is the $-OH$ group and that for alkenes is $C=C$.

▲ Figure 22.6 Alcohols contain the $-OH$ functional group and alkenes contain the $C=C$ functional group.

## STRUCTURAL FORMULAE

A **structural formula** shows *how the atoms in a molecule are joined together*. There are two ways of representing structural formulae: they can be drawn as a **displayed formula** (full structural formula), or they can be written out in condensed form (*condensed structural formula*) by omitting all the carbon–carbon single bonds and carbon–hydrogen single bonds, for example $CH_3CH_2CH_3$. You need to be confident about using either way.

## DISPLAYED FORMULAE

A displayed formula (sometimes called a full structural formula) shows *all* the bonds in the molecule as individual lines. You need to remember that each line represents *a pair of shared electrons in a covalent bond*.

Figure 22.7 shows a model of butane, together with its displayed formula. Notice that the way the displayed formula is drawn is different from the shape of the actual molecule. Displayed formulae are always drawn with the molecule straightened out and flattened. They do, however, show exactly how all the atoms are joined together.

> **REMINDER**
>
> It is important to remember when you draw structures that C always forms four bonds and H always forms one bond. Oxygen will form two bonds and a halogen (chlorine, bromine, etc.) atom one. If you are not sure about why these atoms form this number of bonds, you need to look back at Chapter 8.

▲ Figure 22.7 Butane. The angles between neighbouring C–H bonds in butane are shown as 90° in the two-dimensional displayed formula. In reality they are about 109.5° in a tetrahedral arrangement in three dimensions, similar to the arrangement of C–C bonds in a diamond structure (see Chapter 8, page 93).

## HOW TO DRAW A STRUCTURAL FORMULA

For anything other than the smallest molecules, drawing a fully displayed formula is very time-consuming. You can simplify the formula by writing, for example, $CH_3$ or $CH_2$ instead of showing all the carbon–hydrogen bonds.

The structural formula for butane can be shown as

$CH_3CH_2CH_2CH_3$ or $CH_3-CH_2-CH_2-CH_3$

These both show exactly how the atoms in the molecule are joined together. The corresponding molecular formula, $C_4H_{10}$, doesn't give you the same sort of useful information about the molecule.

All the structures in Figure 22.8 represent butane: even though they look different they are exactly the same molecule.

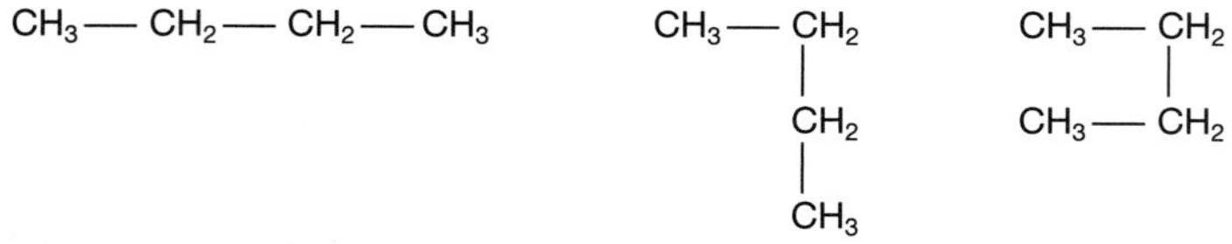

$$CH_3-CH_2-CH_2-CH_3 \qquad CH_3-CH_2 \atop \qquad \qquad \qquad CH_2 \atop \qquad \qquad \qquad CH_3 \qquad CH_3-CH_2 \atop CH_3-CH_2$$

▲ Figure 22.8 All three structures represent butane. The convention is to write the structure with all the carbon atoms in a straight line.

Each structure shows four carbon atoms joined together in a chain, but the chain has simply twisted. This happens in real molecules as well: the atoms can rotate around single carbon–carbon bonds, as shown in Figure 22.9.

▲ Figure 22.9 These diagrams show the shape of a butane molecule better, but you would not draw these in an exam.

A molecule like propene, $C_3H_6$, has a carbon–carbon double bond. This is shown by drawing two lines between the carbon atoms to show *the two pairs of shared electrons*. You would normally write this in a simplified structural formula as $CH_3CH=CH_2$.

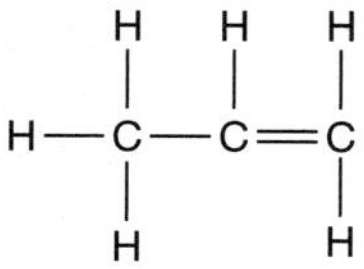

$$\begin{array}{ccccc} & H & & H & & H \\ & | & & | & & | \\ H- & C & - & C & = & C \\ & | & & | & & | \\ & H & & H & & H \end{array}$$

▲ Figure 22.10 The displayed formula for propene

## NAMING ORGANIC COMPOUNDS

Names for organic compounds can appear quite complicated, but they are simply a code that describes the molecule. Each part of a name tells you something specific about the molecule. One part of a name tells you how many carbon atoms there are in the longest chain, another part tells you whether there are any carbon–carbon double bonds, and so on.

## CODING THE CHAIN LENGTH

### HINT

You have to learn these! The first four are the difficult ones because there isn't any pattern. However, a mnemonic can help, something like **M**onkeys **E**at **P**ink **B**ananas. Or you could think of one of your own that you can remember!
'**Pent**' means five (as in **pent**agon) and '**hex**' means six (as in **hex**agon).

Look for the code letters in the name – these are given in Table 22.2.

Table 22.2 Coding the chain length

| Code letters | Number of carbons in chain |
|:---:|:---:|
| meth | 1 |
| eth | 2 |
| prop | 3 |
| but | 4 |
| pent | 5 |
| hex | 6 |

For example, **but**ane has a chain of four carbon atoms. **Prop**ane has a chain of three carbon atoms.

## CODING FOR THE TYPE OF COMPOUND

### ALKANES

**Alkanes** are *a homologous series of similar hydrocarbons (compounds of carbon and hydrogen only) in which all the carbons are joined to each other with single covalent bonds*. Compounds like this are coded with the ending '**ane**'. For example, eth*ane* is a two-carbon chain (because of 'eth') with a carbon–carbon single bond, $CH_3—CH_3$.

### ALKENES

**Alkenes** are *a homologous series of hydrocarbons which contain a carbon–carbon double bond*. This is shown in their name by the ending '**ene**'. For example, eth*ene* is a two-carbon chain containing a carbon–carbon double bond, $CH_2=CH_2$. With longer chains, the position of the double bond could vary in the chain. This is shown by numbering the chain and noting which carbon atom the double bond *starts* from.

### HINT

In more complicated molecules, the presence of the code 'an' in the name again shows that the carbons are joined by single bonds. For example, you can tell that propan-1-ol contains three carbon atoms ('prop') joined together by carbon–carbon single bonds ('an'). The coding on the end gives you more information about the molecule. You will meet this later in the chapter.

Table 22.3 Indicating the position of the double bond in the name of alkenes

| Formula | Name | Description |
|:---:|:---:|:---:|
| $CH_2=CHCH_2CH_3$ | but-1-ene | a four-carbon chain with a double bond starting on the first carbon |
| $CH_3CH=CHCH_3$ | but-2-ene | a four-carbon chain with a double bond starting on the second carbon |

How do you know which end of the chain to number from? The rule is that *you number from the end which produces the smaller numbers in the name*.

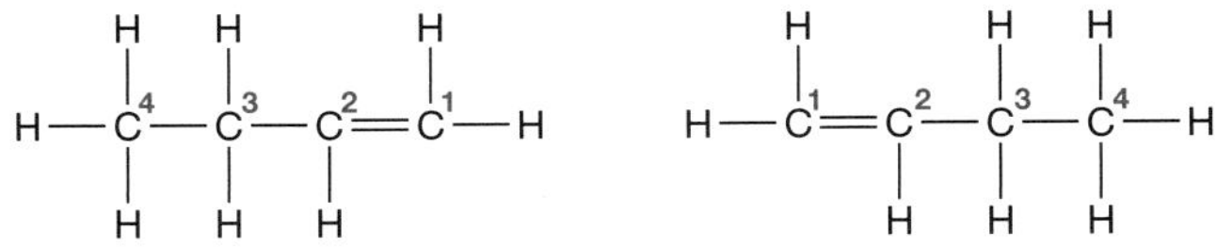

### HINT

Both parts of Figure 22.11 show the same molecule, but in one case it has been turned over so that what was originally on the left is now on the right, and vice versa. It would be silly to change the name every time the molecule moved! Both of the forms in Figure 22.11 are called but-1-ene.

▲ Figure 22.11 Both structures represent but-1-ene.

## CODING FOR BRANCHED CHAINS

Hydrocarbon chains can have side branches on them. You are only likely to come across two small side chains, shown in Table 22.4.

Table 22.4 Coding for branched chains

| Side chain | Coded |
|---|---|
| $CH_3-$ | methyl |
| $CH_3CH_2-$ | ethyl |

### KEY POINT

Notice that the count of the number of carbons in the side chain is coded exactly as before: '**meth**' shows a *one*-carbon side chain, '**eth**' shows *two* carbons. The 'yl' shows that the group is attached to something else.

The name of a molecule is always based on the longest chain you can find in it. *The position of the side chain is shown by numbering the carbon atoms from the end of the longest chain which produces the smaller numbers in the name*, exactly as before with the carbon–carbon double bonds in alkenes.

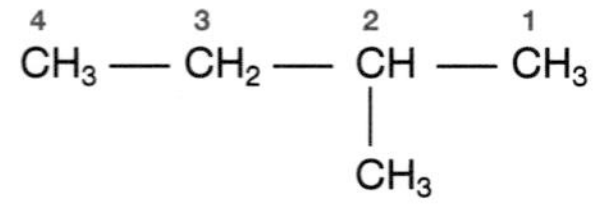

▲ Figure 22.12 Number from the end that produces the smaller number.

The longest chain in the molecule in Figure 22.12 has four carbon atoms ('but') with no double bonds ('ane'). The name is based on butane. There is a methyl group branching from the number 2 carbon. (Remember to number from the end that produces the smaller number.) The compound is called 2-methylbutane.

Where there is more than one side chain, you describe the position of each of them.

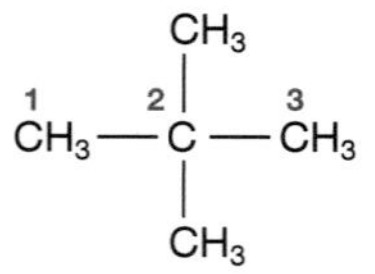

▲ Figure 22.13 Describe the position of each side chain.

The longest chain in the molecule in Figure 22.13 has three carbon atoms and no double bonds. Therefore the name is based on propane.

There are two methyl groups attached to the second carbon. The compound is 2,2-dimethylpropane. The '*di*' in the name shows the presence of the two methyl groups. '2,2' shows that they are both on the second carbon atom.

You can reverse this process and draw a structural formula from a name. All you have to do is decode the name.

For example, what is the structural formula for 2,3-dimethylbut-2-ene?

Start by looking for the code for the longest chain length. 'but' shows a four-carbon chain. 'ene' shows that it contains a carbon–carbon double bond starting on the second carbon atom ('-2-ene'). There are two methyl groups ('dimethyl') attached to the second and third carbon atoms in the chain ('2,3-'). All you have to do now is to fit all this together into a structure.

### HINT

It is much more important to be able to decode names to give structures than the other way around. If you don't know what a teacher or an examiner is talking about, you are completely lost!

Start by drawing the structure without any hydrogen atoms on the main chain. It doesn't matter whether you draw the $CH_3$ groups pointing up or down. Then add enough hydrogens so that each carbon atom forms four bonds. Figure 22.14 shows how your thinking would work.

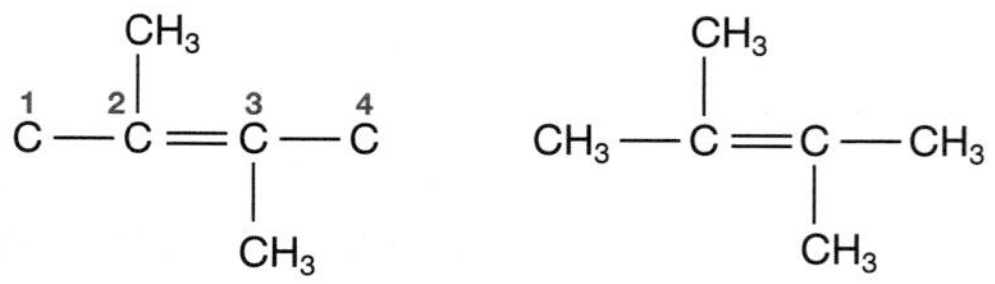

▲ Figure 22.14 First draw the carbon structure, then add hydrogens.

KEY POINT

**Warning!** Don't confuse the word isomer with isotope. *Isotopes* are different atoms of the same element with the same atomic number but different mass numbers.

STRUCTURAL ISOMERISM IN THE ALKANES

# STRUCTURAL ISOMERISM

**Structural isomers** *are molecules with the same molecular formula, but different structural formulae.*

Examples will make this clear.

## ISOMERS OF $C_5H_{12}$

If you had some molecular models and picked out five carbon atoms and 12 hydrogen atoms, you would find it was possible to join them together in more than one way. The different molecules formed are known as isomers. All have the molecular formula $C_5H_{12}$, but different structures.

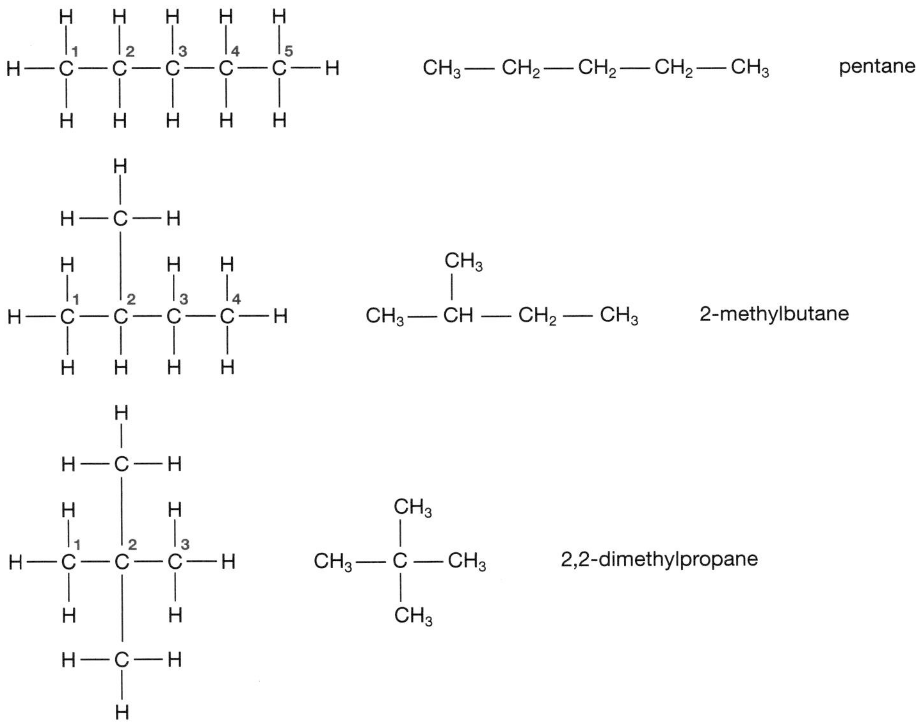

$CH_3 - CH_2 - CH_2 - CH_2 - CH_3$     pentane

$CH_3 - CH - CH_2 - CH_3$     2-methylbutane

$CH_3 - C - CH_3$     2,2-dimethylpropane

▲ Figure 22.15 There are three isomers for $C_5H_{12}$. These can be shown as either displayed formulae or structural formulae.

If you look carefully at the structures in Figure 22.15 you can see that you couldn't change one into the other simply by bending or twisting the molecule. You would have to break some bonds and reconnect the atoms. That's a simple way of telling that you have isomers.

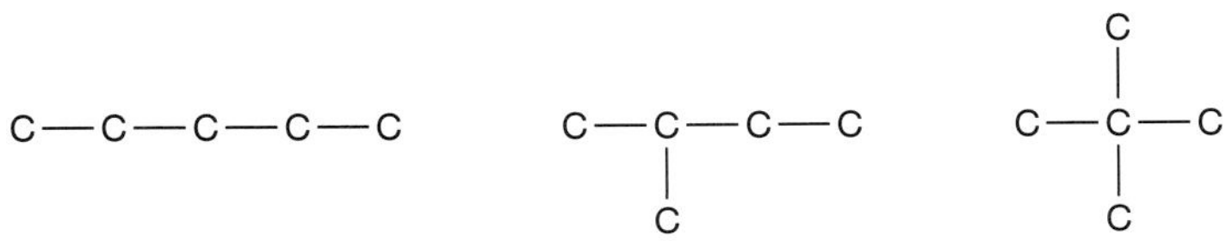

HINT

It is sometimes easier to see that the molecules are different by drawing only the C atoms joined together, without the hydrogens.

But you must show the hydrogen atoms in the exam!

REMINDER

A straight chain is an unbranched chain.

The **straight chain** isomer is called pentane. The *branched chain* isomer in the middle, which has a four-carbon chain ('butane') and a methyl group on the second carbon, is called 2-methylbutane. The final *branched chain* isomer, which has a three-carbon chain ('propane') and two methyl groups on the second carbon, is called 2,2-dimethylpropane. The prefix '*di*' indicates the presence of two identical branches in the molecule.

Students frequently think they can find another isomer as well. If you look closely at this 'fourth' isomer (Figure 22.16) you will see that it is just 2-methylbutane rotated in space.

To avoid this sort of problem, always draw your isomers so that *the longest carbon chain is drawn horizontally* (or draw them out again without the H atoms).

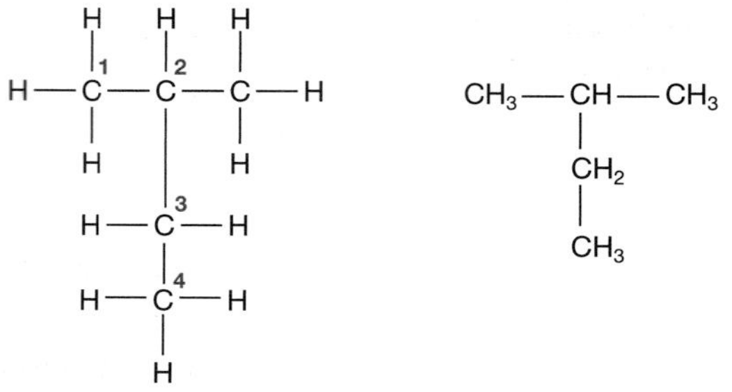

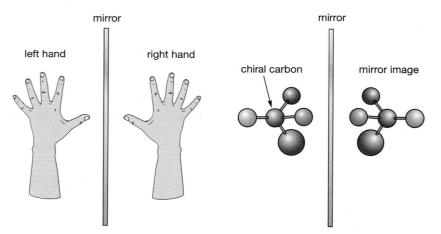

▲ Figure 22.16 There is no fourth isomer for $C_5H_{12}$.

## LOOKING AHEAD

To be precise, the type of isomerism discussed here should be called structural isomerism: molecules have the same molecular formula but different structural formulae. The atoms in the different isomers are joined in a different order. There are also other forms of isomerism and one important type is called optical isomerism. Optical isomers have atoms joined in the same order (i.e. they have the same structural formula), but are arranged differently in three-dimensional space. These isomers are mirror images of each other but cannot be superimposed. A good analogy is to look at your left and right hands. They are exactly the opposite of each other but cannot lie on top of another and align perfectly (Figure 22.17).

▲ Figure 22.17 Optical isomers are mirror images of each other and cannot be superimposed.

Optical isomerism occurs when a molecule has a *chiral centre*, that is to say it has a carbon atom that is connected to four different atoms or groups of atoms. These molecules have a tetrahedral shape around the chiral centre (molecular shape was mentioned in Chapter 8, pages 88–89). Figure 22.17 shows two molecules that are mirror images of each other. However, if you rotate one of the molecules 180° in space, it cannot be put on top of the other.

Optical isomers have identical chemical properties, except when they interact with something that is also chiral. For example, apart from glycine, all naturally occurring amino acids have optical isomers, which means that all the proteins in our body are chiral. This is why we can distinguish the two optical isomers in

**EXTENSION WORK**

The optical isomers are assigned a prefix (*R* or *S*) to show the configuration of atoms around the chiral centre.

Figure 22.18. One smells of spearmint and the other of caraway seed because the two molecules interact differently with the protein receptors in our noses.

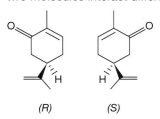

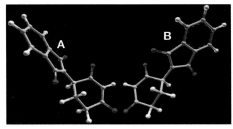

▲ Figure 22.18 Two optical isomers of carvone

▲ Figure 22.19 Optical isomerism in drug molecules

An important application of optical isomerism is in the synthesis of pharmaceutical drugs. You might have heard of the thalidomide tragedy. Back in the 1960s in Germany, the molecule B shown in Figure 22.19 was found to have anti-nausea effects and was therefore used to treat morning sickness in pregnant woman. Unfortunately, molecule B was being manufactured together with its optical isomer A and little was known about the molecule A until many children born in that era were found to have deformed limbs. These days, many drug companies devote a great deal of money to separating optical isomers of new drugs just in case they have side effects.

## STRUCTURAL ISOMERISM IN THE ALKENES

### ETHENE AND PROPENE

Ethene, $CH_2=CH_2$, doesn't have any isomers. Propene, $CH_3CH=CH_2$, doesn't have a structural isomer that is still an alkene.

**EXTENSION WORK**

There is a structural isomer of $C_3H_6$ that doesn't have a carbon–carbon double bond. It has three carbon atoms joined in a ring. It is called cyclopropane, which indicates 'a ring of three carbons with only single bonds between them'.

## ISOMERS OF $C_4H_8$

There are three structural isomers with the molecular formula $C_4H_8$ and containing a carbon–carbon double bond.

$$\overset{4}{C}H_3 - \overset{3}{C}H_2 - \overset{2}{C}H = \overset{1}{C}H_2$$
but-1-ene

$$\overset{1}{C}H_3 - \overset{2}{C}H = \overset{3}{C}H - \overset{4}{C}H_3$$
but-2-ene

$$\overset{3}{C}H_3 - \overset{2}{C} = \overset{1}{C}H_2$$
$$\underset{CH_3}{|}$$
2-methylpropene

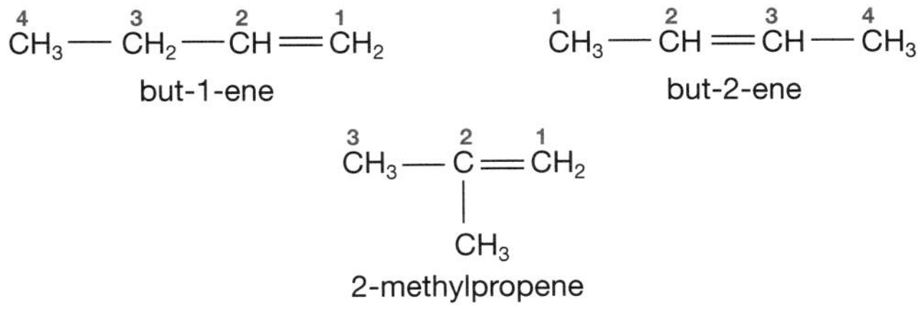

▲ Figure 22.20 Structural isomers with the molecular formula of $C_4H_8$.

Notice the way that you can vary the position of the double bond as well as branching the chain.

## CHEMISTRY ONLY

## A QUICK INTRODUCTORY LOOK AT THE ALCOHOLS

**Alcohols** are *a homologous series of compounds which all contain an –OH functional group attached to a hydrocarbon chain*. This is coded for in the name by the ending '**-ol**'.

**HINT**

If you aren't comfortable with the names, see pages 258–260 on organic names.

Figure 22.21 shows four small alcohols.

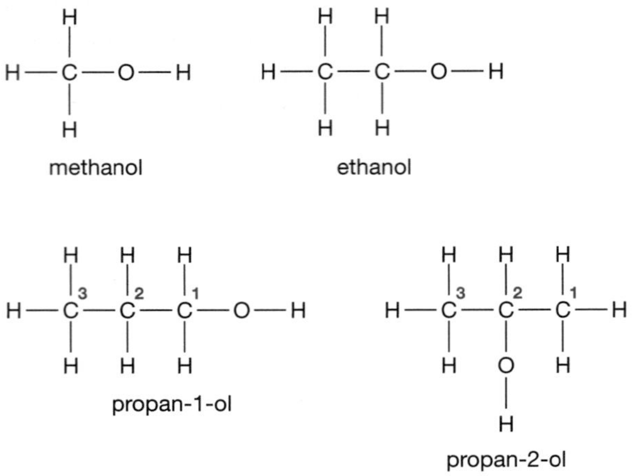

methanol

ethanol

propan-1-ol

propan-2-ol

▲ Figure 22.21 Four small alcohols

The names all end with '-ol', showing there is an –OH functional group in the molecule. Notice the way the number of carbons in the chain is counted exactly as before with 'meth', 'eth' and 'prop'.

There are two different forms of propanol. The –OH group can be attached to the end carbon of the chain (in propan-1-ol) or to the second carbon in the chain (as in propan-2-ol). These are *structural isomers* because there is no way that you can bend or twist one of the molecules to make the other. Bonds would have to be broken to get from one to the other.

**EXTENSION WORK**

It is possible to find a structural isomer of propanol that doesn't have an –OH functional group, for example $CH_3OCH_2CH_3$, which has the oxygen atom bonded between two carbon atoms rather than joined to a hydrogen atom. This compound is named methoxyethane and belongs to a homologous series of compounds called *ethers*.

**END OF CHEMISTRY ONLY**

## SOME CHEMICAL REACTIONS OF ORGANIC COMPOUNDS

In this section we will look at how to classify the reactions of organic compounds. These reactions will be explained much more in the next few chapters.

**COMBUSTION**

▲ Figure 22.22 The burning of a camping stove

All organic compounds that you will meet at International GCSE undergo combustion reactions. Combustion is just another way of saying *burning* and involves a reaction with oxygen.

**Combustion** of hydrocarbons in *excess oxygen* gives rise to *carbon dioxide* and *water*, together with the release of a large amount of heat energy. For example:

propane burning $\quad C_3H_8(g) + 5O_2(g) \rightarrow 3CO_2(g) + 4H_2O(l)$

butene burning $\quad C_4H_8(g) + 6O_2(g) \rightarrow 4CO_2(g) + 4H_2O(l)$

Propane is frequently used in camping stoves outdoors.

Alcohols undergo similar reactions:

ethanol burning $\quad C_2H_5OH(l) + 3O_2(g) \rightarrow 2CO_2(g) + 3H_2O(l)$

## SUBSTITUTION

A **substitution** reaction occurs when an atom or group of atoms is *replaced* by a different atom or group of atoms. For example, the hydrogen atoms in an alkane can be replaced by halogen atoms. We will go into more detail on this reaction in Chapter 24.

ethane with bromine gas

$$CH_3CH_3(g) + Br_2(g) \rightarrow CH_3CH_2Br(g) + HBr(g)$$

ethane            bromine            bromoethane      hydrogen bromide

▲ Figure 22.23 A substitution reaction between an alkane and a halogen

## ADDITION

In an **addition** reaction something is added to a molecule without taking anything away. Alkenes undergo addition reactions. For example, ethene reacts with bromine:

ethene with $Br_2$

$$CH_2CH_2(g) + Br_2(l) \rightarrow CH_2BrCH_2Br(l)$$

ethene            bromine            1,2-dibromoethane

▲ Figure 22.24 An addition reaction between an alkene and a halogen.

## CHAPTER QUESTIONS

Note: the questions below deliberately include some examples which are slightly beyond the level which you would expect to meet in an International GCSE exam.

**SKILLS** CRITICAL THINKING

1 a Write down the names of the following hydrocarbons:

     i $CH_4$

     ii $CH_3CH_2CH_3$

     iii $CH_3CH_2CH_2CH_2CH_2CH_3$

     iv $CH_3CH=CH_2$

     v $CH_2=CH_2$

     vi $CH_2=CHCH_2CH_3$

     vii $CH_3CH_2CH_2OH$

     viii $CH_3CH_2CHOHCH_3$

**SKILLS** INTERPRETATION

b Draw structural and displayed formulae for:

     i ethane               v methanol

     ii propan-1-ol       vi 2-methylpropane

     iii but-2-ene         vii methylpropene

     iv hexane           viii pent-1-ene

SKILLS ▸ CRITICAL THINKING

SKILLS ▸ INTERPRETATION

2 a Explain the term *structural isomerism*.

b There are two structural isomers of $C_4H_{10}$. Draw their structures and name them.

c There are five structural isomers of $C_6H_{14}$. Draw their structures and name them.

SKILLS ▸ CRITICAL THINKING

d How many structural isomers containing a carbon–carbon double bond can you find with a molecular formula $C_4H_8$? Name the straight chain isomers.

SKILLS ▸ INTERPRETATION

e Draw two isomers of $C_4H_8$ that do not contain carbon–carbon double bonds.

3 a Draw the structures for the following structural isomers of $C_4H_9OH$. (Note: this is beyond what you will be asked to do at International GCSE. The answer to the first one is shown to give you hints as to how to work out the others.)

   i   butan-1-ol (Answer: $CH_3CH_2CH_2CH_2OH$)

   ii  butan-2-ol

   iii 2-methylpropan-1-ol

   iv  2-methylpropan-2-ol

SKILLS ▸ CRITICAL THINKING

b Can you find two more structural isomers with the same molecular formula ($C_4H_{10}O$) which don't contain an –OH group?

4 Classify the following reactions as substitution, addition or combustion.

a

b $2C_8H_{18}(l) + 25O_2(g) \rightarrow 16CO_2(g) + 18H_2O(l)$

c $C_4H_8(g) + H_2(g) \rightarrow C_4H_{10}(g)$

d

e $C_2H_5OH(l) + 3O_2(g) \rightarrow 2CO_2(g) + 3H_2O(l)$

f $CH_3CH_2CH_2CH_3(g) + Br_2(g) \rightarrow CH_3CH_2CHBrCH_3(g) + HBr(g)$

SKILLS   CRITICAL THINKING

5 The table below shows the formulae of some organic compounds.

| Compound | Formula |
|----------|---------|
| A | $C_2H_6$ |
| B | $CH_3CH{=}CH_2$ |
| C | $CH_3CH_2CH_2OH$ |
| D | $CH_3CH_2CH_2CH_2CH_3$ |

a   Select one compound from the table which is not a hydrocarbon. Explain your choice.

SKILLS   INTERPRETATION

b   Draw a dot-and-cross diagram to show the bonding in compound **A**.

c   Draw the full displayed formula of compound **B**. Include all bonds in your drawing.

SKILLS   CRITICAL THINKING

d   Compounds **A** and **D** are from the same homologous series. State the general formula for this homologous series.

e   Other than having the same general formula, state *one* other characteristic of members of the same homologous series.

SKILLS   PROBLEM SOLVING

f   Write a balanced chemical equation for the complete combustion of compound **D**.

# 23 CRUDE OIL

The oil industry is at the very heart of modern life. It provides fuels, plastics and the organic chemicals which go to make things as different as solvents, drugs, dyes and explosives. This chapter explores how an unappealing, sticky, black liquid is converted into useful things.

▲ Figure 23.1 The oil industry is BIG business!

## LEARNING OBJECTIVES

- Know that crude oil is a mixture of hydrocarbons
- Describe how the industrial process of fractional distillation separates crude oil into fractions
- Know the names and uses of the main fractions obtained from crude oil: refinery gases, gasoline, kerosene, diesel, fuel oil and bitumen
- Know the trend in colour, boiling point and viscosity of the main fractions
- Know that a fuel is a substance that, when burned, releases heat energy
- Know the possible products of complete and incomplete combustion of hydrocarbons with oxygen in the air
- Know that in car engines the temperature reached is high enough to allow nitrogen and oxygen from air to react, forming oxides of nitrogen

- Explain how the combustion of some impurities in hydrocarbon fuels results in the formation of sulfur dioxide
- Understand how sulfur dioxide and oxides of nitrogen contribute to acid rain
- Describe how long-chain alkanes are converted to alkenes and shorter-chain alkanes by catalytic cracking (using silica or alumina as the catalyst and a temperature in the range of 600–700 °C)
- Explain why cracking is necessary in terms of the balance between supply and demand for different fractions

## WHAT IS CRUDE OIL?

### THE ORIGIN OF CRUDE OIL

Millions of years ago, plants and animals living in the sea died and fell to the bottom. Layers of sediment formed on top of them. Their shells and skeletons formed limestone. The soft tissue was gradually changed by heat and high pressure into **crude oil**. Crude oil is a *finite, non-renewable resource*. Once all the existing supplies have been used, they won't be replaced, or at least not for many millions of years.

▲ Figure 23.2 This sticky black liquid is essential to modern life.

### CRUDE OIL CONTAINS HYDROCARBONS

Crude oil is a mixture of hydrocarbons, *compounds containing carbon and hydrogen only*. There are lots of different hydrocarbons of various sizes in crude oil, ranging from molecules with just a few carbon and hydrogen atoms to molecules containing over 100 atoms.

### HOW THE PHYSICAL PROPERTIES OF HYDROCARBONS CHANGE WITH MOLECULE SIZE

As the number of carbon atoms in hydrocarbon molecules increases, the physical properties of the compounds change. Most of these changes are the result of increasing attractions between neighbouring molecules. *As the molecules become bigger, the intermolecular forces of attraction become stronger* and it becomes more difficult to pull one molecule away from its neighbours.

As the molecules become bigger, the following changes occur.

#### REMINDER

Intermolecular forces are explained in Chapter 8 (page 92).

- *Boiling point increases*: the larger the molecule, the higher the boiling point. This is because large molecules are attracted to each other more strongly than smaller ones. More energy is needed to break these stronger intermolecular forces of attraction to produce the widely separated molecules in the gas.

- *The liquids become less* **volatile**: the bigger the hydrocarbon, the more slowly it evaporates at room temperature. This is again because the bigger molecules are more strongly attracted to their neighbours and so don't turn into a gas so easily.

#### HINT

We usually count a substance as being volatile if it turns to a vapour easily at room temperature. That means it will evaporate quickly at that temperature.

- *The liquids become more* **viscous** *and flow less easily*: liquids containing small hydrocarbon molecules are runny. Those containing large molecules flow less easily because of the stronger forces of attraction between their molecules.

- The liquids become darker in colour.

- Bigger hydrocarbons do not burn as easily as smaller ones. This limits the use of the bigger ones as fuels.

## SEPARATING CRUDE OIL

Crude oil itself has no uses and it has to be separated into *fractions* before it can be used. These fractions are all mixtures, but each one contains a narrow range of sizes of hydrocarbons with similar boiling points. We use **fractional distillation** to separate crude oil into fractions. This is carried out in an oil refinery.

### FRACTIONAL DISTILLATION

Crude oil is heated until it boils and the vapours pass into a **fractionating column**, which is cooler at the top and hotter at the bottom. The vapours rise up the column. How far up the column a particular hydrocarbon moves depends on its boiling point.

### KEY POINT

The boiling point is the same as the temperature at which a gas condenses to form a liquid; it could also be called the *condensation point*.

Suppose a hydrocarbon boils at 120 °C. At the bottom of the column, the temperature is much higher than 120 °C and so the hydrocarbon remains as a gas. As it travels up the column, the temperature of the column becomes lower. When the temperature falls to 120 °C, that hydrocarbon will turn into a liquid, it *condenses* and can be removed.

Smaller molecules have lower boiling points and get further up the column before they condense. Longer chain hydrocarbons have higher boiling points and condense lower down in the column. This way, the crude oil is split into various *fractions*.

### HINT

You will find a lot of disagreement, from various sources, about exactly what fractions are produced in fractional distillation. Figure 23.3 matches the requirements of the Edexcel International GCSE specification, but this is a major simplification of what really goes on.

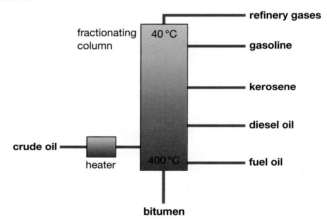

▲ Figure 23.3 Fractional distillation of crude oil

The hydrocarbons in the refinery gas fraction (see Figure 23.3) have very small masses and their boiling points are so low that the temperature of the column never falls low enough for them to condense to liquids. The temperature of the column isn't hot enough to boil some of the very large hydrocarbons found in the crude oil, and so they remain as a liquid. These are removed as a residue from the bottom of the column. Bitumen (see below), which is used in road making, is made from this.

## USES OF THE FRACTIONS

### AS FUELS

All hydrocarbons burn in air (oxygen) to form *carbon dioxide and water*, and release a lot of heat in the process. The various fractions can therefore be used as **fuels**.

*A fuel is a substance which, when burned, releases heat energy.*

## HINT

Don't try to learn these equations as there are too many possible hydrocarbons you could be asked about. Provided you know (or are told) the formula of the fuel and remember the products of the combustion reaction, you can balance the equation yourself.

▲ Figure 23.4 As well as all the other poisonous or cancer-causing compounds, cigarette smoke contains carbon monoxide due to incomplete combustion.

For example, burning methane (the major constituent of natural gas):

$$CH_4(g) + 2O_2(g) \rightarrow CO_2(g) + 2H_2O(l)$$

or burning octane (one of the hydrocarbons present in gasoline, petrol):

$$2C_8H_{18}(l) + 25O_2(g) \rightarrow 16CO_2(g) + 18H_2O(l)$$

If there isn't enough air (or oxygen), you get **incomplete combustion**. This leads to the formation of *carbon (soot)* or *carbon monoxide* instead of carbon dioxide.

For example, if methane burns in a badly maintained gas appliance, there may not be enough oxygen available to produce carbon dioxide, and so you get toxic carbon monoxide instead:

$$2CH_4(g) + 3O_2(g) \rightarrow 2CO(g) + 4H_2O(l)$$

The formation of carbon monoxide from the incomplete combustion of hydrocarbons is very dangerous. Carbon monoxide is colourless and odourless, and is very poisonous. *Carbon monoxide is poisonous because it reduces the ability of the blood to carry oxygen around the body.* This will make you ill, or you may even die, because not enough oxygen gets to the cells in your body for respiration to provide energy.

### DID YOU KNOW?

Carbon monoxide combines with *haemoglobin* (the molecule that carries oxygen in the red blood cells), preventing it from carrying the oxygen. It binds more strongly to the haemoglobin than oxygen does.

### EXTENSION WORK

If you try to draw a dot-and-cross diagram for carbon monoxide, you will see that it is a very strange molecule because carbon does not form four bonds.

## REFINERY GASES

**Refinery gases** are *a mixture of methane, ethane, propane and butane*, which can be separated into individual gases if required. These gases are commonly used as liquefied petroleum gas (LPG) for domestic heating and cooking.

## GASOLINE (PETROL)

As with all the other fractions, petrol is a mixture of hydrocarbons with similar boiling points. It is used as a fuel in cars.

## KEROSENE

Kerosene is used as a fuel for jet aircraft, as domestic heating oil and as 'paraffin' for small heaters and lamps.

▲ Figure 23.5 Kerosene is used as aviation fuel.

**DIESEL**

This is used as a fuel for buses, lorries, some cars, and some railway engines. Some is also converted to other more useful organic chemicals, including petrol, in a process called *cracking* (see page 273–275 for more details on cracking).

▲ Figure 23.6 A train powered by diesel

**FUEL OIL**

This is used as a fuel for ships and for industrial heating.

▲ Figure 23.7 Ships' boilers burn fuel oil.

**BITUMEN**

Bitumen is a thick, black material, which is melted and mixed with small pieces of rock to make the top surface of roads.

▲ Figure 23.8 Bitumen is used in road construction.

## ENVIRONMENTAL PROBLEMS ASSOCIATED WITH THE BURNING OF FOSSIL FUELS FROM CRUDE OIL

**KEY POINT**

**Fossil fuels** include coal, gas and fuels derived from crude oil. These all come from things that were once alive.

There are major environmental problems associated with the burning of fossil fuels derived from crude oil. First of all, the carbon dioxide produced when hydrocarbons are burned is a greenhouse gas. Greenhouse gases *trap the heat radiated from the Earth's surface (originally from the Sun)* and many people believe that this could lead to climate change. For more discussion on the greenhouse effect, see Chapter 13.

## ACID RAIN: SULFUR DIOXIDE AND OXIDES OF NITROGEN

▲ Figure 23.9 Use of very low-sulfur fuels limits the production of sulfur dioxide, but the spark in a petrol engine causes oxygen and nitrogen from the air to combine to make oxides of nitrogen, $NO_x$.

▲ Figure 23.10 Trees dying from the effects of acid rain.

Rain is naturally slightly acidic (pH = ~5.6) because of dissolved carbon dioxide. Acid rain is rain with a pH lower than this (pH < 5.6) because of the presence of various pollutants. The pH of acid rain is often about 4.

*Acid rain is formed when water and oxygen in the atmosphere react with sulfur dioxide to produce sulfuric acid ($H_2SO_4$), or with various oxides of nitrogen, $NO_x$, to give nitric acid ($HNO_3$).* $SO_2$ and $NO_x$ come mainly from power stations and factories burning fossil fuels, or from motor vehicles.

Fossil fuels contain a small amount of sulfur. When the fuel is burned, the sulfur reacts with oxygen, producing sulfur dioxide:

$$S(s) + O_2(g) \rightarrow SO_2(g)$$

Reactions in the atmosphere with oxygen and water can convert this to *sulfuric acid ($H_2SO_4$)*, a strong acid and an important component of acid rain:

$$2SO_2(g) + 2H_2O(l) + O_2(g) \rightarrow 2H_2SO_4(aq)$$

Note: when sulfur dioxide reacts with water, a weaker acid called *sulfurous acid ($H_2SO_3$)* is formed:

$$SO_2(g) + H_2O(l) \rightarrow H_2SO_3(aq)$$

In petrol engines, sparks are used to ignite the petrol–air mixture to power the car. The temperature reached in the engine is high enough to allow nitrogen and oxygen in the air to combine to produce oxides of nitrogen. For example:

$$N_2(g) + O_2(g) \rightarrow 2NO(g)$$

These nitrogen oxides can be converted to *nitric acid ($HNO_3$)* in the atmosphere and therefore contribute to acid rain.

Acid rain is a major problem, mainly because of its devastating effect on trees and on life in lakes. Acid rain kills trees and fish in lakes. In some lakes the water is so acidic that it won't support life at all. Limestone buildings and marble statues (both made of *calcium carbonate*) and some metals such as iron are also attacked by acid rain.

Reaction between limestone and sulfuric acid:

$$CaCO_3(s) + H_2SO_4(aq) \rightarrow CaSO_4(s) + H_2O(l) + CO_2(g)$$

The solution to acid rain involves removing sulfur from fuels (this is usually done for petrol used in cars), 'scrubbing' the gases from power stations and factories to remove $SO_2$ and $NO_x$, and using catalytic converters in cars.

## CRACKING

Although most of the fractions from the fractional distillation of crude oil are useful as fuels, some fractions are more useful and more profitable to sell than others.

The amounts of each fraction obtained will depend on the proportions of the various hydrocarbons in the original crude oil, not the amounts in which they are needed. The problem is:

■ there are far too many long-chain hydrocarbons, which are not in such high demand and are not as profitable to sell

■ there are not enough shorter-chain hydrocarbons that can be used as fuel for cars.

Far more petrol is needed than can be supplied by just separating crude oil into its fractions.

**Cracking** is a process in which *long-chain alkanes are converted to alkenes and shorter-chain alkanes*. The big hydrocarbon molecules in fuel oil, for example, can be broken down into the smaller ones needed for petrol. A very simple equation could be:

$$C_{14}H_{30} \rightarrow C_{10}H_{22} + C_4H_8$$
   alkane       alkane     alkene

**HOW CATALYTIC CRACKING WORKS**

The fuel oil fraction is heated to give a gas and then passed over a catalyst of *silicon dioxide* (also called silica) and *aluminium oxide* (also called alumina) at about *600–700 °C*. Cracking can also be carried out at higher temperatures without a catalyst (*thermal cracking*).

Cracking is just an example of *thermal decomposition*: a big molecule splitting into smaller ones on heating. The majority of the hydrocarbons found in crude oil have single bonds between the carbon atoms. During the cracking process, C–C single bonds are broken and new C=C double bonds are formed.

The molecules are broken up in a fairly random way. One possibility is shown in Figure 23.11.

**REMINDER**

In Figure 23.11, the molecules have been drawn to show the various covalent bonds. They have also been 'straightened out' with bond angles of 90°.

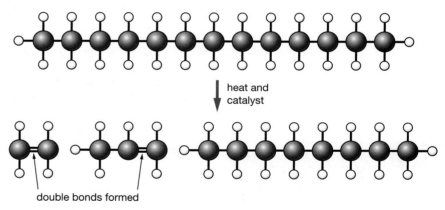

heat and catalyst

double bonds formed

▲ Figure 23.11 How cracking works.

As an equation, this would read:

$$C_{13}H_{28}(l) \rightarrow C_2H_4(g) + C_3H_6(g) + C_8H_{18}(l)$$

Cracking produces a mixture of alkanes and alkenes. In this case, two different alkenes are produced: ethene and propene. Octane, an alkane, is also formed. This particular cracking reaction has therefore produced two types of useful molecules for the chemical industry. Ethene and propene are both used to make important polymers, as you will find in Chapter 29. Both are also used in making other organic chemicals. The reaction also produces octane, which is a component of petrol (gasoline).

**HINT**

Don't worry too much if you can't remember which are gases and which are liquids. State symbols are very commonly missed out in organic chemistry.

There are therefore two important reasons why oil companies carry out cracking:

■ to produce more petrol

■ to produce more alkenes that can be used for making polymers (plastics) (alkenes are more reactive than alkanes and also have other uses).

The molecule might have split quite differently. All sorts of reactions are happening in a catalytic cracker. Two other possibilities might be:

$$C_{13}H_{28}(l) \rightarrow 2C_2H_4(g) + C_9H_{20}(l)$$

$$C_{13}H_{28}(l) \rightarrow 2C_2H_4(g) + C_3H_6(g) + C_6H_{14}(l)$$

Some reactions might even produce a small percentage of free hydrogen. For example:

$$C_{13}H_{28}(l) \rightarrow 2C_2H_4(g) + C_3H_6(g) + C_6H_{12}(l) + H_2(g)$$

In this case, all the hydrocarbons formed will have double bonds. They are all alkenes.

There will also be very many other ways in which this particular molecule might have split up. Also, remember that the fraction being cracked contains a complex mixture of hydrocarbons, not just one. At the end of the cracking process you will have an equally complex mixture of smaller hydrocarbons, both alkanes and alkenes. This mixture will have to go through a lot of further processing (including further fractional distillation) to separate everything out into pure compounds.

**EXTENSION WORK**

In industry, catalytic cracking is done by passing hydrocarbons through a bed of zeolite (aluminosilicate) catalyst at about 500 °C and moderate pressures. Apart from breaking down large straight-chain molecules into smaller ones, isomerisation can occur during the process to produce branched-chain alkanes and alkanes with ring structures. The presence of these isomers in a fuel helps to increase the *octane number* of a fuel, which is a measure of the tendency of a fuel to not undergo auto-ignition in an engine. *Auto-ignition* is when fuel burns spontaneously out of control. It leads to wear in the engine and wastage of petrol. The higher the octane number, the lower the tendency for a fuel to undergo auto-ignition.

## CHAPTER QUESTIONS

**SKILLS** CRITICAL THINKING

**1** Crude oil is a mixture of hydrocarbons.

  **a** State which two elements are present in the compounds in crude oil.

  **b** Crude oil is separated into fractions by heating and passing the vapour into a fractionating column. Explain how the fractions separate in the column.

  **c** Two of the fractions are gasoline and diesel. State one use of each.

  **d** Name two fractions formed in the fractional distillation of crude oil, other than gasoline and diesel.

  **e** Describe the differences between gasoline and diesel. In your answer you should refer to the average size of the molecules in the two liquids, the colour of the two liquids and the viscosities of the two liquids.

  **f** One of the hydrocarbons found in the diesel fraction is $C_{18}H_{38}$. Suggest the formula of a hydrocarbon that could be found in the gasoline fraction.

  **g**   **i** Write a balanced chemical equation for the complete combustion of butane, $C_4H_{10}$.

    **ii** Incomplete combustion of hydrocarbons produces carbon monoxide. Explain why carbon monoxide is harmful to humans.

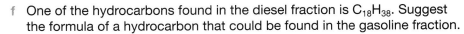

**SKILLS** PROBLEM SOLVING

**SKILLS** REASONING

**SKILLS** REASONING

2 Explain why:

  a burning a fuel containing sulfur as an impurity causes acid rain

  b petrol engines produce oxides of nitrogen

  c the presence of too much sulfur dioxide, $SO_2$, in the atmosphere has a damaging effect on many buildings made of marble, a form of calcium carbonate, $CaCO_3$.

**SKILLS** CRITICAL THINKING

3 Cracking is a process that splits larger hydrocarbons into smaller ones.

  a Give two reasons why an oil company might want to crack a hydrocarbon.

  b Give the conditions under which cracking is carried out.

**SKILLS** PROBLEM SOLVING

  c A molecule of the hydrocarbon $C_{11}H_{24}$ was cracked to give two molecules of ethene, $C_2H_4$, and one other molecule. Write a balanced chemical equation for the reaction which took place. (You can omit the state symbols from your equation.)

  d Write an equation for an alternative cracking reaction involving the same hydrocarbon, $C_{11}H_{24}$.

**SKILLS** CREATIVITY

4 Imagine a world in which fossil fuels such as coal, natural gas and oil had never formed. This would have effects other than the immediately obvious ones. For example, the iron and steel industry depends on coke made from coal, although in the past (on a much smaller scale) it used charcoal made from wood. Choose one aspect of the modern world and explain, in no more than 300 words, how it would be different in the absence of fossil fuels. You could choose from transport, use of materials, landscape, disease prevention, and power generation, for example, but you don't have to restrict yourself to this list.

Try not to be too simplistic about this, human beings are inventive! For example, it is possible to obtain organic chemicals from alcohol (from fermented sugar) and fuels from plant oils.

# 24 ALKANES

The hydrocarbons separated from the fractional distillation of crude oil are mainly alkanes. This is the simplest family of hydrocarbons, containing only carbon–carbon and carbon–hydrogen single bonds. Here we look into the structures and chemical reactions of these compounds in more detail. It is assumed that you have already read Chapter 22 (Introduction to Organic Chemistry).

## LEARNING OBJECTIVES

- Know the general formula for alkanes
- Explain why alkanes are classified as saturated hydrocarbons

- Understand how to draw the structural and displayed formulae for alkanes with up to five carbon atoms in the molecule and to name the unbranched-chain isomers
- Describe the reactions of alkanes with halogens in the presence of ultraviolet radiation, limited to mono-substitution

▲ Figure 24.1 The small alkanes are gases and are burned as fuels.

The alkanes are the simplest family of hydrocarbons. Hydrocarbons are compounds that contain carbon and hydrogen only.

Many of the alkanes are used as fuels. Methane is the major component of natural gas. Ethane and propane are also present in small quantities in natural gas, and are important constituents of liquefied petroleum gases (LPG) from crude oil distillation (discussed in Chapter 23).

The first five members of the alkanes series are shown in Table 24.1.

Table 24.1 Some alkanes

| Name | Molecular formula | Structural formula | Displayed formula |
|---|---|---|---|
| methane | $CH_4$ | $CH_4$ | |
| ethane | $C_2H_6$ | $CH_3CH_3$ | |
| propane | $C_3H_8$ | $CH_3CH_2CH_3$ | |
| butane | $C_4H_{10}$ | $CH_3CH_2CH_2CH_3$ | |
| pentane | $C_5H_{12}$ | $CH_3CH_2CH_2CH_2CH_3$ | |

### KEY POINT

The alkanes are described as *saturated* hydrocarbons because they contain only C–C single bonds and have no double (C=C) or triple (C≡C) bonds. A way of thinking about this is that the alkanes are saturated with hydrogen, in other words they have the maximum number of hydrogen atoms possible for that number of carbon atoms. But do not write this definition down in the exam!

## ISOMERS OF ALKANES

Isomers are *compounds which have the same molecular formula but different structural formulae*. The alkanes shown in Table 24.1 can all be described as *straight-chain alkanes* or as having an unbranched chain. There are also isomers of butane and pentane that have branched chains. These are shown in Figure 24.2.

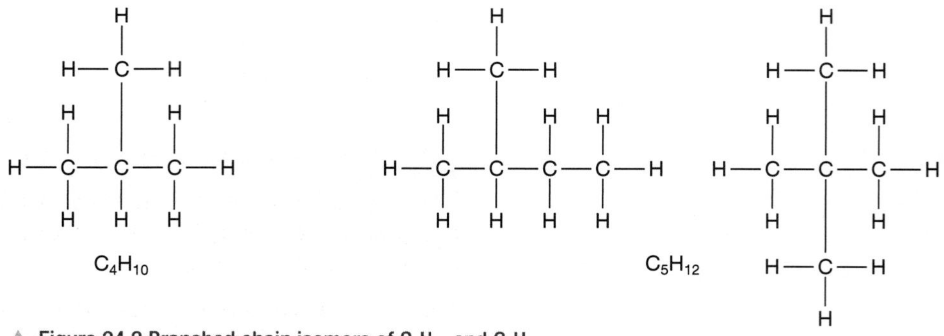

$C_4H_{10}$
$C_5H_{12}$

▲ Figure 24.2 Branched chain isomers of $C_4H_{10}$ and $C_5H_{12}$.

## HOMOLOGOUS SERIES

**HINT**

When asked about a homologous series in the exam, it is probably best to only write about the properties (the second and third bullet points). Sometimes you do not get separate marks for 'the same functional group' and 'similar chemical properties', and perhaps a better way of writing this is 'they have similar chemical properties because they have the same functional group'.

The alkanes form a homologous series. A homologous series is a series of compounds that:

- have the same functional group
- *have similar chemical properties*
- *show a trend (gradation) in physical properties*
- can be described by the same general formula
- differ from the next by a $-CH_2-$ unit.

The alkanes form the simplest homologous series. The alkanes do not really have a functional group as they just contain single C–C and C–H bonds, which are the basis of all other organic compounds.

### MEMBERS OF A HOMOLOGOUS SERIES HAVE THE SAME GENERAL FORMULA

In the case of the alkanes, if there are $n$ carbons, there are $2n + 2$ hydrogens.

The general formula for the alkanes is $\mathbf{C_nH_{2n+2}}$.

So, for example, if there are three carbons, there are $(2 \times 3) + 2 = 8$ hydrogens. The formula for propane is $C_3H_8$.

If you want the formula for an alkane with 15 carbons, you could work out that it is $C_{15}H_{32}$ and so on.

The reason that the alkanes and other homologous series can be described by a general formula is that each member differs from the next by a $-CH_2-$ unit.

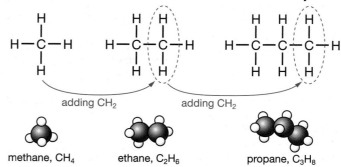

adding $CH_2$
adding $CH_2$
methane, $CH_4$
ethane, $C_2H_6$
propane, $C_3H_8$

▲ Figure 24.3 The three smallest alkanes. Each member differs from the next by a $-CH_2-$ unit.

## MEMBERS OF A HOMOLOGOUS SERIES SHOW A TREND (GRADATION) IN PHYSICAL PROPERTIES

Figure 24.4 shows the boiling points of the first eight alkanes.

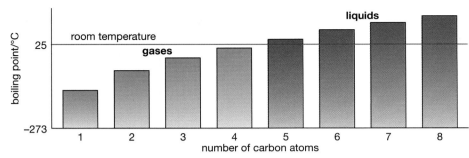

▲ Figure 24.4 Boiling points of the first eight alkanes

The first four alkanes are gases at room temperature (~25 °C). All the other alkanes you are likely to see at International GCSE are liquids. Solids start to appear at about $C_{18}H_{38}$.

The molecules of the members of a homologous series increase in size in a regular way and you can see that the *boiling points* also *increase* in a regular way.

As the molecules become bigger, the strength of the *intermolecular forces of attraction* between them *increases*. This means that *more energy* has to be put in to *break* the attractions between one molecule and its neighbours. One effect of this is that the boiling points increase in a regular way.

## MEMBERS OF A HOMOLOGOUS SERIES HAVE SIMILAR CHEMICAL PROPERTIES

Chemical properties are dependent on the functional groups and bonding within the molecules. Because alkanes only contain carbon–carbon single bonds and carbon–hydrogen bonds, they are all going to behave in the same way. These are strong bonds, therefore alkanes are fairly unreactive organic compounds and are often thought of as being quite inert.

### KEY POINT

The alkanes are not inert in the sense that they don't react with anything, like neon for example, but they are not very reactive for organic compounds and only really undergo two reactions.

# TWO REACTIONS OF THE ALKANES

## COMBUSTION

### HINT

Don't try to learn these equations! Practise working them out for a wide range of different hydrocarbons. You will find guidance on how to work out the ethane equation in Chapter 5, page 41.

### HINT

Remember, in combustion reactions it is highly unlikely that hydrogen, $H_2$, will be formed as a product because hydrogen is very flammable. You always obtain water, $H_2O$, even if there is insufficient oxygen.

All alkanes burn in air or oxygen. If there is enough oxygen, they burn *completely* to give *carbon dioxide* and *water*, for example:

$$CH_4(g) + 2O_2(g) \rightarrow CO_2(g) + 2H_2O(l)$$

$$2C_2H_6(g) + 7O_2(g) \rightarrow 4CO_2(g) + 6H_2O(l)$$

If there isn't enough oxygen, there is *incomplete* combustion of the hydrocarbon, and you obtain *carbon monoxide* or *carbon (soot)* instead of carbon dioxide, for example:

$$2C_2H_6(g) + 5O_2(g) \rightarrow 4CO(g) + 6H_2O(l)$$

or

$$2C_2H_6(g) + 3O_2(g) \rightarrow 4C(s) + 6H_2O(l)$$

**SUBSTITUTION**

Alkanes react with halogens in the presence of **ultraviolet radiation** (UV light), for example from sunlight. A hydrogen atom in the alkane is replaced by a halogen atom. This is known as a *substitution* reaction because *one atom has been replaced (or substituted) by a different one*.

**EXTENSION WORK**

UV light is required for substitution reactions to split halogen molecules into atoms. Energy is needed to break the covalent bonds between halogen atoms. The atoms formed are called *free radicals* due to the presence of unpaired electrons on them. These are very reactive species which then go on to react with the usually unreactive alkanes.

A mixture of methane and bromine gas is orange because of the presence of the bromine. If it is exposed to sunlight, it loses its colour, and a mixture of bromomethane and hydrogen bromide gases is formed.

$$CH_4(g) \ + \ Br_2(g) \ \rightarrow \ CH_3Br(g) \ + \ HBr(g)$$

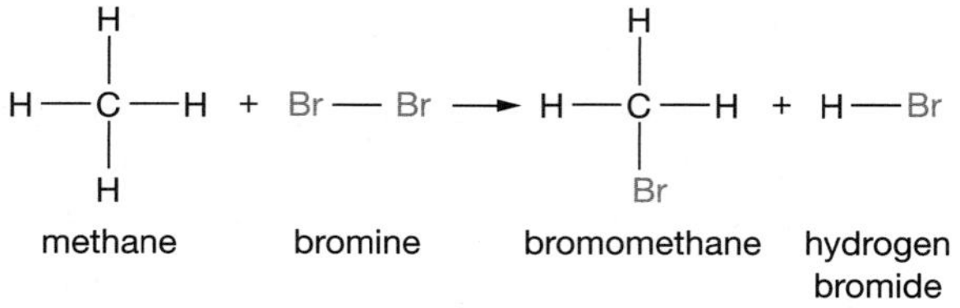

▲ Figure 24.5 A substitution reaction between methane and bromine.

**EXTENSION WORK**

If bromine is in excess, this reaction can go on to substitute more of the hydrogens in the methane. What you obtain is a mixture of $CH_3Br$, $CH_2Br_2$, $CHBr_3$ and $CBr_4$, plus HBr of course. You will find out more about *multi-substitution* if you go on to do chemistry at a more advanced level.

An exactly similar reaction happens between ethane and chlorine exposed to UV light. In this case, you get a mixture of chloroethane and hydrogen chloride gases:

$$CH_3CH_3(g) + Cl_2(g) \rightarrow CH_3CH_2Cl(g) + HCl(g)$$

**Mono-substitution** occurs when *only one hydrogen atom in the alkanes is replaced by a halogen atom*.

When propane reacts with bromine, it is possible to form two organic products even when only mono-substitution occurs. The two products are structural isomers of each other, they have the same molecular formula but different structural formulae.

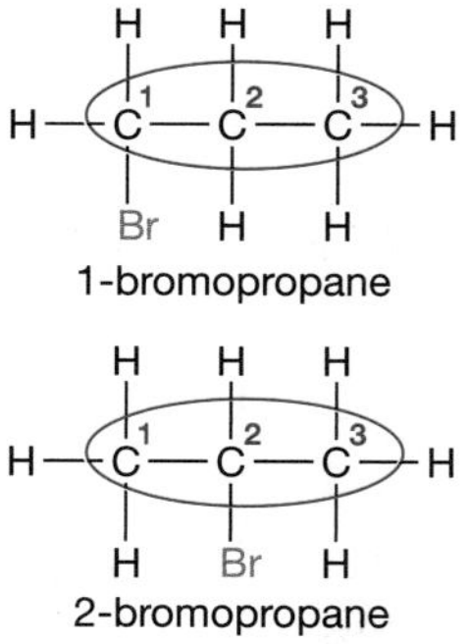

**HINT**

Notice the HCl(g) formed in this reaction is called hydrogen chloride rather than hydrochloric acid. To form an acid, hydrogen chloride gas must dissolve in water and dissociate to form H$^+$ ions (see Chapter 16).

▲ Figure 24.6 Two isomers of bromopropane, formed from the mono-substitution of propane with bromine.

The products formed from substitution reactions of alkanes with halogens are called *halogenoalkanes*. They are named according to the format *x-haloalkane*, where *x* indicates the position of the halogen atom in the longest carbon chain. So, when bromine reacts with propane we get 1-bromopropane and 2-bromopropane.

## CHAPTER QUESTIONS

SKILLS CRITICAL THINKING

1 a Alkanes are *saturated* hydrocarbons. Explain the term *saturated*.

   b Undecane is an alkane with 11 carbon atoms.

      i Write down the molecular formula for undecane.

SKILLS REASONING
      ii What physical state (solid, liquid or gas) would you expect undecane to be in at room temperature (25 °C)?

SKILLS PROBLEM SOLVING
      iii Write a balanced chemical equation for the complete combustion of undecane.

      iv Write a balanced chemical equation for the incomplete combustion of undecane to produce carbon monoxide and explain why this reaction is harmful to humans.

SKILLS ANALYSIS

2 Write the molecular formulae of the following compounds.

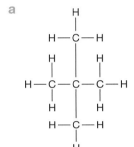
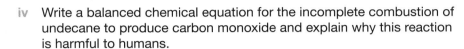
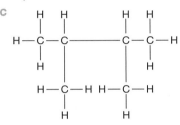

3 Draw two isomers of each of the following compounds.

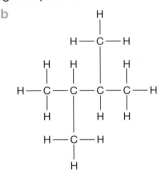

SKILLS CRITICAL THINKING

4 Ethane reacts with bromine under certain conditions.

   a State the conditions needed for this reaction to occur.

   b What is the type of reaction that occurs?

SKILLS PROBLEM SOLVING
   c Write a balanced chemical equation for the reaction using displayed formulae and name the organic product produced.

SKILLS INTERPRETATION

5 Butane reacts with chlorine to form two organic compounds with the formula $C_4H_9Cl$.

   a Write a balanced chemical equation for the formation of one of these compounds.

   b Draw the structures and give the names of the two possible *organic* products of this reaction.

# 25 ALKENES

The alkenes are another family of hydrocarbons. They all contain a carbon–carbon double bond. Alkenes can be produced from alkanes by cracking and they are much more reactive than the alkanes.

▶ Figure 25.1 Sunflower oil is polyunsaturated and contains lots of C=C groups.

## LEARNING OBJECTIVES

- Know that alkenes contain the functional group >C=C<
- Know the general formula for alkenes
- Explain why alkenes are classified as unsaturated hydrocarbons

- Describe the reactions of alkenes with bromine to produce dibromoalkanes
- Describe how bromine water can be used to distinguish between an alkane and an alkene

The alkenes are a homologous series of hydrocarbons which contain a carbon–carbon double bond. The **C=C** bond is the functional group of the alkenes.

The first three members of the alkene series are shown in Table 25.1.

Table 25.1 The first three alkenes

| Name | Molecular formula | Empirical formula | Structural formula | Displayed formula |
|---|---|---|---|---|
| ethene | $C_2H_4$ | $CH_2$ | $CH_2=CH_2$ | |
| propene | $C_3H_6$ | $CH_2$ | $CH_2=CHCH_3$ | |
| but-1-ene | $C_4H_8$ | $CH_2$ | $CH_2=CHCH_2CH_3$ | |

**KEY POINT**

It is not possible to have an alkene with one carbon atom because alkenes must contain a C=C group so you need a minimum of two carbon atoms.

## ISOMERS

There are two straight-chain isomers with the formula $C_4H_8$:

but-1-ene

but-2-ene

and one branched-chain isomer:

## UNSATURATED HYDROCARBONS

Alkenes are *unsaturated* compounds because they contain a C=C bond. *A saturated compound contains single C–C bonds only whereas an unsaturated compound contains one or more double or triple C–C bonds.*

### THE GENERAL FORMULA

Alkenes have the general formula $C_nH_{2n}$. The number of hydrogen atoms is twice the number of carbon atoms. This means that all alkenes have the same empirical formula, $CH_2$.

The alkene with 11 carbon atoms will have 22 hydrogen atoms and the molecular formula $C_{11}H_{22}$.

### PHYSICAL PROPERTIES

These are very similar to those of the alkanes. Remember that the small alkanes with up to four carbon atoms are gases at room temperature. The same is true for the alkenes. They are gases up to $C_4H_8$, and the next dozen or so are liquids. Again, the members of the homologous series *show a trend in physical properties*.

### CHEMICAL REACTIONS OF THE ALKENES

#### COMBUSTION

In common with all hydrocarbons, alkenes burn in air or oxygen to give carbon dioxide and water, for example:

$$C_2H_4(g) + 3O_2(g) \rightarrow 2CO_2(g) + 2H_2O(l)$$

This isn't a reaction anybody would ever choose to do. Alkenes are much too useful to waste by burning them.

#### ADDITION

The C=C double bond is the functional group. The functional group determines the chemical properties of a compound. Alkenes have characteristic chemical properties which are different from those of alkanes. Alkenes are much more reactive than alkanes.

*Alkenes undergo addition reactions.* Part of the double bond breaks to become a single C–C bond and the electrons are used to join other atoms onto the two carbon atoms.

## THE ADDITION OF BROMINE

Bromine adds to alkenes without any need for heat, light or a catalyst. The reaction is often carried out using bromine water (aqueous bromine solution) as, for example, in Figure 25.2, with ethene.

You know a reaction has happened because bromine water is *orange* but the product, called 1,2-dibromoethane, is a *colourless* liquid.

You can write this as an equation in two ways: first, using displayed formulae and second, using structural formulae.

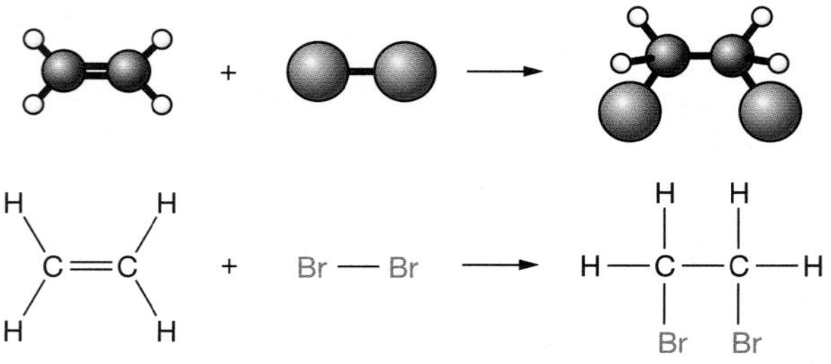

$CH_2{=}CH_2(g) + Br_2(aq) \rightarrow CH_2BrCH_2Br(l)$.

ethene + bromine → 1,2-dibromoethane

▲ Figure 25.2 An addition reaction: an alkene plus bromine

### HINT

Decode the name and make sure you can see how it works. 'eth' shows a two-carbon chain. 'ane' shows only carbon–carbon single bonds. The beginning of the name, '1,2-dibromo', shows two bromine atoms attached to the first and second carbon atoms.

### HINT

Make sure you can see how these formulae relate to the full structures in the equation above.

## THE TEST FOR UNSATURATED COMPOUNDS

Any compound with a carbon–carbon double bond will react with bromine in a similar way to ethene. This is used to *test for a carbon–carbon double bond*. If you shake an unknown organic compound with bromine water and the orange bromine water is *decolourised* (the colour changes from orange to colourless), the compound contains a carbon–carbon double bond. If your unknown compound is a gas, you can simply bubble it through bromine water with the same effect.

The left-hand tube in Figure 25.3 shows the effect of shaking a liquid alkene with bromine water. The organic layer (containing the alkene) is on top. You can see that the bromine water has been completely decolourised, showing the presence of the carbon–carbon double bond.

The right-hand tube in Figure 25.3 shows what happens if you use a liquid alkane, which doesn't have a carbon–carbon double bond. The colour of the bromine is still there. Alkanes do not decolourise bromine water because they do not contain a carbon–carbon double bond.

The addition reaction between bromine and propene is very similar to that of ethene. Part of the carbon–carbon double bond breaks and two bromine atoms add across the bond, as shown in Figure 25.4.

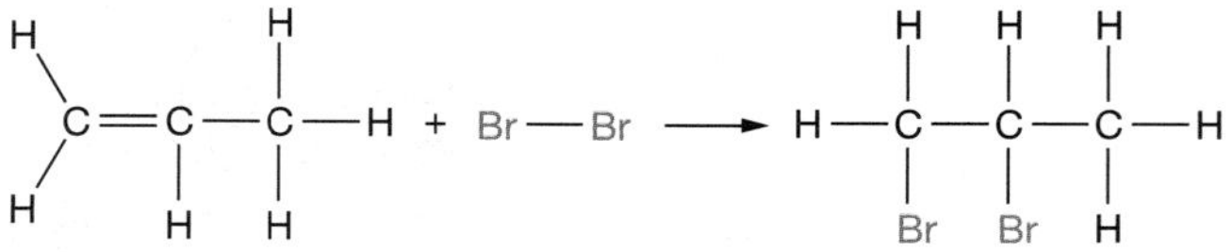

$CH_2{=}CHCH_3 + Br_2 \longrightarrow CH_2BrCHBrCH_3$

propene + bromine → 1,2-dibromopropane

▲ Figure 25.4 An addition reaction of propene with bromine

The product is called 1,2-dibromopropane. Its structure is based on propane and the prefix '1,2-dibromo' shows there are two bromine atoms, on the first and the second carbon atoms.

### HINT

You must be careful with the words you use here: the final mixture is *colourless* and not clear. Do not use the word clear. Clear means not cloudy, you can see through it. The bromine water is clear and any other solution can also be described as clear.

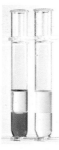

▲ Figure 25.3 The result of shaking a liquid alkene (left) or alkane (right) with bromine water.

### EXTENSION WORK

In Figure 25.3, interestingly, most of the colour in the right-hand tube is now in the top organic layer. That is because the covalent bromine, which has a simple molecular structure, is more soluble in the organic compound than it is in water.

## OTHER IMPORTANT ADDITION REACTIONS OF ALKENES

▲ Figure 25.5 Production of margarine

**EXTENSION WORK**

### Hydrogenation

A **hydrogenation** reaction involves the *addition of hydrogen*. An alkene reacts with hydrogen to form an alkane, for example:

$$CH_2{=}CH_2 \ + \ H_2 \ \longrightarrow \ CH_3CH_3$$

This reaction is used for the production of margarine, which is made by hydrogenation of polyunsaturated (i.e. many C=C bonds) plant oils.

## CHEMISTRY ONLY

### HYDRATION

A **hydration** reaction involves the *addition of water*. An alkene reacts with water to form an alcohol, for example:

$$CH_2{=}CH_2 \ + \ H_2O \ \longrightarrow \ CH_3CH_2OH$$

| | |
|---|---|
| Starting materials: | ethene and steam |
| Temperature: | 300 °C |
| Pressure: | 60–70 atmospheres (the atmosphere is a unit of pressure, equal to 101325 pascals) |
| Catalyst: | phosphoric acid |

This reaction is used for the production of high-purity ethanol in industry (see Chapter 26 for more detail on this reaction).

**END OF CHEMISTRY ONLY**

## CHAPTER QUESTIONS

**SKILLS**   CRITICAL THINKING    **6**

**SKILLS**   PROBLEM SOLVING, INTERPRETATION    **7**

    **5**

1 The alkenes are a *homologous series* of *unsaturated* hydrocarbons.

    a   i   State two characteristics of the members of a *homologous series*.
       ii   Explain the term *unsaturated*.

    b   Ethene is the simplest member of the alkenes. Bromine water can be used to distinguish alkenes from alkanes.
       i   Bromine water is added to ethene. State the starting colour of the bromine water and the finishing colour of the reaction mixture.
       ii   Complete the equation by drawing and naming the displayed formula of the product.

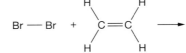

Name of the product:

**SKILLS** INTERPRETATION

c Isomers are compounds that have the same molecular formula but different displayed formulae. Draw the displayed formulae of three alkene isomers that have the molecular formula $C_4H_8$.

d Can you draw any other isomers of $C_4H_8$ which are *not* alkenes?

**SKILLS** CRITICAL THINKING

2 A gaseous hydrocarbon, **P**, with three carbon atoms, decolourises bromine water.

a Explain what the decolorisation of the bromine water tests for.

b Write the displayed formula for the hydrocarbon.

**SKILLS** PROBLEM SOLVING

c Write a balanced chemical equation for the reaction between the hydrocarbon and bromine, using displayed formulae.

d The reaction between ethane and bromine gas in the presence of UV light also causes the bromine to lose colour.

  i Write a balanced chemical equation for this reaction.

**SKILLS** REASONING

  ii This reaction is described as a substitution reaction, whereas the reaction between **P** and bromine is an addition reaction. Using the equations you have already written, explain the difference between addition and substitution.

**SKILLS** CRITICAL THINKING

3 Alkenes undergo addition reactions.

a Predict the structural formula of the compound formed when chlorine reacts with propene. Name the product formed.

b Ethene reacts with hydrogen. Give the name and structural formula of the product.

4 Propene can be made by heating propane and sulfur.

a Name another method of making propene and state two conditions for making alkenes from alkanes.

**SKILLS** EXECUTIVE FUNCTION

b Describe a chemical test which can be used to distinguish between propane and propene.

Test:

Result with propane:

Result with propene:

**SKILLS** CRITICAL THINKING

5 Give the structural formulae and names of the product when but-1-ene reacts with each of the following:

a steam

b hydrogen chloride (HCl)

c bromine.

# 26 ALCOHOLS

Alcohols are an important class of organic compounds. In this chapter we will study some of their reactions.

▶ Figure 26.1 Alcohols are very useful solvents.

## LEARNING OBJECTIVES

■ Know that alcohols contain the functional group –OH

■ Understand how to draw structural and displayed formulae for methanol, ethanol, propanol (propan-1-ol only) and butanol (butan-1-ol only), and name each compound

■ Know that ethanol can be oxidised by:
  • burning in air or oxygen (complete combustion)
  • reaction with oxygen in the air to form ethanoic acid (microbial oxidation)
  • heating with potassium dichromate(VI) in dilute sulfuric acid to form ethanoic acid

■ Know that ethanol can be manufactured by:
  • reacting ethene with steam in the presence of a phosphoric acid catalyst at a temperature of about 300 °C and a pressure of about 60–70 atm
  • the fermentation of glucose, in the absence of air, at an optimum temperature of about 30 °C and using the enzymes in yeast

■ Understand the reasons for fermentation in the absence of air and at an optimum temperature

What we commonly refer to as 'alcohol' is just one member of a large family (homologous series) of similar compounds. Alcohols all contain the functional group **–OH** covalently bonded to a carbon chain.

The familiar alcohol in drinks is $C_2H_5OH$ (or, better, showing the structure, $CH_3CH_2OH$) and should properly be called *ethanol*.

## DRAWING AND NAMING THE ALCOHOLS

We are only going to look at the four simplest alcohols: methanol, ethanol, propan-1-ol and butan-1-ol. These are shown in Table 26.1.

Table 26.1 Formulae and names of the first four members of the alcohol homologous series

| Name | Molecular formula | Structural formula | Displayed formula |
|------|-------------------|--------------------|-------------------|
| methanol | $CH_4O$ | $CH_3OH$ | |
| ethanol | $C_2H_6O$ | $CH_3CH_2OH$ | |
| propan-1-ol | $C_3H_8O$ | $CH_3CH_2CH_2OH$ | |
| butan-1-ol | $C_4H_{10}O$ | $CH_3CH_2CH_2CH_2OH$ | |

**HINT**

When drawing the displayed formula for alcohols you must show the bond between the O and the H, so O–H.

Look carefully at the names. The 'meth', 'eth', 'prop' and 'but' parts count *the number of carbon atoms*. The 'an' part tells you that the molecules are *saturated* and there aren't any carbon–carbon double bonds. The 'ol' tells you there is *an –OH functional group*.

Why the number 1 in propan-1-ol and butan-1-ol? The number tells you *where the –OH group is in the carbon chain*. You count the numbers from the end which gives you the smallest possible number in the name (see Table 26.1). Here, because the –OH group is attached to the end carbon, it is on carbon number 1. It is also possible to have propan-2-ol or butan-2-ol if the –OH is attached to a middle carbon. Propan-1-ol and propan-2-ol are structural isomers of each other, the same holds for butan-1-ol and butan-2-ol.

For the rest of this topic, we are just going to use ethanol as a representation of alcohols.

**HINT**

The names propanol and butanol are acceptable to use in the exam.

## THE OXIDATION OF ETHANOL

**ETHANOL BURNS IN AIR**

All alcohols burn to form carbon dioxide and water. With ethanol:

$$C_2H_5OH(l) + 3O_2(g) \rightarrow 2CO_2(g) + 3H_2O(l)$$

Ethanol is a **biofuel**, that is, *a fuel that is made from biological sources, such as sugar cane or corn*. Mixtures of petrol with ethanol are increasingly used in countries such as Brazil, which have little or no oil industry to produce their own petrol. On the other hand, they often have a climate which is good for growing sugar cane. Other countries are introducing biofuels such as ethanol to reduce dependence on fossil fuels, which are finite **non-renewable resources**.

▲ Figure 26.2 Ethanol being used as a fuel in Brazil.

## ETHANOL IS OXIDISED BY THE AIR IN THE PRESENCE OF MICROBES (MICROBIAL OXIDATION)

A bottle of wine left open to the air turns sour. The French for sour wine is *vin aigre*, which has been distorted into *vinegar*. The ethanol in the wine is oxidised by air with the help of microorganisms such as bacteria or yeast to form ethanoic acid, $CH_3COOH$.

### DID YOU KNOW?

The old name for ethanoic acid was acetic acid.

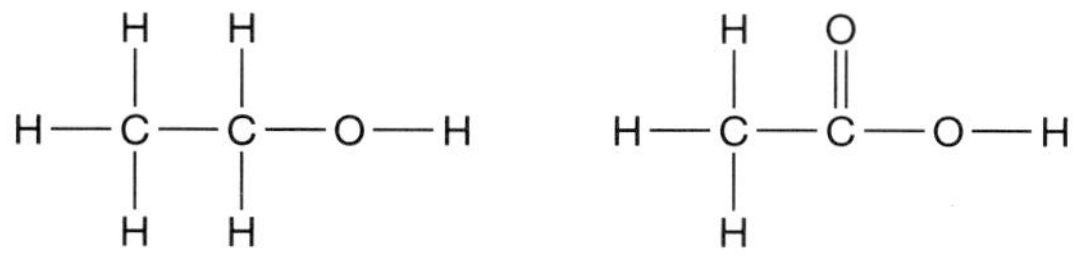

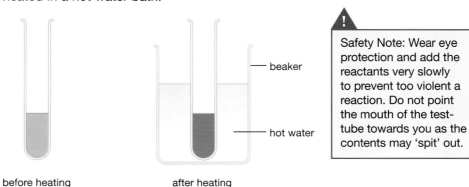

ethanol                    ethanoic acid

▲ Figure 26.3 Displayed formulae of ethanol and ethanoic acid

## ETHANOL CAN BE OXIDISED BY HEATING WITH POTASSIUM DICHROMATE(VI) IN DILUTE SULFURIC ACID

### EXTENSION WORK

In the name potassium dichromate(VI), the 'di-' tells you that there are two chromium atoms in the compound, and the roman numeral (VI) shows the *oxidation state* of chromium in this compound. We have seen these symbols before in Chapter 7, where they simply represent the charges on the ions, i.e. Cu(II) is $Cu^{2+}$ and Fe(III) is $Fe^{3+}$. The situation in the dichromate ion, $Cr_2O_7^{2-}$, is a little more complicated as there is covalent bonding between the chromium and oxygen atoms within the ion. You do not need to worry about the concept of oxidation state at International GCSE. However, it is important to add the (VI) after the name of potassium dichromate as there are other types of chromates which cannot oxidise alcohols.

This is the usual way of oxidising ethanol (and other alcohols) in the lab. The oxidising agent is a mixture of potassium dichromate(VI) and dilute sulfuric acid. Potassium dichromate(VI) is a strong oxidising agent and has the formula $K_2Cr_2O_7$. The dilute sulfuric acid is important for the potassium dichromate(VI) to act as an oxidising agent. Without the $H^+$ ions from the acid, no redox reaction will occur.

A few drops of ethanol are added to a solution containing the orange mixture of potassium dichromate(VI) and dilute sulfuric acid in a test-tube. The tube is heated in a hot water bath.

— beaker

— hot water

before heating          after heating

**!** Safety Note: Wear eye protection and add the reactants very slowly to prevent too violent a reaction. Do not point the mouth of the test-tube towards you as the contents may 'spit' out.

▲ Figure 26.4 Oxidation of ethanol by heating with potassium dichromate(VI) and dilute sulfuric acid.

The solution turns green and now contains a very dilute solution of ethanoic acid together with other products. The green colour indicates the presence of $Cr^{3+}$ ions, which are formed when the potassium dichromate(VI) is reduced. The equation for this reaction is very complicated, and so in organic chemistry we frequently show an oxidising agent in equations as [O].

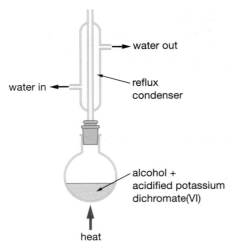

▲ Figure 26.6 Ethanol is refluxed with acidified potassium dichromate(VI) to form ethanoic acid.

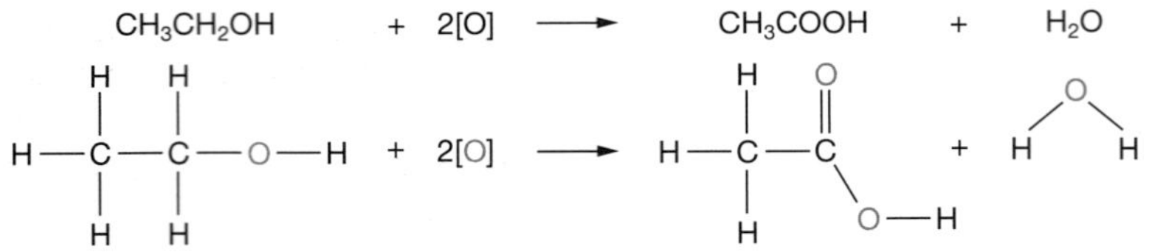

$$CH_3CH_2OH \quad + \quad 2[O] \quad \longrightarrow \quad CH_3COOH \quad + \quad H_2O$$

**Starting materials:** ethanol and potassium dichromate(VI) in dilute sulfuric acid

**Conditions:** heat under reflux

▲ Figure 26.5 Oxidation of ethanol to ethanoic acid

### EXTENSION WORK

You can use this reaction to make a sample of ethanoic acid by heating ethanol with an excess of the oxidising agent in a flask with a condenser placed vertically in the top (Figure 26.6). This is known as *heating under reflux* and condenses any vapours so that they run back into the flask and don't escape.

When the reaction is complete, the condenser is rearranged for a simple distillation. Ethanoic acid and water are both distilled and a solution of ethanoic acid is collected.

## THE PRODUCTION OF ETHANOL

Here we will look at two different process that are used industrially to make ethanol.

### MAKING ETHANOL BY FERMENTATION

▲ Figure 26.7 Sugar cane: one of the possible raw materials for making ethanol. Others include maize (called corn in the USA), wheat and other starchy materials.

Yeast is added to a sugar (or starch) solution and left in the warm (about 30 °C) for several days in the absence of air (anaerobic conditions). **Enzymes** (biological catalysts) in the yeast convert the sugar into ethanol and carbon dioxide. This process is known as **fermentation**.

The *absence of air* and the *temperature* are both important. In the presence of air (aerobic conditions), enzymes in the yeast produce carbon dioxide and water instead of ethanol. The enzymes are protein catalysts and if the temperature is increased much above 40 °C they lose their structure and don't work any longer. The proteins are said to be **denatured**. At a lower temperature, the reaction is too slow. Somewhere between 30 and 40 °C is the *optimum temperature* for the reaction.

### EXTENSION WORK

The biochemistry is very complicated. Assuming you are starting from ordinary sugar (sucrose), the sucrose is split into two smaller sugars, glucose and fructose. Glucose and fructose have the same molecular formulae, but different structures. They are structural isomers.

$$C_{12}H_{22}O_{11}(aq) + H_2O(l) \rightarrow C_6H_{12}O_6(aq) + C_6H_{12}O_6(aq)$$
sucrose          glucose     fructose

Enzymes in the yeast convert glucose into ethanol and carbon dioxide in a large number of steps. The overall equation for that conversion is

$$C_6H_{12}O_6(aq) \rightarrow 2C_2H_5OH(aq) + 2CO_2(g)$$
glucose        ethanol       carbon dioxide

Starting materials: glucose (from sugar cane)
Temperature: 30 °C
Catalyst: enzymes in yeast
Other conditions: anaerobic

**REMINDER**

You will find a diagram and discussion for a lab-based fractional distillation of an ethanol/water mixture in Chapter 2, page 18.

Yeast is killed by more than about 15% of alcohol in the mixture, and so it is impossible to make pure alcohol by fermentation. The alcohol is purified by fractional distillation. This takes advantage of the difference in boiling point between ethanol and water. Water boils at 100 °C whereas ethanol boils at 78 °C.

## MAKING ETHANOL BY THE HYDRATION OF ETHENE

Ethanol is also made by reacting ethene with steam, a process known as hydration.

$$CH_2{=}CH_2(g) + H_2O(g) \rightarrow CH_3CH_2OH(g)$$

Starting materials:   ethene and steam
Temperature:   300 °C
Pressure:   60–70 atmospheres
Catalyst:   phosphoric acid ($H_3PO_4$)

Only a small proportion of the ethene reacts. The ethanol produced is condensed as a liquid and the unreacted ethene is recycled through the process.

**KEY POINT**

This is an *addition reaction*: the water is added to the ethene without removing anything.

## COMPARING THE TWO METHODS OF PRODUCING ETHANOL

| | Fermentation | Hydration of ethene |
|---|---|---|
| Use of resources | Uses renewable resources: sugar beet or sugar cane, corn and other starchy materials | Uses finite, non-renewable resources: once all the oil has been consumed, there won't be any more |
| Type of process | A batch process: everything is mixed together in a reaction vessel and then left for several days<br><br>That batch is then removed and a new reaction is set up<br><br>This is inefficient | A continuous flow process: a stream of reactants is constantly passed over the catalyst<br><br>This is more efficient than a batch process |
| Rate of reaction | Slow, taking several days for each batch | Quick |
| Quality of product | Produces very impure ethanol which needs further processing | Produces much purer ethanol |
| Reaction conditions | Uses gentle temperatures and atmospheric pressure | Uses high temperatures and pressures, requiring a high input of energy |

At the moment, countries which have easy access to crude oil produce ethanol mainly from ethene, but one day there will be no oil remaining. At that point, the production of ethanol from sugar and starch will provide an alternative route to many of the organic chemicals we need.

**CHAPTER QUESTIONS**

SKILLS > INTERPRETATION     **7**

**1 a** Draw the structural formulae for the following compounds:

    **i** methanol             **ii** propan-1-ol

   **b** Draw the displayed formulae for these compounds:

    **i** ethanol               **ii** butan-1-ol

SKILLS > CRITICAL THINKING

   **c** Name the following compounds:

    **i** $CH_3CH_2CH_2CH_2OH$        **ii** $CH_3CH_2OH$

SKILLS > INTERPRETATION     **8**

**2 a** Draw a dot-and-cross diagram showing the bonding in a molecule of ethanol.

SKILLS > CRITICAL THINKING    **6**

   **b** Ethanol can be produced by the catalytic hydration of ethene. Describe two conditions used in this conversion.

   **c** In some countries ethanol is manufactured by a different method called fermentation.

**5**

    **i** State the raw material used to manufacture ethanol in this way.

SKILLS > REASONING    **7**

    **ii** Explain why the following conditions are important for making ethanol by the fermentation of glucose.

     ▪ Yeast is added into the reaction mixture.

     ▪ A temperature between 30 and 40 °C is used.

     ▪ Fermentation needs to be carried out in the absence of air.

**6**

    **iii** Explain why some countries, such as Brazil, manufacture ethanol in this way.

SKILLS > PROBLEM SOLVING    **9**

   **d** Ethanol is used as a biofuel in Brazil. Write a balanced chemical equation for the complete combustion of ethanol.

**3** Glucose can be converted to ethanoic acid in the following sequence of reactions:

$$\text{glucose} \xrightarrow{\text{reaction 1}} \text{ethanol} \xrightarrow{\text{reaction 2}} \text{ethanoic acid}$$

   **a** Write a balanced chemical equation for reaction 1 (the formula for glucose is $C_6H_{12}O_6$).

SKILLS > CRITICAL THINKING

   **b** Ethanoic acid can be made by the oxidation of ethanol in a lab. The ethanoic acid can be distilled off from the resulting mixtures and collected as a solution in water.

**5**

    **i** Give suitable reagents for reaction 2.

    **ii** Describe the colour change during reaction 2.

SKILLS > PROBLEM SOLVING    **9**

    **iii** Write a balanced chemical equation for this process, using [O] to represent the oxidising agent.

**8**

   **c** Ethanol can be converted to ethene and another product in a single step. Complete the equation below.

$$C_2H_5OH \rightarrow C_2H_4 + \underline{\hspace{2cm}}$$

**END OF CHEMISTRY ONLY**

# 27 CARBOXYLIC ACIDS

Previously you have seen how alcohols can be oxidised to form a class of organic compounds called carboxylic acids. Here we look at their structures and some of their reactions in more detail.

▲ Figure 27.1 Stinging nettles inject methanoic acid.

## LEARNING OBJECTIVES

■ Know that carboxylic acids contain the functional group

$$\begin{array}{c} \text{O} \\ \| \\ \text{—C—OH} \end{array}$$

■ Understand how to draw structural and displayed formulae for unbranched-chain carboxylic acids with up to four carbon atoms in the molecule, and name each compound

■ Describe the reactions of aqueous solutions of carboxylic acids with metals and metal carbonates

■ Know that vinegar is an aqueous solution containing ethanoic acid

Acids such as ethanoic acid are known as carboxylic acids and they all contain the functional group **–COOH**

$$\begin{array}{c} \text{O} \\ \| \\ \text{—C—OH} \end{array}$$

The most familiar one is ethanoic acid (old name *acetic acid*). Vinegar is *an aqueous solution containing ethanoic acid*.

Carboxylic acids are formed by oxidation of the corresponding alcohols, for example ethanoic acid is formed when ethanol is left in the air or oxidised using potassium dichromate(VI) in dilute sulfuric acid. If you need a reminder, have at look at Chapter 26, pages 289–290.

## DRAWING AND NAMING THE CARBOXYLIC ACIDS

Again, we are only going to be looking at the four simplest carboxylic acids: methanoic acid, ethanoic acid, propanoic acid and butanoic acid. These are shown in Table 27.1.

Table 27.1 Formulae and names of the first four members of the carboxylic acid homologous series

| Name | Molecular formula | Structural formula | Displayed formula |
|------|-------------------|--------------------|-------------------|
| methanoic acid | $CH_2O_2$ | HCOOH | 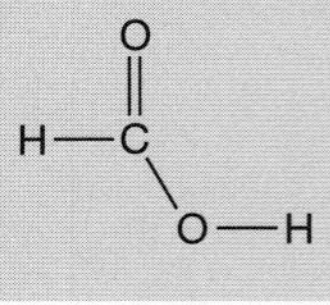 |
| ethanoic acid | $C_2H_4O_2$ | $CH_3COOH$ | 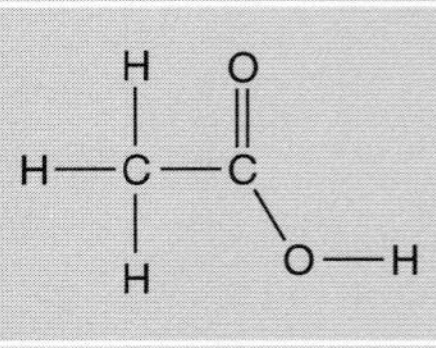 |
| propanoic acid | $C_3H_6O_2$ | $CH_3CH_2COOH$ | 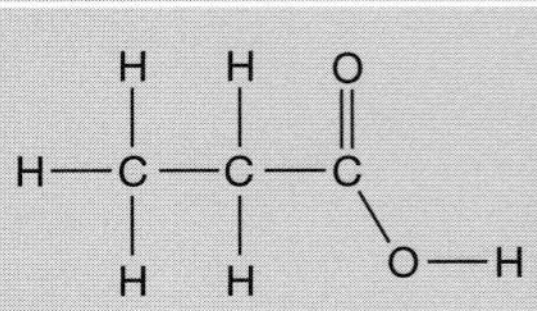 |
| butanoic acid | $C_4H_8O_2$ | $CH_3CH_2CH_2COOH$ | 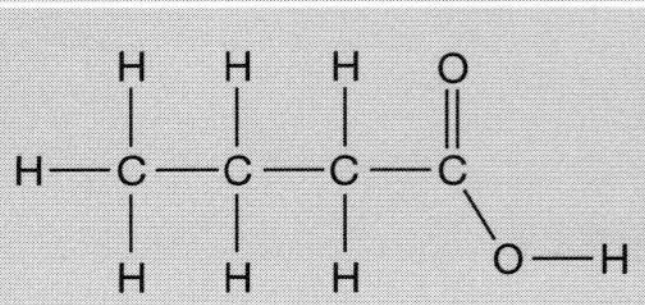 |

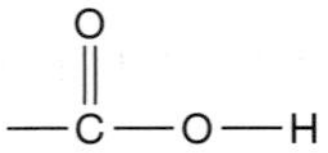

▲ Figure 27.2 The carboxylic acid functional group

## ACID PROPERTIES OF THE CARBOXYLIC ACIDS

Ethanoic acid and the other carboxylic acids are weak acids with a pH of about 3–5, depending on the concentration of the solution. They will turn blue litmus paper red, and react with all the things you expect acids to react with.

In the reactions described here, we are taking ethanoic acid as an example because it is the one you are most likely to use. All the others will behave similarly, and if necessary all you need to do is change the formula from $CH_3COOH$ to, for example, $CH_3CH_2COOH$.

### REACTION WITH METALS

▲ Figure 27.3 Reaction between magnesium ribbon and ethanoic acid.

Dilute ethanoic acid reacts with metals in the same way as other dilute acids such as hydrochloric acid or sulfuric acid, only *more slowly*. Remember *metals react with acids to give a salt and hydrogen* (see Chapter 17, page 174).

For example, dilute ethanoic acid reacts with magnesium with a lot of fizzing to produce a salt and hydrogen. The salt formed is magnesium ethanoate, which is soluble in water and so you get a colourless solution:

$$Mg(s) + 2CH_3COOH(aq) \rightarrow (CH_3COO)_2Mg(aq) + H_2(g)$$

Students frequently have problems with the formulae of salts like magnesium ethanoate. You have to remember that this is an ionic compound, and so contains magnesium ions, $Mg^{2+}$, and ethanoate ions, $CH_3COO^-$. That means that you need two ethanoate ions for every magnesium ion to make the charges balance. The metal is written after the ethanoate because that reflects the structure; the magnesium ion is attracted to the negative charge on the oxygen atoms in the ethanoate ions.

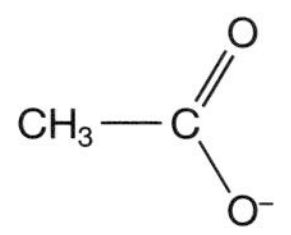

$$CH_3 - C \begin{matrix} O \\ \\ O^- \end{matrix}$$

▲ Figure 27.4 The structure of an ethanoate ion

Other metals behave as you would expect. Sodium is dangerously reactive; zinc produces bubbles of hydrogen more slowly than it does with hydrochloric acid, for example; metals below hydrogen in the reactivity series won't react.

Wherever a reaction takes place, a metal ethanoate will be formed as well as hydrogen. Zinc ethanoate is $(CH_3COO)_2Zn$.

### EXTENSION WORK

Ethanoic acid is a weak acid, whereas the acids we have met before, hydrochloric acid or sulfuric acid, are called strong acids. Students can have the misconception that strong acids are equivalent to concentrated acids. Whether an acid is strong or weak has, in fact, nothing to do with its concentration. Strong acids dissociate completely into their constituent ions in water, for example in hydrochloric acid you find mostly $H^+$ and $Cl^-$ ions and hardly any HCl molecules. In weak acids, the dissociation is only partial. About 1 in ~75 $CH_3COOH$ molecules will dissociate into $H^+$ and $CH_3COO^-$ in water. This is why if you have the same concentration of ethanoic acid and hydrochloric acid, the ethanoic acid reacts with metals or metal carbonates much more slowly than the hydrochloric acid as the concentration of the $H^+$ ions is much less.

## REACTION WITH CARBONATES

*Carbonates react with acids to give a salt, carbon dioxide and water* (Chapter 17, page 177), and ethanoic acid behaves like any other acid. For example, with sodium carbonate you get a lot of fizzing and a colourless solution of sodium ethanoate:

$$Na_2CO_3(s) + 2CH_3COOH(aq) \rightarrow 2CH_3COONa(aq) + CO_2(g) + H_2O(l)$$

With calcium carbonate you again get lots of fizzing and a colourless solution – this time containing calcium ethanoate:

$$2CH_3COOH(aq) + CaCO_3(s) \rightarrow (CH_3COO)_2Ca(aq) + CO_2(g) + H_2O(l)$$

## CHAPTER QUESTIONS

SKILLS   INTERPRETATION

1 a Draw the structural formulae for the following compounds:
    i ethanoic acid
    ii butanoic acid

  b Draw the displayed formulae for these compounds:
    i methanoic acid
    ii propanoic acid

  c Name the following compounds:
    i $CH_3CH_2CH_2COOH$
    ii $CH_3COOH$

2 What are the functional groups shown in the compound below (choose one answer from the table below)?

| | Alkene | Alcohol | Carboxylic acid |
|---|---|---|---|
| A | ✓ | ✓ | ✓ |
| B | ✓ | ✓ | ✗ |
| C | ✓ | ✗ | ✓ |
| D | ✗ | ✓ | ✓ |

3 Propan-1-ol can be oxidised by heating it under reflux with potassium dichromate(VI) in dilute sulfuric acid.

  a Draw the full displayed formula of the product of this reaction.

  b Describe the colour change during the reaction.

  c A student suggested the following apparatus for the oxidation of propan-1-ol. Explain why this method might be too dangerous.

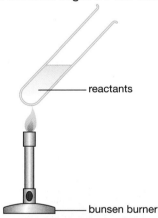

reactants

bunsen burner

  d Describe what you would see and write balanced chemical equations for the reactions if you added some of the product you collected to test-tubes containing:

   i magnesium ribbon          ii sodium carbonate solution.

4 Vinegar can react with baking soda, which contains sodium hydrogencarbonate ($NaHCO_3$), to give a salt, carbon dioxide and water.

  a Balance the following equation for the reaction between vinegar and baking soda.

$$\underline{\hspace{2cm}} + NaHCO_3 \rightarrow \underline{\hspace{2cm}} + CO_2 + H_2O$$

   vinegar    baking soda         a salt

  b Suggest the type of reaction occurring.

  c Suggest an observation that could be made during this reaction.

**END OF CHEMISTRY ONLY**

# 28 ESTERS

Alcohols, carboxylic acids and esters are organic compounds that contain oxygen as well as carbon and hydrogen. They are closely related. Carboxylic acids are made from oxidation of alcohols. In this chapter, you will learn how we can make esters by reacting alcohols with carboxylic acids. Esters are frequently used for perfumes and artificial flavours.

▲ Figure 28.1 Fats and oils contain the ester functional group.

## LEARNING OBJECTIVES

■ Know that esters contain the functional group

$$\begin{matrix} & O \\ & \| \\ -C & -O- \end{matrix}$$

■ Know that ethyl ethanoate is the ester produced when ethanol and ethanoic acid react in the presence of an acid catalyst

■ Understand how to write the structural and displayed formulae of ethyl ethanoate

■ Understand how to write the structural and displayed formulae of an ester, given the name or formula of the alcohol and carboxylic acid from which it is formed and vice versa

■ Know that esters are volatile compounds with distinctive smells and are used as food flavourings and in perfumes

■ Practical: Prepare a sample of an ester such as ethyl ethanoate

Esters are organic compounds formed by the reaction of an alcohol with a carboxylic acid. They have the functional group

$$\begin{matrix} & O \\ & \| \\ -C & -O- \end{matrix}$$

This can be written in condensed form as **–COO–**.

**DID YOU KNOW?**

**Ethyl ethanoate has quite a strong smell. It can be used as a solvent in nail polish and nail polish remover.**

## MAKING A SIMPLE ESTER: ETHYL ETHANOATE

Heating a mixture of ethanoic acid and ethanol with a few drops of concentrated sulfuric acid produces a liquid called ethyl ethanoate.

$$\underset{\text{ethanoic acid}}{CH_3COOH(l)} + \underset{\text{ethanol}}{CH_3CH_2OH(l)} \rightleftharpoons \underset{\text{ethyl ethanoate}}{CH_3COOCH_2CH_3(l)} + \underset{\text{water}}{H_2O(l)}$$

| | |
|---|---|
| Starting materials: | ethanol and ethanoic acid |
| Catalyst: | concentrated sulfuric acid |
| Conditions: | heat |

The concentrated sulfuric acid isn't written into the equation because it is a *catalyst* and it isn't consumed in the reaction.

The reaction is called **esterification**. It can also be described as a **condensation** reaction *because water is made when two molecules are joined together*. We will talk more about condensation reactions in the next chapter.

This reaction is *reversible* so to maximise the yield of ethyl ethanoate we often use pure ethanol and ethanoic acid (pure ethanoic acid is called *glacial ethanoic acid*). In Chapter 21 we discussed how the position of equilibrium always shifts to counteract any changes you make. If water is added to the reactants, the system would want to shift to the left to consume the water and therefore decrease the amount of ethyl ethanoate made. That is not what we want if we plan to make an ester!

## DRAWING AND NAMING ESTERS

In order to work out the structure of the ester formed when an alcohol and a carboxylic acid react together, it is easier to start by drawing the alcohol and carboxylic acid so that their OH groups are next to each other. Now remove $H_2O$ and join together what is left.

**EXTENSION WORK**

In an esterification reaction, an H from the alcohol and an OH from the acid are removed to form water. However, this distinction is not very important at International GCSE level. If you remove OH from the alcohol and H from the acid, you will obtain the same structure if you just join together what is left over.

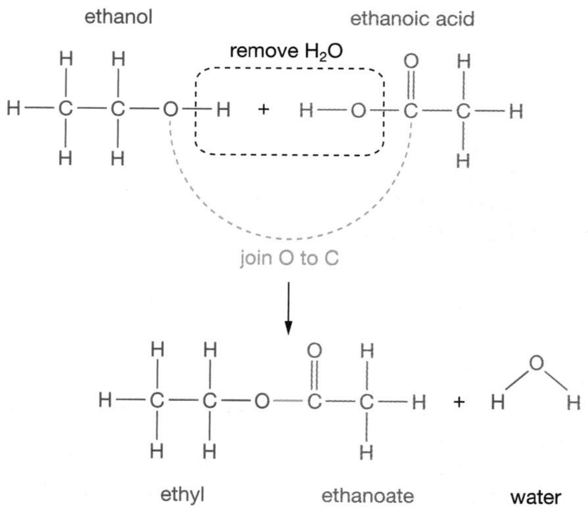


▲ **Figure 28.2 Formation of an ester**

An ester can be drawn either way round (see Figure 28.3) but the name is always written the same way round: ethyl ethanoate.

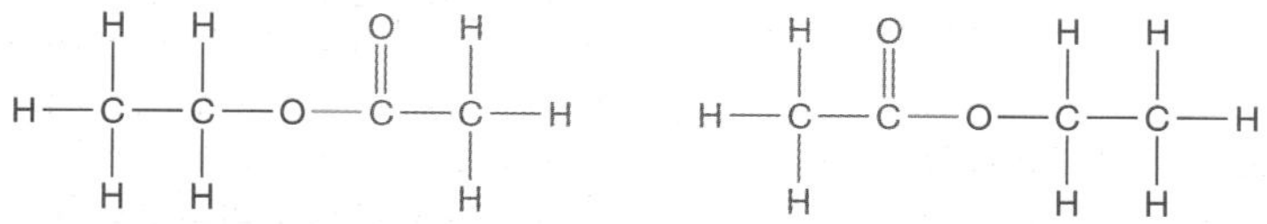

▲ **Figure 28.3 Different ways of drawing ethyl ethanoate**

**KEY POINT**

When naming esters the part with the C=O is always the 'oate' part.

Some other examples are shown in Table 28.1.

Table 28.1 Examples of esters

The colours in Table 28.1 are used to show how the esters are made. The parts in red come from the alcohols and the parts in blue come from the carboxylic acids.

It isn't actually important which C–O bond you break, as long as you add H to one side and OH to the other to give two OH groups you will get the same answer.

When working out the alcohol and carboxylic acid that an ester comes from, the 'yl' bit tells you the alcohol and the 'oate' bit tells you the carboxylic acid. So, propyl butanoate is made from propanol and butanoic acid.

Esters are often described as having a sweet, fruity smell.

You also need to be able to reverse this and work out which carboxylic acid and alcohol you would have to react together to produce a given ester. To do this, break apart the molecule between the C and O and add OH to one side and H to the other so that you have two OH groups.

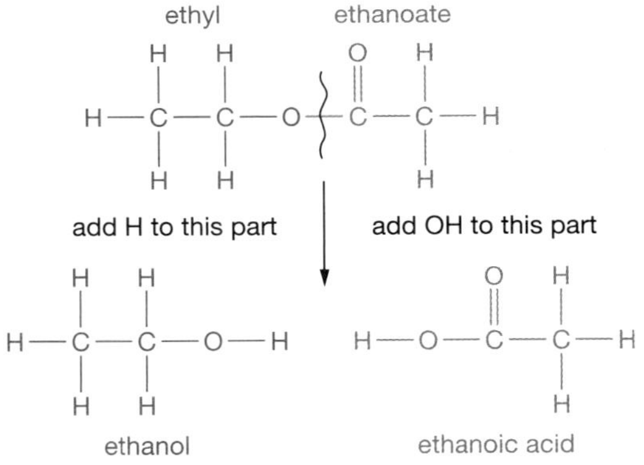

▲ Figure 28.4 Splitting an ester

$$CH_3COOCH_2CH_3(l) + H_2O(l) \rightleftharpoons CH_3COOH(l) + CH_3CH_2OH(l)$$

ethyl ethanoate          water          ethanoic acid          ethanol

## USES OF ESTERS

The small esters like ethyl ethanoate are commonly used in solvents, but most esters are used in food flavourings and perfumes. They are *volatile* liquids (*they turn to vapour easily*) with distinctive smells. The typical smell of bananas, raspberries, pears and other fruit is due in part to naturally occurring esters. Chemists create artificial flavourings and perfumes using mixtures of esters and other organic compounds.

**!**

Safety Note: Wear goggles and dispense the acids in a fume cupboard. Avoid all skin contact with the acids; wash off immediately with plenty of cold water if you get some on your skin. Carefully follow the instructions as to how to smell the esters produced.

### ACTIVITY 1

▼ **PRACTICAL: MAKING AND TESTING THE SMELL OF SOME ESTERS IN THE LAB**

▲ Figure 28.5 Cosmetics researcher smelling a perfume under development.

It is easy to produce esters in the lab on a test-tube scale by reacting various combinations of carboxylic acids and alcohols in the presence of a few drops of concentrated sulfuric acid. The mixture has to be heated, but not by a direct Bunsen flame because the contents of the tubes are flammable. The following procedure could be used:

■ Put 1 cm³ of ethanoic acid and 1 cm³ of ethanol into a boiling tube.

■ Mix well and carefully add a few drops of concentrated sulfuric acid.

■ Place the boiling tube in a beaker of hot water at about 80 °C for 5 minutes.

■ Allow the contents of the tube to cool.

■ When cool, pour the mixture into a beaker half-full of 0.5 mol/dm³ sodium carbonate solution. There will be some effervescence as the excess acid reacts with the metal carbonate. The reason for doing this is to be able to smell the ester better. Not much ester is likely to be formed in such a short time, and the smell of the ester in the tube is going to be masked by the smell of the carboxylic acid.

■ A layer of ester will separate and float on top of the water. The acid and alcohol will mostly dissolve in the water, but the ester won't. The small esters that you are likely to make in the lab float on the water, which makes them easy to smell.

■ Smell the product by gently wafting the odour towards your nose with your hand. Do not put your nose near the top of the beaker!

■ Repeat this procedure for other combinations of acids and alcohols. For example, you can make butyl butanoate, which should smell like pineapple, and propyl ethanoate, which smells like pear.

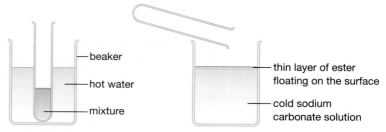

beaker

hot water

mixture

thin layer of ester floating on the surface

cold sodium carbonate solution

▲ Figure 28.6 Making esters on a test-tube scale. The ester is likely to be colourless but colour has been used in the diagram to make it easier to see.

## CHAPTER QUESTIONS

**SKILLS** ▶ INTERPRETATION

1 a Draw the structural formulae for the following compounds:

     i   ethyl ethanoate

     ii   butyl butanoate

   b Draw the displayed formulae for the following compounds:

     i   methyl butanoate      ii   ethyl propanoate

**SKILLS** ▶ CRITICAL THINKING

   c Name the following compounds:

     i   $CH_3CH_2COOCH_3$      ii   $HCOOCH_2CH_3$

**SKILLS** ▶ INTERPRETATION

2 Suppose you have access to bottles of ethanol, propan-1-ol, ethanoic acid, propanoic acid and concentrated sulfuric acid.

   a Draw the structural formulae of all the esters you could make from combinations of the alcohols and carboxylic acids.

**SKILLS** ▶ EXECUTIVE FUNCTION

   b Describe how you would make small samples of each of these esters.

   c Explain how you could detect the formation of the esters in these reactions.

**SKILLS** ▶ CRITICAL THINKING

   d Give two common uses for esters.

**SKILLS** ▶ CRITICAL THINKING, INTERPRETATION

3 Give the names and draw the structural formulae of the alcohols and carboxylic acids that could be used to make the following esters:

   a methyl propanoate

   b

```
        H    H    H         O
        |    |    |         ||
  H —— C —— C —— C —— O —— C —— H
        |    |    |
        H    H    H
```

   c butyl ethanoate

**SKILLS** ▶ INTERPRETATION

4 Ethyl ethanoate can be made from ethanol as the only organic starting material in a number of steps.

   a Draw the displayed formula for ethyl ethanoate.

**SKILLS** ▶ CRITICAL THINKING

   b Identify the carboxylic acid that is required for the esterification of ethanol to form ethyl ethanoate.

   c Describe how you could make the carboxylic acid in b from ethanol.

**SKILLS** ▶ PROBLEM SOLVING

   d Write a balanced chemical equation for the formation of ethyl ethanoate from ethanol and the carboxylic acid in b.

**SKILLS** ▶ CRITICAL THINKING

   e Write down the conditions required for the esterification step.

**SKILLS** ▶ EXECUTIVE FUNCTION

5 If you were given three unlabelled samples of ethanol, ethanoic acid and ethyl ethanoate, but weren't allowed to smell them, suggest how you might identify which one was which, with as much certainty as possible.

**END OF CHEMISTRY ONLY**

# 29 SYNTHETIC POLYMERS

Polymers are everywhere, ranging from naturally occurring polymers, for example proteins (polypeptides) and sugar (polysaccharides), which you probably have met before, to synthetic polymers used for plastic bags and clothing. Polymers are large molecules consisting of many repeating units. This chapter looks at two different ways that polymers can be made: addition polymerisation for polymers such as poly(ethene) and PVC, and condensation polymerisation for polymers such as polyester.

▲ Figure 29.1 Synthetic polymers are extremely useful for making a variety of products, from the coating for electrical wires to plastic bottles and intelligent label chips.

## LEARNING OBJECTIVES

- Know that an addition polymer is formed by joining up many small molecules called monomers

- Understand how to draw the repeat unit of an addition polymer, including poly(ethene), poly(propene), poly(chloroethene) and poly(tetrafluoroethene)

- Understand how to deduce the structure of a monomer from the repeat unit of an addition polymer and vice versa

- Explain problems in the disposal of addition polymers, including:
  - their inertness and inability to biodegrade
  - the production of toxic gases when they are burned

### CHEMISTRY ONLY

- Know that condensation polymerisation, in which a dicarboxylic acid reacts with a diol, produces a polyester and water.

- Understand how to write the structural and displayed formula of a polyester, showing the repeat unit, given the formulae of the monomers from which it is formed, including the reaction of ethanedioic acid and ethanediol:

$$n\text{H—O—C(=O)—C(=O)—O—H} + n\text{H—O—CH}_2\text{CH}_2\text{—O—H} \longrightarrow \left[\text{C(=O)—C(=O)—O—CH}_2\text{CH}_2\text{—O}\right]_n + 2n\text{H}_2\text{O}$$

- Know that some polyesters, known as biopolyesters, are biodegradable.

## ADDITION POLYMERISATION

**THE POLYMERISATION OF ETHENE**

Ethene is the smallest alkene and can be produced by cracking. It is the smallest hydrocarbon containing a carbon–carbon double bond. Figure 29.2 shows different ways of representing an ethene molecule.

$$C_2H_4 \qquad CH_2{=}CH_2$$

▲ Figure 29.2 Ethene

Under the right conditions, molecules containing carbon–carbon double bonds can join together to produce very long chains. Part of the double bond is broken to become a single bond, and the electrons in it are used to join to neighbouring molecules.

*Polymerisation is the joining up of lots of little molecules (monomers) to make one big molecule (polymer).* In the case of ethene, lots of ethene molecules join together to make *poly(ethene)*, which is more usually called *polythene. Molecules simply add onto each other without anything else being formed.* This is called **addition polymerisation**.

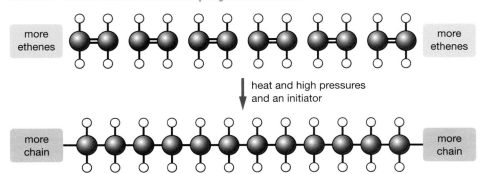

▲ Figure 29.3 The polymerisation of ethene

The chain length can vary from about 4000 to 40 000 carbon atoms.

For normal purposes, the polymerisation reaction is written using displayed formulae.

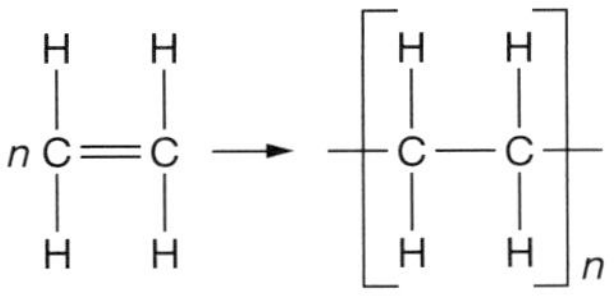

$$n\ \overset{H}{\underset{H}{C}}=\overset{H}{\underset{H}{C}} \longrightarrow \left[\overset{H}{\underset{H}{C}}-\overset{H}{\underset{H}{C}}\right]_n$$

▲ Figure 29.4 Formation of poly(ethene)

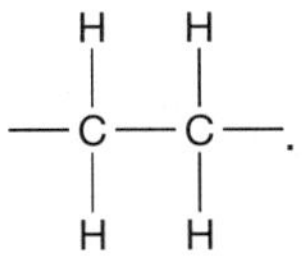

**KEY POINT**

An *initiator* is used to start the process. You mustn't call it a catalyst because it is consumed in the reaction.

**EXTENSION WORK**

People occasionally wonder what happens at the ends of the chains. They don't end tidily! Bits of the initiator are bonded on at either end. You don't need to worry about that for International GCSE.

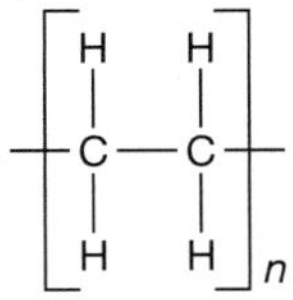

**KEY POINT**

In this structure for poly(ethene) $n$ represents a large but variable number. It simply means that the structure in the brackets (called the *repeat unit*) repeats itself many times in the molecule. When representing a polymer, the bonds on the two sides of the repeat unit should extend outside the square brackets to show that the carbon atoms are joined to the next ones in a chain, as shown in Figure 29.4.

**HINT**

Make sure you can distinguish between the different terms monomer, polymer and repeat unit. In the equation in Figure 29.4, the monomer is $\overset{H}{\underset{H}{C}}=\overset{H}{\underset{H}{C}}$,

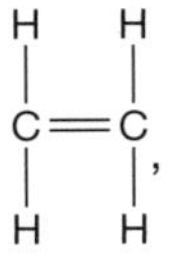

the polymer is $\left[\overset{H}{\underset{H}{C}}-\overset{H}{\underset{H}{C}}\right]_n$ and the repeat unit is $-\overset{H}{\underset{H}{C}}-\overset{H}{\underset{H}{C}}-$.

**USES FOR POLY(ETHENE)**

**KEY POINT**

Although the polymer is called *poly(ethene)* it is actually an *alkane*. The polymer is saturated and will not decolourise bromine water.

Poly(ethene) comes in two types: low-density poly(ethene) (LDPE) and high-density poly(ethene) (HDPE). Low-density poly(ethene) is mainly used as a thin film to make polythene bags. It is very flexible and not very strong.

▲ Figure 29.5 Plastic bags made from poly(ethene)

High-density poly(ethene) is used where greater strength and rigidity are needed, for example to make plastic bottles such as milk bottles. If you can find a recycling symbol with the letters HDPE next to it, then the bottle is made of high-density poly(ethene). If it has some other letters there, then it is made of a different polymer.

## HOW TO DEDUCE THE POLYMERISATION REACTION FOR ANY ALKENE

Figure 29.6 shows how to work out a polymerisation reaction starting from any alkene. The key thing is to always draw the alkene as shown with just the C=C in the middle.

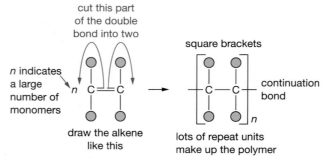

▲ Figure 29.6 How to work out a polymerisation reaction

**THE POLYMERISATION OF PROPENE**

Propene is another alkene, this time with three carbon atoms in each molecule. Its formula is normally written as $CH_3CH=CH_2$. Think of it as a modified ethene molecule, with a $CH_3$ group attached in place of one of the hydrogen atoms.

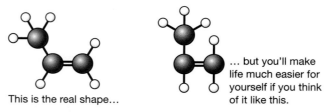

▲ Figure 29.7 Propene

When propene is polymerised you obtain *poly(propene).* This used to be called *polypropylene.*

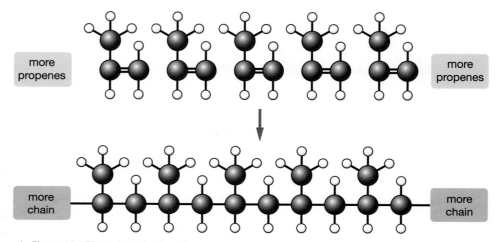

▲ Figure 29.8 The polymerisation of propene

Write this as:

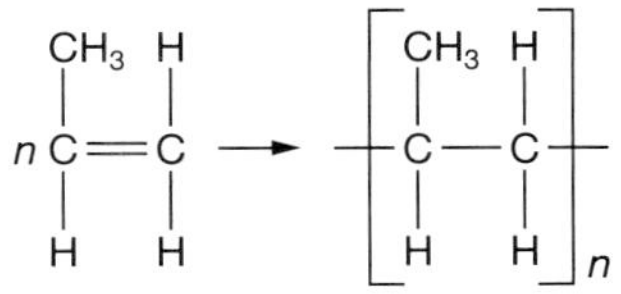

▲ Figure 29.9 Formation of poly(propene)

## USES OF POLY(PROPENE)

Poly(propene) is somewhat stronger than poly(ethene). It is used to make ropes and crates (among many other things). If an item has a recycling mark with PP inside it or near it, it is made of poly(propene).

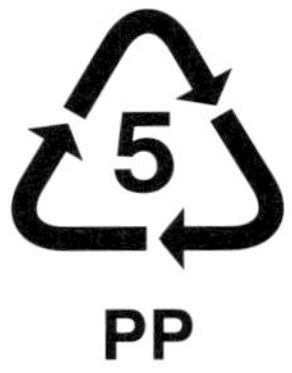

**PP**

▲ Figure 29.10 PP recycling symbol

▲ Figure 29.11 Poly(propene) is used to make crates . . .

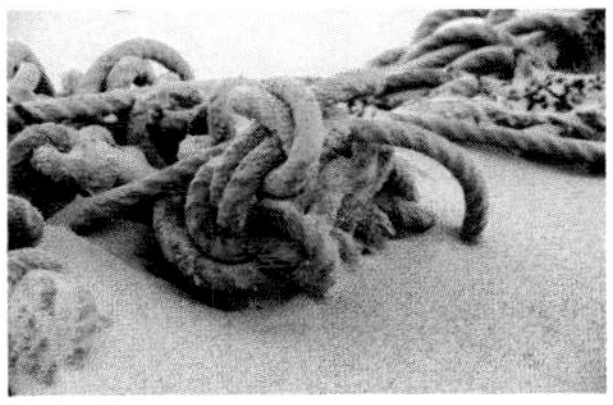

▲ Figure 29.12 . . . and ropes.

## THE POLYMERISATION OF CHLOROETHENE

Chloroethene is a molecule in which one of the hydrogen atoms in ethene is replaced by a chlorine. Its formula is $CH_2=CHCl$. In the past, it was called vinyl chloride. Polymerising chloroethene gives you poly(chloroethene). This is usually known by its old name, polyvinylchloride or PVC.

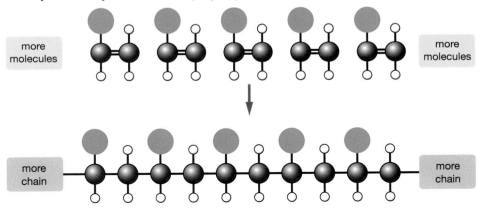

▲ Figure 29.13 The polymerisation of chloroethene

Write this as:

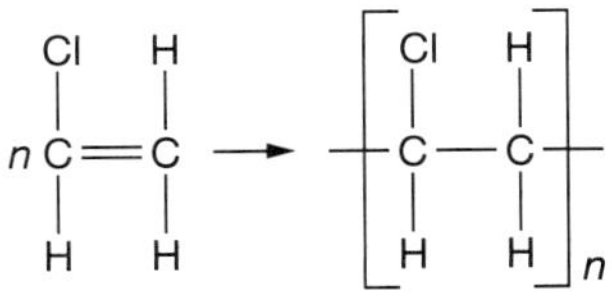

▲ Figure 29.14 Formation of poly(chloroethene)

## USES OF POLY(CHLOROETHENE)

Poly(chloroethene), known as PVC, has a lot of uses. It is quite strong and rigid, and so can be used for water pipes or replacement windows. It can also be made flexible by adding *plasticisers*. This makes it useful for sheet floor coverings, and even clothing. These polymers don't conduct electricity, therefore PVC can be used for electrical insulation.

▲ Figure 29.15 PVC has a variety of uses.

## THE POLYMERISATION OF TETRAFLUOROETHENE

Tetrafluoroethene is another molecule derived from ethene in which all four hydrogen atoms are replaced by fluorine. Its formula is $CF_2=CF_2$. Polymerising tetrafluoroethene gives you poly(tetrafluoroethene) or PTFE, more commonly known as Teflon®.

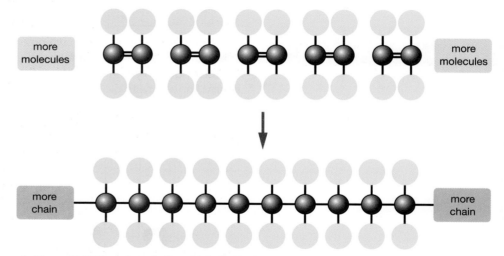

more molecules       more molecules

more chain       more chain

▲ Figure 29.16 The polymerisation of tetrafluoroethene

Write this as:

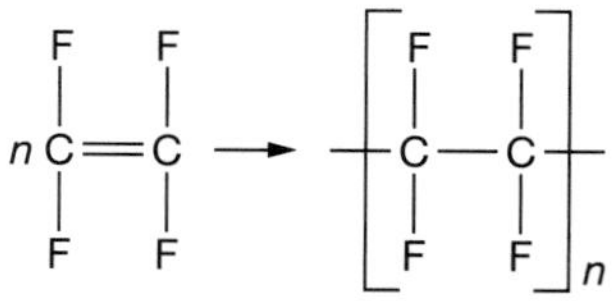

$$n\,C{=}C \longrightarrow \left[ \begin{array}{c} F \quad F \\ | \quad\ | \\ C{-}C \\ | \quad\ | \\ F \quad F \end{array} \right]_n$$

▲ Figure 29.17 Formation of poly(tetrafluoroethene)

## USES OF POLY(TETRAFLUOROETHENE)

Poly(tetrafluoroethene) is often used as a non-stick coating for pots and pans. It is very unreactive due to the strong carbon–fluorine bonds, and can be found lining containers for corrosive chemicals.

▲ Figure 29.18 PTFE is used for non-stick coatings.

## WORKING OUT THE MONOMER FOR A GIVEN ADDITION POLYMER

In an exam you may find that you are given the structure of a polymer and asked to deduce what the repeat unit is or what monomer it was made from. Figure 29.19 shows you how to do this.

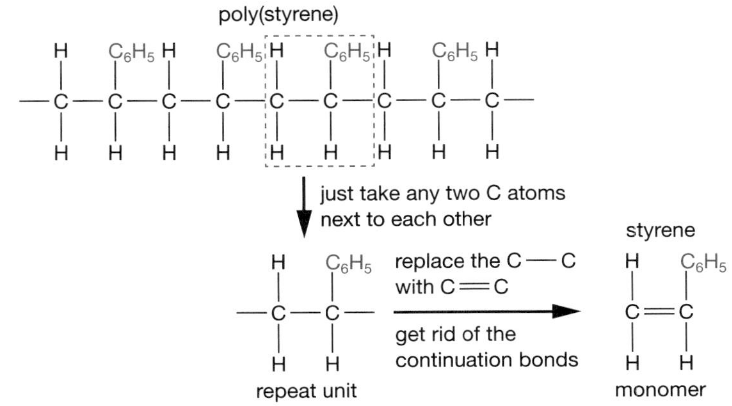

poly(styrene)

just take any two C atoms next to each other

replace the C — C with C = C

get rid of the continuation bonds

repeat unit

styrene

monomer

▲ Figure 29.19 How to work out the repeat unit and monomer for poly(styrene)

The $C_6H_5$ group is complicated, but we don't need to worry about that when we write the structure of the monomer; work from the structure you are given. First find the repeat unit, which can be done simply by taking any two adjacent carbon atoms in the main polymer chain. To derive the monomer, a double bond needs to be put back between the two middle carbon atoms of the repeat unit.

## DISPOSAL OF ADDITION POLYMERS

Recycling of plastics is important, not just to save raw materials, but because it takes a very long time for addition polymers like poly(ethene) and poly(chloroethene) to break down in the environment. They contain strong covalent bonds, making them essentially *inert* at ordinary temperatures. They are **non-biodegradable**, meaning *they cannot be broken down by bacteria in the environment*.

One solution to the problem of their disposal is to *bury them in landfill sites*, which are basically just big holes in the ground, where they will remain unchanged for thousands of years.

▲ Figure 29.20 Plastics (and other rubbish) can be disposed of in landfill sites.

Some countries, including Denmark and Japan, *incinerate (burn) plastics* in order to tackle the problem of disposal. This releases a lot of heat energy, which can be used to generate electricity. However, there are problems associated with this. For example, carbon dioxide is produced, which contributes to **global warming**. Toxic gases are also released, including carbon monoxide and hydrogen chloride.

There are advantages and disadvantages for using either method to dispose of plastics.

| Disposal method | Advantages | Disadvantages |
|---|---|---|
| Landfill | No greenhouse gases or toxic gases produced from plastics<br>Cheap | Ugly, smelly and noisy; no one wants to live next to a landfill site<br>Uses large areas of land<br>The waste will be there for thousands of years |
| Incineration | Requires little space<br>Can produce heat for local homes/offices and/or produce electricity | It is expensive to build and maintain the plant<br>Produces greenhouse gases<br>Releases toxic gases<br>The ash produced must still be disposed of in landfill sites |

## CHEMISTRY ONLY

## CONDENSATION POLYMERISATION

### MAKING A POLYESTER

When a lot of ethene molecules combine to make a chain, the double bonds break and the monomer molecules just add onto each other to make a polymer. *Nothing is lost*. That is why it is called *addition polymerisation*.

Making a polyester is different. Instead of one type of monomer, you often have two, joining together alternately. Each time two monomers combine, a small molecule such as water or hydrogen chloride is lost. The elimination of water gives the name *condensation* to this type of polymerisation.

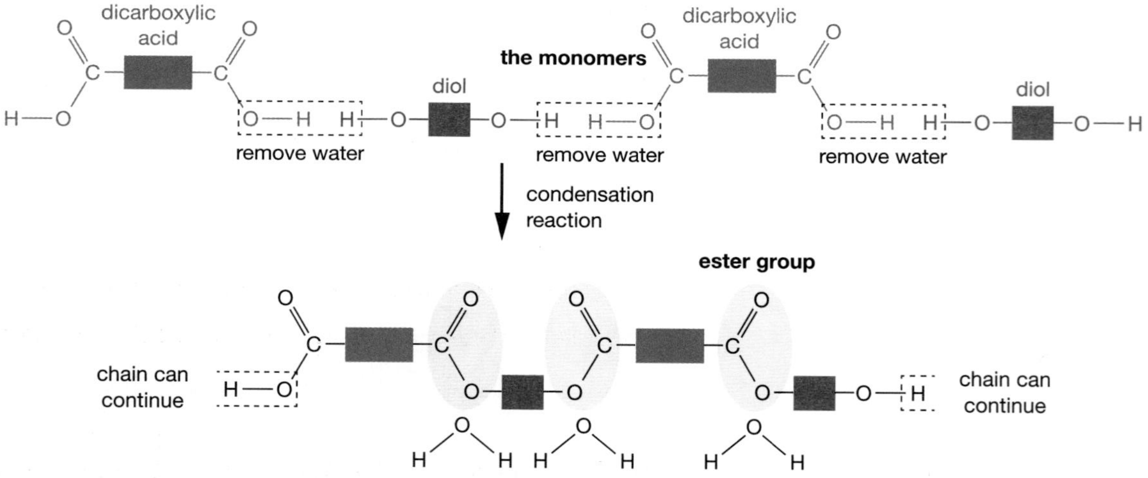


▲ Figure 29.21 Making a polyester

### THE MONOMERS

**KEY POINT**

It is not a mistake, there is an extra 'e' in each of these names. We talk about propan-1-ol but propan*e*-1,3-diol.

One of the monomers is a *diol*, an alcohol molecule with one –OH at each end. For example, ethane-1,2-diol has the structural formula $CH_2OHCH_2OH$. The prefix 'eth' indicates the presence of two carbon atoms in the molecule. The code 'di' in the name shows that there are two –OH groups present. The number '1,2' tells you the –OH groups are attached to carbons number 1 and 2, one on each carbon. Propane-1,3-diol, which has one more carbon in the carbon chain, has the structural formula $CH_2OHCH_2CH_2OH$.

The second type of monomer is a *dicarboxylic acid*, a carboxylic molecule with one –COOH at each end. For example, ethanedioic acid has the structural formula HOOCCOOH. A more complicated example of a dicarboxylic acid is hexane-1,6-dioic acid, with the formula $HOOCCH_2CH_2CH_2CH_2COOH$, where 'hex' indicates six carbon atoms in the main chain of the molecule and '1,6-dioic acid' means there are two carboxylic acid functional groups at either end of the molecule, on carbons 1 and 6.

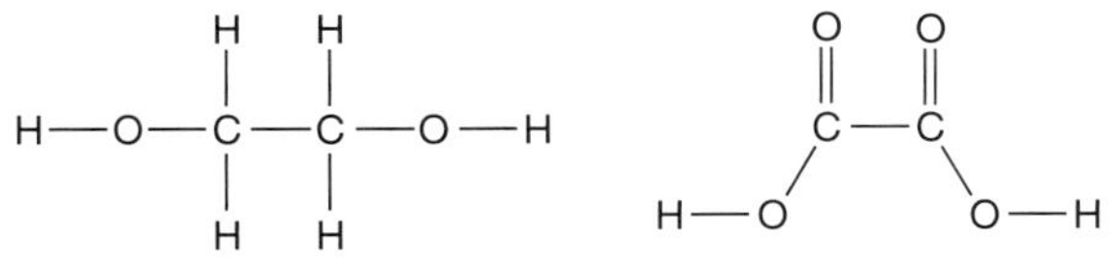

ethane-1,2-diol      ethanedoic acid

▲ Figure 29.22 Displayed formulae for ethane-1,2-diol and ethanedoic acid, monomers for a condensation polymerisation reaction to form a polyester.

## FORMING A POLYESTER

### HINT

It does not matter whether you take the OH from the dicarboxylic acid and the H from the alcohol or the other way around – the polymer formed will still be the same.

Under the right conditions, diol and dicarboxylic molecules can join together with the loss of a molecule of water each time a new bond is formed.

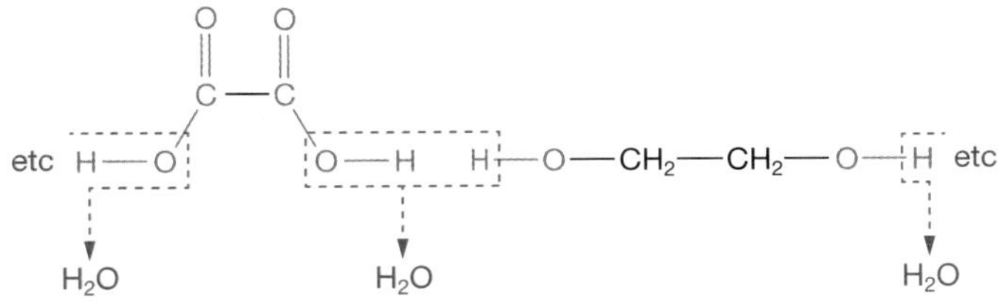

▲ Figure 29.23 Joining the monomers together...

Both monomers react at each end, joining together to form a long-chain polymer held together by a large number of ester groups, as shown in Figure 29.24.

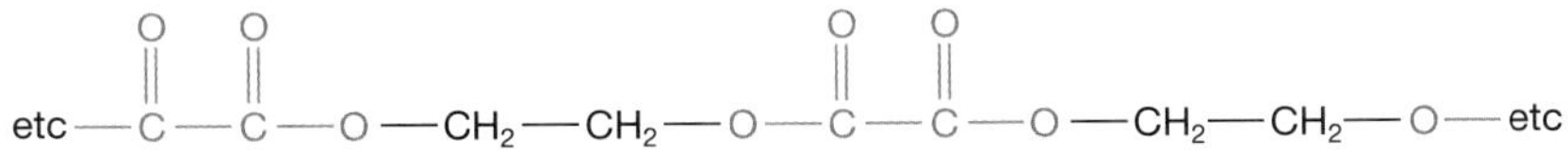

▲ Figure 29.24 ... to make a chain.

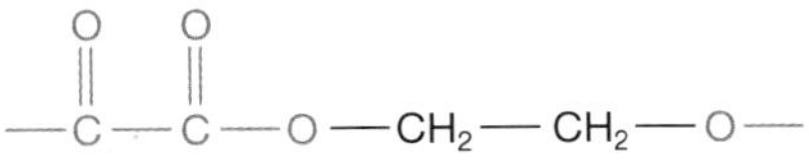

▲ Figure 29.25 Repeat unit of a polyester made from ethane-1,2-diol and ethanedoic acid.

The balanced chemical equation for this polymerisation is:

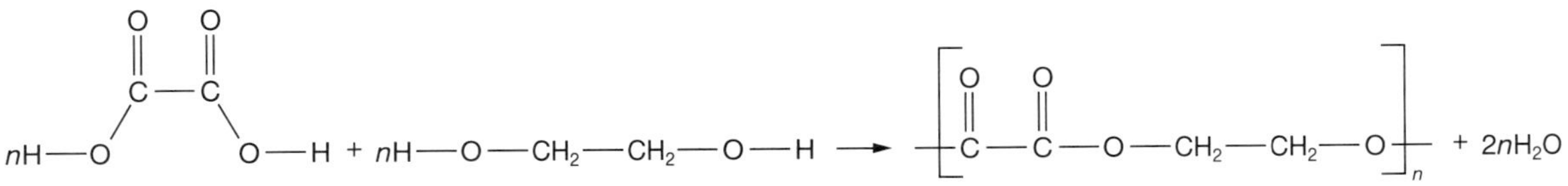

### EXTENSION WORK

The equation we have shown balances with $2n$ water molecules but in reality there will only be $(2n - 1)$ water molecules formed because there will still be one –OH and one –COOH left at the ends of the polyester molecule.

We have already seen one *block diagram* at the beginning of this section which shows the general condensation polymerisation reaction (Figure 29.21). A block diagram is a useful way of learning this because you can use the box to represent any hydrocarbon chain. For example, if we put in some carbons and hydrogens to make the reaction between hexane-1,6-dioic acid and ethane-1,2-diol, we obtain:

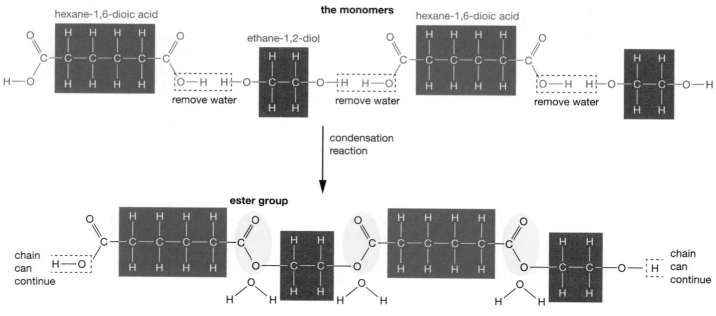

**the monomers**

hexane-1,6-dioic acid    ethane-1,2-diol    hexane-1,6-dioic acid

remove water    remove water    remove water

condensation reaction

ester group

chain can continue    chain can continue

**the polymer (and water molecules)**

▲ Figure 29.26 Using a block diagram to work out the structure of any polyester.

**BIODEGRADABLE POLYESTERS**

▲ Figure 29.28 A biodegradable plastic bag

As with addition polymers there are environmental issues with the disposal of condensation polymers. These are more reactive because of the ester linkage and will all break down eventually, although this could take hundreds of years. Scientists have developed some *biodegradable polyesters* which break down much more quickly. These are called **biopolyesters**.

Biodegradable polymers can be made from lactic acid. Lactic acid can be obtained from corn starch, and when it undergoes polymerisation it forms a biodegradable polyester (polylactic acid, or PLA), which can be used for making biodegradable plastic bags or in surgery for internal stitches.

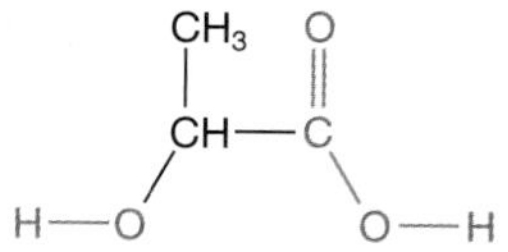

▲ Figure 29.27 Lactic acid monomer for making polylactic acid (PLA)

**END OF CHEMISTRY ONLY**

## CHAPTER QUESTIONS

1 Ethene, $C_2H_4$, is an important molecule used in industry. It is an *unsaturated hydrocarbon*.

   a Explain the terms *unsaturated* and *hydrocarbon*.

   b Ethene burns easily and releases a large amount of heat during combustion. Write a balanced chemical equation for the complete combustion of ethene.

   c Explain why in industry ethene is considered too valuable to be used as a fuel.

2 Propene, $C_3H_6$, can be polymerised to make poly(propene).

   a Explain the term *polymerisation*.

   b Draw a displayed formula for propene.

   c Draw a diagram to represent the structure of a poly(propene) chain showing three repeat units.

   d The formation of poly(propene) is an example of *addition* polymerisation. Explain what is meant by the word *addition*.

   e Styrene has the formula $C_6H_5CH=CH_2$. Write a balanced chemical equation to show what happens when styrene is polymerised to make polystyrene. Your equation should clearly show the structure of the polystyrene. (Show the $C_6H_5$ group as a whole.)

   f A small part of a Perspex molecule looks like this:

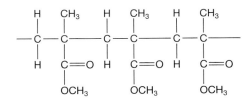

   Draw the structure of the monomer from which Perspex is made.

   g Many addition polymers, for example those used for making plastic bags, are not biodegradable. Used plastic bags can be either buried underground in landfill sites or incinerated. State which method of disposal you think is better and give two reasons for your choice.

## CHEMISTRY ONLY

3 A polymer has the structure shown below:

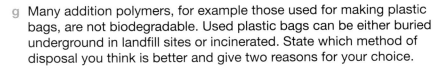

   a Identify the type of the polymer.

   b Draw the structures of the monomers that are used to make this polymer.

4 Polyesters such as Terylene (for clothes) or PET (commonly used to make drinks bottles) are made by condensation polymerisation from ethane-1,2-diol and terephthalic acid (properly known as benzene-1,4-dicarboxylic acid). The structures of these are:

Ethane-1,2-diol: $HO-CH_2CH_2-OH$

Terephthalic acid: $HO-\underset{\underset{\displaystyle O}{\|}}{C}-C_6H_4-\underset{\underset{\displaystyle O}{\|}}{C}-OH$

a Draw a chain of the polymer with two repeat units.

b Write a balanced chemical equation for the formation of PET.

*(The structure of the $C_6H_4$ group is complicated, and you can write it simply as $C_6H_4$. You can draw this as a block diagram if you wish, but it is more satisfying (and no more difficult in this case) to draw the structure properly.)*

5 *(This question contains new material and is designed to look difficult. It is actually not too difficult as long as you understand about polyester.)* Nylon-6,6 is made by a condensation polymerisation of the monomers 1,6-diaminohexane ($H_2NCH_2CH_2CH_2CH_2CH_2CH_2NH_2$) and hexane-1,6-dioic acid. The amine group $-NH_2$ reacts in a very similar manner to the $-OH$ group in an alcohol during condensation polymerisation with carboxylic acids, forming amide functional groups which join the monomers together into a polymer chain.

$$-\underset{\underset{\displaystyle H}{|}}{\underset{\displaystyle N}{}}-\quad\text{with}\quad-\underset{\underset{\displaystyle O}{\|}}{C}-$$

a i Explain what is meant by *condensation polymerisation* and how it differs from *addition polymerisation*.

ii Using a block diagram, draw one repeat unit for nylon-6,6.

b Nylon-6,10 is made from 1,6-diaminohexane and a longer chain acid, decanedioic acid, containing a total of 10 carbon atoms: $HOOCCH_2CH_2CH_2CH_2CH_2CH_2CH_2CH_2COOH$.

i How will a chain of nylon-6,10 differ from one of nylon-6,6? Refer to the diagram you drew in a ii.

ii In what way(s) will the two chains be the same? Again, refer to the diagram you drew in a ii.

**END OF CHEMISTRY ONLY**

# UNIT QUESTIONS

SKILLS ANALYSIS

**1** Crude oil is a complex mixture of hydrocarbons. The diagram shows the separation of crude oil into simpler mixtures called fractions.

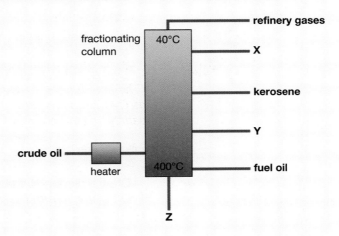

a What could X, Y and Z represent (choose one answer)? **(1)**

|   | X | Y | Z |
|---|---|---|---|
| **A** | gasoline | bitumen | diesel |
| **B** | diesel | gasoline | bitumen |
| **C** | bitumen | gasoline | diesel |
| **D** | gasoline | diesel | bitumen |

SKILLS CRITICAL THINKING

b State a use for the refinery gas fraction. **(1)**

c Name the liquid that leaves the fractionating column at the lowest temperature. **(1)**

d Describe how crude oil is separated into fractions in industry. **(4)**

e Write down and explain the relationship between the number of carbon atoms in a hydrocarbon and its boiling point. **(2)**

SKILLS PROBLEM SOLVING

f One of the hydrocarbons, $C_{15}H_{32}$, called pentadecane, is present in the kerosene fraction. It could be used as a fuel in jet engines. Write an equation for the incomplete combustion of pentadecane to produce carbon monoxide. **(2)**

g The complete combustion of alkanes produces carbon dioxide, $CO_2$, which can dissolve in water to form a weakly acidic solution.

SKILLS INTERPRETATION

i Draw a dot-and-cross diagram to show the bonding in $CO_2$. Show the outer electrons only. **(2)**

SKILLS REASONING

ii Predict the most likely pH of a solution of $CO_2$ (choose one answer). **(1)**

**A** 1 **B** 5 **C** 7 **D** 9

h $C_{15}H_{32}$ is a member of the alkane family. Long-chain alkanes are often cracked to form alkenes.

   i Write down two conditions used for cracking alkanes in industry. **(2)**

   ii Give two reasons why cracking is necessary in the oil industry. **(2)**

   iii During cracking, one molecule of $C_{15}H_{32}$ is converted into one molecule of ethene, one molecule of another alkene $C_4H_8$ and one molecule of hydrocarbon **W**. State the molecular formula of **W**. **(1)**

   iv Alkenes are unsaturated hydrocarbons. Write down what is meant by the term *unsaturated*. **(1)**

   v One possible structure for $C_4H_8$ is

$$H_2C=CH-CH_2-CH_3$$

Draw a displayed formula for each of the other two isomers of $C_4H_8$, which are also alkenes. **(2)**

**(Total 22 marks)**

**2** The table shows the formulae of some organic compounds.

| A | B | C |
|---|---|---|
| $CH_2=CHCl$ | $CH_3CH_2CH_3$ | $CH_2=CHCH_2CH_3$ |
| D | E | F |
| $C_2H_6$ |  CH₃CH₂OH displayed | $CH_2BrCHBrCH_2$... displayed |

a Give one reason why compound **E** is *not* a hydrocarbon. **(1)**

b i State the name of compound **F**. **(1)**

   ii Draw the displayed formula of an isomer of compound **F**. **(2)**

c Compound **C** can undergo an addition reaction with bromine. Complete the equation below showing the structural formula of the product(s). **(1)**

$$CH_2=CHCH_2CH_3 + Br_2 \rightarrow$$

d Two of the compounds shown belong to the same homologous series.

   i Give the letters of these two compounds. **(1)**

   ii Write down the general formula of this homologous series. **(1)**

   iii Write down *two* other characteristics of the compounds in a homologous series. **(2)**

e Compound **A** can undergo addition polymerisation.

   i Give the name of the addition polymer formed by compound **A**. **(1)**

   ii Draw the repeat unit of the addition polymer of compound **A**. **(1)**

SKILLS ▶ REASONING

iii Explain the problems in the disposal of the addition polymer formed by compound **A**. **(2)**

iv Write down and explain the observations you would make if bromine water was added to the addition polymer of **A**. **(2)**

SKILLS ▶ CRITICAL THINKING

f Compound **B** undergoes a reaction with chlorine.

i Write down the condition required for this reaction to take place. **(1)**

ii Write down the type of reaction that occurs. **(1)**

SKILLS ▶ INTERPRETATION

iii Draw the displayed formulae for the two possible organic products from the reaction between compound **B** and chlorine. **(2)**

**(Total 19 marks)**

SKILLS ▶ PROBLEM SOLVING  **3**

Compound **X** contains 55% carbon, 9% hydrogen and 36% oxygen by mass.

a Calculate the empirical formula of **X**. **(3)**

b The relative molecular mass ($M_r$) of **X** is 88. Deduce the molecular formula of **X**. **(1)**

## CHEMISTRY ONLY

SKILLS ▶ INTERPRETATION

c **X** belongs to a class of compound called the carboxylic acids. Draw the displayed formula of **X** and name the compound. **(2)**

SKILLS ▶ PROBLEM SOLVING

d Write a balanced equation for the reaction between **X** and methanol. Name the organic product formed. **(3)**

### END OF CHEMISTRY ONLY

**(Total 9 marks)**

SKILLS ▶ EXECUTIVE FUNCTION  **4**

Below are two hydrocarbons with the molecular formula $C_3H_6$.

compound 1          compound 2

a Describe a chemical test for distinguishing between the two hydrocarbons and give the results for this test. **(3)**

Test:

Result with compound 1:

Result with compound 2:

SKILLS ▶ CRITICAL THINKING

b Compound 3 is in the same homologous series as compound 1. Compound 3 can form an addition polymer. Draw the repeat unit of this polymer. **(2)**

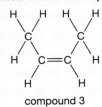

compound 3

SKILLS ▶ CRITICAL THINKING

c Name the addition polymer formed by compound 3. **(1)**

**(Total 6 marks)**

# CHEMISTRY ONLY

SKILLS ▶ INTERPRETATION  **5**

Esters occur naturally, but can be made in the laboratory from an alcohol and a carboxylic acid. Esters have many uses.

a Butyl ethanoate is an important industrial solvent.

   i Draw the displayed formula of butyl ethanoate. **(2)**

SKILLS ▶ CRITICAL THINKING

   ii Butyl ethanoate can be made from butanol and ethanoic acid. Give the conditions for making ethanoic acid from ethanol in a lab. **(3)**

SKILLS ▶ INTERPRETATION

b Esters can be used in perfumes. One of them, allyl hexanoate, is used because it smells of pineapple. Draw out the structural formulae of the alcohol and the carboxylic acid used to make allyl hexanoate. **(2)**

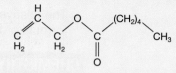

SKILLS ▶ CRITICAL THINKING

c Imagine that you wish to prepare ethyl ethanoate according to the reaction:

$$CH_3COOH(l) + CH_3CH_2OH(l) \rightleftharpoons CH_3COOCH_2CH_3(l) + H_2O(l)$$

   i Write down the meaning of the sign $\rightleftharpoons$. **(1)**

   ii The reaction is catalysed in order to increase the rate of the reaction. State the name of the catalyst used for this reaction in a laboratory. **(1)**

d Write down and explain the effect of the catalyst on the position of equilibrium in the reaction in **c**. **(2)**

SKILLS ▶ PROBLEM SOLVING

e Fats are naturally occurring esters. The diagram below shows one type of fat in our body, a triglyceride.

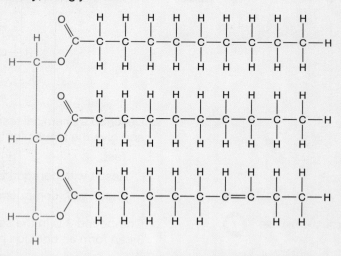

Triglycerides are made from the alcohol glycerol and different long-chain carboxylic acids. The structure of glycerol is shown below. Complete and balance the equation for the reaction between glycerol and ethanoic acid. **(3)**

glycerol           ethanoic acid

**(Total 14 marks)**

**END OF CHEMISTRY ONLY**

**SKILLS** ▸ INTERPRETATION   8   **6** ▸ This question is about synthetic polymers. Synthetic polymers can be either addition polymers or condensation polymers.

a Poly(propene) is an addition polymer made from propene $CH_2=CHCH_3$. Draw a length of the poly(propene) chain showing two repeat units. **(2)**

b Another addition polymer, polyacrylamide, is used to make hydrogel in contact lenses, nappies and drug delivery systems. The structure of the polymer is shown in the diagram. Draw the structural formula of the monomer used to make polyacrylamide.

$$\left[ CH_2-CH \atop \quad\quad\quad C=O \atop \quad\quad\quad NH_2 \right]_n$$

**(1)**

**CHEMISTRY ONLY**

c Polyesters are condensation polymers.

**SKILLS** ▸ CRITICAL THINKING   7

i Describe how condensation polymerisation is different from addition polymerisation. **(2)**

**SKILLS** ▸ INTERPRETATION   8

ii The structures of two monomers that are used to make a polyester are:

$$H-O-\overset{O}{\underset{\|}{C}}-CH_2-\overset{O}{\underset{\|}{C}}-O-H \qquad HO-CH_2-CH_2-OH$$

Draw the structure of the repeat unit of the polyester formed from these two monomers. **(2)**

d Polylactic acid is a biopolyester. Lactic acid has the structure shown below.

**SKILLS** ▸ CRITICAL THINKING

$$H-O-\overset{\textcircled{1}\,\,H}{\underset{\underset{H}{\overset{|}{\underset{|}{C}}-H}}{\overset{|}{C}}}-\overset{O}{\underset{\|}{C}}-O-H\,\,\textcircled{2}$$

7

i Explain the meaning of *biopolyester*. **(1)**

ii Which of the two –OH groups labelled can react with aqueous sodium hydroxide (choose one answer)? **(1)**

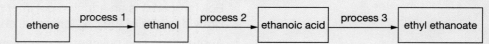

|   | 1 | 2 |
|---|---|---|
| A | ✓ | ✓ |
| B | ✓ | ✗ |
| C | ✗ | ✓ |
| D | ✗ | ✗ |

**(Total 9 marks)**

**7** The diagram shows one possible synthetic route to make ethyl ethanoate, an important solvent used in glues and nail polish removers.

ethene → process 1 → ethanol → process 2 → ethanoic acid → process 3 → ethyl ethanoate

a Process 1 is called catalytic hydration.

i Name the catalyst and state one more condition required for this reaction in industry. **(2)**

ii The equation for the conversion of ethene to ethanol can be written using displayed formulae.

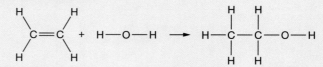

The table below gives some average bond energies.

| Bond | C–C | C=C | C–H | C–O | O–H |
|---|---|---|---|---|---|
| Average bond energy/kJ/mol | 348 | 612 | 412 | 360 | 463 |

Use the information in the table to calculate the enthalpy change for this reaction. **(3)**

iii Complete and label the energy level diagram for this reaction. Show clearly the enthalpy change, ΔH, of the reaction. **(2)**

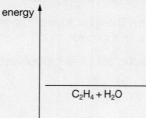

energy

$C_2H_4 + H_2O$

b Ethanol can also be made from plants, including sugar cane. Yeast is added to a sugar solution and left in the warm for several days in anaerobic conditions.

i Name the process in which ethanol is produced from glucose. **(1)**

ii What is the purpose of the yeast? **(1)**

iii Explain why anaerobic conditions are required for this reaction to occur. **(1)**

iv Why do countries such as Brazil use this process to produce ethanol rather than the catalytic hydration of ethene? **(2)**

c In process 2, ethanol is oxidised to form ethanoic acid using potassium dichromate(VI).

   i Write down two other conditions required for this reaction to occur in a laboratory. **(2)**

   ii Write a balanced chemical equation for this reaction.
(You may use [O] to represent the oxidising agent.) **(2)**

   iii What colour change do you observe in this reaction? **(2)**

   iv Ethanoic acid can react with metals to produce hydrogen gas.
Write a balanced equation for the reaction between ethanoic acid and magnesium. **(2)**

d In process 3, ethanoic acid reacts with ethanol to form ethyl ethanoate in a condensation reaction.

   i Explain why this is called a condensation reaction. **(1)**

   ii How could you separate the product from the reaction mixture? **(2)**

**(Total 23 marks)**

**END OF CHEMISTRY ONLY**

# APPENDIX A: PERIODIC TABLE

**Group**

| Period | 1 | 2 | | | | | | | | | | | | 3 | 4 | 5 | 6 | 7 | 0 |
|---|---|---|---|---|---|---|---|---|---|---|---|---|---|---|---|---|---|---|---|
| 1 | 1 H Hydrogen 1 | | | | | | | | | | | | | | | | | | 4 He Helium 2 |
| 2 | 7 Li Lithium 3 | 9 Be Beryllium 4 | | | | | | | | | | | | 11 B Boron 5 | 12 C Carbon 6 | 14 N Nitrogen 7 | 16 O Oxygen 8 | 19 F Fluorine 9 | 20 Ne Neon 10 |
| 3 | 23 Na Sodium 11 | 24 Mg Magnesium 12 | | | | | | | | | | | | 27 Al Aluminium 13 | 28 Si Silicon 14 | 31 P Phosphorus 15 | 32 S Sulfur 16 | 35.5 Cl Chlorine 17 | 40 Ar Argon 18 |
| 4 | 39 K Potassium 19 | 40 Ca Calcium 20 | 45 Sc Scandium 21 | 48 Ti Titanium 22 | 51 V Vanadium 23 | 52 Cr Chromium 24 | 55 Mn Manganese 25 | 56 Fe Iron 26 | 59 Co Cobalt 27 | 59 Ni Nickel 28 | 63.5 Cu Copper 29 | 65 Zn Zinc 30 | | 70 Ga Gallium 31 | 73 Ge Germanium 32 | 75 As Arsenic 33 | 79 Se Selenium 34 | 80 Br Bromine 35 | 84 Kr Krypton 36 |
| 5 | 85 Rb Rubidium 37 | 88 Sr Strontium 38 | 89 Y Yttrium 39 | 91 Zr Zirconium 40 | 93 Nb Niobium 41 | 96 Mo Molybdenum 42 | (98) Tc Technetium 43 | 101 Ru Ruthenium 44 | 103 Rh Rhodium 45 | 106 Pd Palladium 46 | 108 Ag Silver 47 | 112 Cd Cadmium 48 | | 115 In Indium 49 | 119 Sn Tin 50 | 122 Sb Antimony 51 | 128 Te Tellurium 52 | 127 I Iodine 53 | 131 Xe Xenon 54 |
| 6 | 133 Cs Caesium 55 | 137 Ba Barium 56 | ■ | 178 Hf Hafnium 72 | 181 Ta Tantalum 73 | 184 W Tungsten 74 | 186 Re Rhenium 75 | 190 Os Osmium 76 | 192 Ir Iridium 77 | 195 Pt Platinum 78 | 197 Au Gold 79 | 201 Hg Mercury 80 | | 204 Tl Thallium 81 | 207 Pb Lead 82 | 209 Bi Bismuth 83 | 209 Po Polonium 84 | (210) At Astatine 85 | (222) Rn Radon 86 |
| 7 | (223) Fr Francium 87 | (226) Ra Radium 88 | ■ ■ | (267) Rf Rutherfordium 104 | (268) Db Dubnium 105 | (269) Sg Seaborgium 106 | (270) Bh Bohrium 107 | (277) Hs Hassium 108 | (278) Mt Meitnerium 109 | (281) Ds Darmstadtium 110 | (282) Rg Roentgenium 111 | (285) Cn Copernicium 112 | | (286) Nh Nihonium 113 | (289) Fl Flerovium 114 | (290) Mc Moscovium 115 | (293) Lv Livermorium 116 | (294) Ts Tennessine 117 | (294) Og Oganesson 118 |

| 139 La Lanthanum 57 | 140 Ce Cerium 58 | 141 Pr Praseodymium 59 | 144 Nd Neodymium 60 | (145) Pm Promethium 61 | 150 Sm Samarium 62 | 152 Eu Europium 63 | 157 Gd Gadolinium 64 | 159 Tb Terbium 65 | 163 Dy Dysprosium 66 | 165 Ho Holmium 67 | 167 Er Erbium 68 | 169 Tm Thulium 69 | 173 Yb Ytterbium 70 | 175 Lu Lutetium 71 |
|---|---|---|---|---|---|---|---|---|---|---|---|---|---|---|
| (227) Ac Actinium 89 | 232 Th Thorium 90 | 231 Pa Protoactinium 91 | 238 U Uranium 92 | (237) Np Neptunium 93 | (244) Pu Plutonium 94 | (243) Am Americium 95 | (247) Cm Curium 96 | (247) Bk Berkelium 97 | (251) Cf Californium 98 | (252) Es Einsteinium 99 | (257) Fm Fermium 100 | (258) Md Mendelevium 101 | (259) No Nobelium 102 | (266) Lr Lawrencium 103 |

| a |
| X |
| Name |
| b |

a = relative atomic mass
X = atomic symbol
b = atomic number

(Masses in brackets are the mass numbers of the longest-lived isotope)

# APPENDIX B: COMMAND WORDS

| Command word | Definition |
|---|---|
| Add/label | Requires the addition or labelling of a stimulus material given in the question, for example labelling a diagram or adding units to a table. |
| Calculate | Obtain a numerical answer, showing relevant working. |
| Comment on | Requires the synthesis of a number of variables from data/information to form a judgement. |
| Complete | Requires the completion of a table/diagram. |
| Deduce | Draw/reach conclusion(s) from the information provided. |
| Describe | Give an account of something. Statements in the response need to be developed, as they are often linked, but do not need to include a justification or reason. |
| Design | Plan or invent a procedure from existing principles/ideas. |
| Determine | The answer must have an element that is quantitative from the stimulus provided, or must show how the answer can be reached quantitatively. To gain maximum marks, there must be a quantitative element to the answer. |
| Discuss | ▪ Identify the issue/situation/problem/argument that is being assessed within the question.<br>▪ Explore all aspects of an issue/situation/problem/ argument.<br>▪ Investigate the issue/situation etc. by reasoning or argument. |
| Draw | Produce a diagram either using a ruler or freehand. |
| Estimate | Find an approximate value, number or quantity from a diagram/given data or through a calculation. |
| Evaluate | Review information (e.g. data, methods) then bring it together to form a conclusion, drawing on evidence including strengths, weaknesses, alternative actions, relevant data or information. Come to a supported judgement of a subject's quality and relate it to its context. |

| Command word | Definition |
|---|---|
| Explain | An explanation requires a justification/exemplification of a point. The answer must contain some element of reasoning/justification, which can include mathematical explanations. |
| Give/state/name | All of these command words are really synonyms. They generally all require recall of one or more pieces of information. |
| Give a reason/reasons | When a statement has been made and the requirement is only to give the reason(s) why. |
| Identify | Usually requires some key information to be selected from a given stimulus/resource. |
| Justify | Give evidence to support (either the statement given in the question or an earlier answer). |
| Plot | Produce a graph by marking points accurately on a grid from data that is provided and then draw a line of best fit through these points. A suitable scale and appropriately labelled axes must be included if these are not provided in the question. |
| Predict | Give an expected result. |
| Show that | Verify the statement given in the question. |
| Sketch | Produce a freehand drawing. For a graph, this would need a line and labelled axes with important features indicated. The axes are not scaled. |
| State what is meant by | When the meaning of a term is expected but there are different ways for how these can be described. |
| Suggest | Use your knowledge to propose a solution to a problem in a novel context. |
| Verb proceeding a command word | |
| Analyse the data/graph to explain | Examine the data/graph in detail to provide an explanation. |
| Multiple-choice questions | |
| What, why | Direct command words used for multiple-choice questions. |

# APPENDIX C: A GUIDE TO PRACTICAL QUESTIONS

This appendix is designed to help you better understand the practical work in your study, and also to give you guidance on how to answer the practical-based questions in your International GCSE Chemistry papers.

## WHY IS THIS APPENDIX SO IMPORTANT?

Chemistry is a practical subject. Through experiments, you develop better understanding of the theoretical concepts in the subject. This is a guide to things you should consider when planning and carrying out an experiment, and how to process your results.

Of the possible marks in your chemistry exams, about 20% will be given to questions designed to find out whether you have some important laboratory skills. This appendix offers suggestions on how to approach these questions.

## SAFETY

Before starting a particular experiment, you need to consider what risks are involved in the experiment and which safety precautions are needed.

The obvious precaution is to *wear eye protection*. That's true of *all* practical work in chemistry. However, you might be asked a question that says 'Apart from wearing eye protection, explain a safety precaution that you would need to take in doing this experiment. **(2 marks)**'.

Notice the word 'explain' in the last sentence. 'Explain' means that, as well as giving a precaution, you have to give a reason for it. Look at the number of marks available. If there is more than 1 mark, you have to give more than one piece of information.

Table A1 Other possible precautions

| Presence of: | Precaution and reason (marks) |
|---|---|
| poisonous gas, e.g. chlorine | Do the experiment in a fume cupboard (1) because chlorine is poisonous (1) |
| corrosive liquid, e.g. acid or sodium hydroxide solution | Wear gloves (1) because the acid (or whatever) is corrosive (1) |
| flammable liquid, e.g. ethanol | Keep away from naked flames (1) because the ethanol might catch fire (1) |
| any hot liquid or apparatus | Take care not to touch (1) in case you burn yourself (for solids) or scald yourself (for liquids) (1) |
| any especially fragile apparatus, e.g. pipette or thermometer | Be careful not to break the fragile pipette (1) because you might cut yourself (1) |

## CHOOSING AND USING MEASURING EQUIPMENT

If you want to measure the mass of something, you should use a balance. If you want to measure a temperature, you should use a thermometer. When it comes to volumes, however, you are faced with a number of choices of equipment: measuring cylinders of various sizes, burettes, pipettes and gas syringes. It is important to pick the right item.

## MEASURING VOLUMES OF LIQUID

### Measuring cylinders

Measuring cylinders are acceptable for approximate volumes. Choose the smallest measuring cylinder that will take the volume you need. Trying to measure, for example, $8\,cm^3$ in a $100\,cm^3$ measuring cylinder is going to be fairly imprecise. Use a $10\,cm^3$ measuring cylinder instead.

### Pipettes

Pipettes will measure the volume printed on them very precisely (typically to within $0.05\,cm^3$). Common pipettes have volumes of 25 or $10\,cm^3$, but you might also come across other sizes, such as 5, 20 or $50\,cm^3$.

### Burettes

The most common size of burette will measure variable volumes up to $50\,cm^3$, and you can take measurements to the nearest $0.05\,cm^3$.

Note you will never use a beaker or a conical flask to measure the volume of a liquid. They are too imprecise! The precision of a measurement depends on how big the divisions are on an instrument. The smaller the division, the more precise the data.

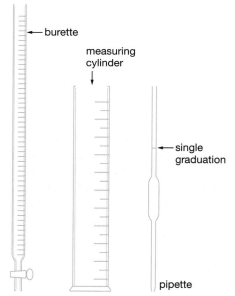

▲ Figure A1 Devices for measuring volumes of liquid.

## MEASURING VOLUMES OF GAS

To measure the volume of a gas you are collecting, you can either use a gas syringe or collect the gas into an inverted measuring cylinder over water. You will find diagrams for both methods on page 229. A gas syringe is more precise and has to be used if the gas is soluble in water.

## TAKING READINGS

You take your measurement from the bottom of the meniscus (if you have a liquid), ignoring any liquid that has 'curled up' at the sides (see Figure A2). Edexcel examiners expect you to read a scale to a precision of *half the smallest divisions* marked on whatever piece of apparatus you are using. If the smallest division is $0.1\,cm^3$, you are expected to estimate the volume to the nearest $0.05\,cm^3$ if the meniscus falls between two marks.

The diagram shows a burette reading of $5.85\,cm^3$. Be very careful to look closely at which way the scale runs. Students might make the mistake of reading this as $6.15\,cm^3$ because they don't notice that the meniscus is somewhere between 5 and 6 on the scale. Don't simply read from the nearest number, think about what you are doing!

▲ Figure A2 Measuring from the bottom of the meniscus.

## RECORDING READINGS

The number you write down should show the degree of precision used. For example, if a burette reading falls exactly on the $11\,cm^3$ line, you should still record it as $11.00\,cm^3$ because it was possible to read it to $0.05\,cm^3$ and your answer wasn't $11.05$ or $10.95\,cm^3$. Recording it as $11\,cm^3$ implies that you couldn't read it more precisely than to the nearest $1\,cm^3$.

# TABLES OF RESULTS

## DRAWING UP A TABLE

Table A2 contains some real measurements taken during a piece of practical work to measure the heat evolved when magnesium reacted with dilute hydrochloric acid. Ignore the bottom row for the moment.

Notice that the figure for volume of acid shows how precisely it was measured.

Volume of acid used = $50.0\,cm^3$

Table A2 Results from an experiment to measure the heat evolved when magnesium reacts with dilute hydrochloric acid

|  | Experiment | | |
|---|---|---|---|
|  | 1 | 2 | 3 |
| Mass of weighing bottle plus Mg/g | 10.81 | 10.81 | 10.82 |
| Mass of weighing bottle afterwards/g | 10.69 | 10.69 | 10.69 |
| Mass of Mg used/g | 0.12 | 0.12 | 0.13 |
| Initial temperature/°C | 17.4 | 17.5 | 17.4 |
| Maximum temperature/°C | 27.5 | 27.4 | 28.4 |
| Temperature rise/°C | 10.1 | 9.9 | 11.0 |
| Precision check – temperature rise per gram of Mg/°C/g | 84.2 | (82.5) | 84.6 |

Notice the figure of 28.4 °C in the third column of Table A2. This shows that the temperature was measured to a precision of 0.1 of a degree celsius.

Notice that Table A2 contains both measured and calculated results (the mass of Mg used and the temperature rise). Notice that every row and column is properly labelled, including units, wherever appropriate. Column or row headings for things such as mass, temperature, volume and time are useless unless the units are included, and you will lose marks for leaving them out.

If you have measured time, for example, in a variety of units while you were doing the experiment, you must change them to one consistent unit before you draw up your table. For example, if you have measured times of 15 s, 30 s, 45 s, 1 min, 1 min 30 s, 2 min and 5 min, you must convert them all to seconds before you can make use of them.

There are various ways that you can show the units in a table (or on the axes of a graph). Some people put the units in brackets, for example (g). The most correct way is to divide by the units. So, for example, if we have a mass of 10.81 g and we divide this by the unit we obtain

$$\frac{10.81\,g}{g} = 10.81,$$

which is the number we have put in the table.

# RELIABILITY OF RESULTS

If you just produce one set of results, you can't be sure that you haven't made a mistake somewhere. Experiments have to be repeated to check their reliability. *Reliable results are those that are in close agreement, but not necessarily the same.* All practical work has built-in errors, which you can't avoid. If you have two results exactly the same, that's just luck!

Sometimes the word 'concordant' is used to represent data that are consistent with each other.

How can you tell whether your results are consistent and reliable? In Table A2, the last row calculates the temperature rise per gram of magnesium used. In this experiment, that result should be consistent if your experiment is reliable.

Notice that in Table A2 two values for temperature rise per gram of Mg are in close agreement, but the other one is slightly different. If you were doing calculations from these results, you would omit the results from experiment 2 because they aren't likely to be quite as reliable as the others.

If you are asked how you would improve the reliability of a set of results, say that you would repeat the experiment (possibly more than once), and check the new results for consistency with the old results. If two sets of results agree closely and a third doesn't, then you discard the third set of results.

## PROCESSING THE RESULTS
### GRAPHS
### Drawing good graphs

It may sound obvious, but the most important tool you need is a *very* sharp pencil. You can't draw a good graph with a pen, and, of course, it is impossible to rub out a mistake.

In an exam, you are likely to be given a graph grid, possibly with at least one axis unlabelled. Choose a scale that uses as much of the graph paper as possible, but without making it really difficult to plot the points. Label each axis clearly (in ink) and don't forget the units. The *independent variable* – what you vary – goes on the *x*-axis and the *dependent variable* – what you measure – goes on the *y*-axis. Plot the points carefully and if any don't seem to be following the general pattern (towards a straight line or a smooth curve), double-check them.

### Anomalous results

Sometimes you will see one result that is clearly wrong because it falls well away from the pattern of the others. This is called an **anomalous result**. You might be asked to explain what might have caused it. When you are drawing your straight line or curve, you ignore the anomalous point completely.

Your explanation should be as precise as you can make it. Decide whether the point is too high or too low on the graph, then try to think of an experimental reason why that might have happened. It isn't enough to say simply that 'wrong measurements were taken or human error', you have to be much more specific, for example 'too much calcium carbonate was added by mistake, and that produced more carbon dioxide than expected'.

The best way of being sure what the examiners want in this sort of case is to look at mark schemes and Examiners' Reports from past papers. You will find more about this at the end of this appendix.

### Straight-line graphs

Most (but not all) straight-line graphs that you will get at International GCSE will go through the origin (0, 0). Does your graph look as if it is going to go through the origin? If so, is it *reasonable* that it should go through the origin?

For example, if you were plotting results from a reaction between calcium carbonate and hydrochloric acid, it would be reasonable that if you didn't use any calcium carbonate, you wouldn't obtain any carbon dioxide. On the other hand, if you were plotting rates of reaction against temperature, it isn't reasonable to assume that a reaction rate would be zero if the temperature was 0 °C.

Suppose you were plotting a graph from an experiment reacting small amounts of calcium carbonate with dilute hydrochloric acid (Table A3). You are trying to find the relationship between the volume of carbon dioxide produced and the mass of calcium carbonate used.

Table A3 Volume of carbon dioxide produced and mass of calcium carbonate used when reacting calcium carbonate with dilute hydrochloric acid

| Mass of $CaCO_3$/g | 0.020 | 0.040 | 0.060 | 0.080 | 0.100 | 0.120 |
|---|---|---|---|---|---|---|
| Volume of $CO_2$/cm³ | 3 | 10 | 13 | 13 | 24 | 31 |

The graph is plotted with volume of $CO_2$ (what you measure – dependent variable) on the *y*-axis and mass of $CaCO_3$ (what you change – independent variable) on the *x*-axis. It must also go through (0, 0) as explained above.

Figure A3 is just a sketch graph of these figures. You would need to draw this yourself on graph paper to see exactly how the points should lie.

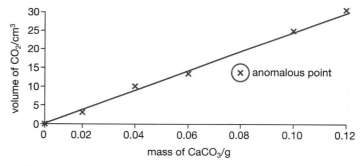

▲ Figure A3 Volume of carbon dioxide produced versus mass of calcium carbonate used.

There is a best-fit line, and the anomalous point hasn't been included. *It mustn't be – it is obviously wrong!* When you draw a best-fit line, make sure that you have an even spread of points on each side of the line you have drawn. Don't worry if it only goes through one or two of your points. In this case, the only point you are certain about is (0, 0). Your line *must* go through (0, 0) in this particular experiment.

Straight lines *must be drawn with a ruler* rather than freehand.

## Curved graphs

Exactly the same principles apply, except that it is more difficult to draw a smooth curve than a straight line. The line should still be a best-fit curve and, again, you shouldn't include any anomalous point(s) when you draw it.

When you are asked to draw a line of best fit, this can be a straight line or a curve.

If you are going to be drawing a lot of graphs in the future (because you are going to do maths or science subjects at a higher level, for example), it is really worth buying a flexible rubber ruler designed for the job. Do an internet search for flexible ruler or 'flexicurve'. These are rulers that can be bent to allow you to draw smooth curves.

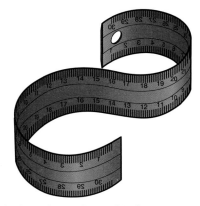

▲ Figure A4 A flexible ruler can be used to draw a curve of best fit.

Figure A5 shows how the rate at which oxygen was given off during a reaction ($y$-axis) varied with temperature ($x$-axis). These are imaginary results just to illustrate a point. They don't relate to any real experiment. Don't assume that the curve must go through the origin. In Figure A5, it won't.

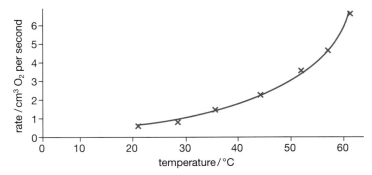

▲ Figure A5 The rate of an imaginary reaction against temperature.

In a case like this, you might be asked how you would modify the experiment to obtain more results between 0 and 20 °C. The obvious thing to do would be to cool the substances you are going to react together before you mix them, by surrounding them with ice. More than once in the past, examiners have commented that students often suggested moving the experiment to a colder climate. You won't get credit for that!

This is an excellent example of why it is essential to look at mark schemes for past papers and read Examiners' Reports. They point out the type of answers they won't accept. If you have read it, you won't make the same mistake yourself.

## Graphs with unusual shapes

Occasionally, you might be given results that increase for a bit, and then decrease again. The final graph might be two intersecting straight lines or a curve, as shown in Figure A6.

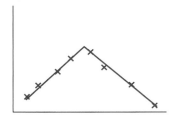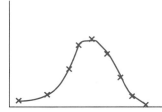

▲ Figure A6 Plotting unusual curves.

You need to look carefully at how the points fall to see which of these you need to draw. You may well find a clue in the question, for example the use of the word 'curve' at some point or 'draw two straight lines'.

After drawing graphs of this kind, you might be asked how you could improve their accuracy. The answer is to take more readings around what is known as the 'turning point' of the graph. With either of these graphs (but especially the curve), what you draw would be more accurate if you had more points around the peak in each case.

## Bar charts

Bar charts are easy to draw. Don't forget to label them. These types of graphs are used when the independent variables are not continuous and can only take certain values. For example, in Figure 19.12 on page 214, the number of carbon atoms in an alcohol molecule is an example of a non-continuous variable.

## DESCRIBING THE RELATIONSHIPS SHOWN BY A GRAPH

Look back at the straight line graph in Figure A3. What does it show? The correct expression is that 'the volume of carbon dioxide is *directly proportional to* the mass of calcium carbonate used'. If you just say that 'as the mass of calcium carbonate increases, the volume of carbon dioxide increases', that isn't precise enough, and you may not get the full mark. That faulty statement could equally well describe a curve.

The phrase *directly proportional* means if you double the mass of calcium carbonate, the volume of carbon dioxide produced will double as well. If you triple the mass of calcium carbonate, the volume of carbon dioxide produced also triples. For the relationship between two variables to be directly proportional, the graph must be a *straight line passing through the origin*. If a straight line does not pass through the origin, you should say there is a *linear relationship* between the two variables. If the line has a *positive gradient*, you could say there is a *positive and linear correlation* between the two variables.

The words *proportional* and *directly proportional* mean the same thing, there is no difference between them.

The graph in Figure A5 is more difficult to describe. You can start by saying that as the temperature increases, the rate gets faster, but you have to make the point that the line is curved. The examiners commented on a similar question by saying that they were looking for words like 'non-linear' or 'exponential'. You could say something like 'The rate increases exponentially with temperature'. That suggests a strong upward curve. Alternatively you would have to use more words and say something like 'the amount that the rate increases for each 10 °C rise in temperature is not constant but increases'. This is a bit more complicated, but it explains what is happening.

Technically, the term 'exponential increase' has a precise mathematical meaning, which won't necessarily apply to all curves that look like this. However, you do not need to worry about this at International GCSE level.

# CALCULATING RESULTS

You will find that the calculations involved with these questions are usually easy, often with a simple formula that you have to slot numbers into. Two things need a comment.

### CALCULATING AVERAGES

If you are calculating an average from a number of results that include one or more inconsistent ones, you would normally just take the average of the consistent ones and ignore the anomalous result.

For example, with good titration technique you should get results that are consistent to within $0.10 \, cm^3$. Suppose your actual results were

21.20    21.45    21.30 $cm^3$

You have two consistent results (21.20 and 21.30), and one that is outside the usual limits of consistency for a titration. Take the average of the two consistent ones (21.25 $cm^3$) and ignore the other one. You will sometimes be told in these questions what constitutes consistent results, for example 'concordant results are those within 0.10 $cm^3$ of each other'.

Important! If a question just gives you some numbers to find an average of, without any comment about consistency, just take a simple average. For example, one past question asked for an average of 95, 96 and 91. Strictly speaking, you should ignore the 91 value because it is out of line with the other two. The answer given in the mark scheme, however, was 94 (an average of all three). You need to look at a lot of mark schemes and Examiners' Reports to judge how to decide what to do in cases like this! (See the end of this appendix.)

### SIGNIFICANT FIGURES

You shouldn't quote an answer to more significant figures than the least precise piece of information you are using in the calculation.

For example, if you are measuring a volume of gas as 8.00 $cm^3$ and a mass of solid of 1.325 g, and use both those figures in your calculation, you can't quote your answer to more than three significant figures because that's all the least precise number (the volume) is quoted to.

When counting significant figures, you start counting from the left with the first non-zero number. For example, 0.0053 has two significant figures because we do not count the first three zeros. It has four *decimal places*, however. 0.6720 has four significant figures and four decimal places. The last zero is significant here.

With zeroes that come before a decimal point, it is more difficult to know how many significant figures there are, for example 500 could be one, two or three significant figures. You would only know which if you took the measurement yourself. Sometimes standard form is used to show this:

$5 \times 10^2$ one significant figure

$5.0 \times 10^2$ two significant figures

$5.00 \times 10^2$ three significant figures

Be careful when you round numbers to the correct number of significant figures. Remember that you are rounding them, not just removing the unwanted figures from your number. So, for example:

12.382651 on your calculator rounds to 12.4 to three significant figures, *not* 12.3.

Remember that you round a digit followed by a 5 or more upwards; a digit followed by less than 5, you leave the same. For example:

2.465 rounds to 2.47 to three significant figures;

2.464 rounds to 2.46 to three significant figures.

# MODIFYING EXPERIMENTS

Suppose you were given a question involving rates of reaction, for example the decomposition of hydrogen peroxide in the presence of manganese(IV) oxide as a catalyst. The results you are given are designed to show how the temperature of the reaction affects the rate. At the end of the question, you might then be asked to suggest how you would find out how changing the concentration of the hydrogen peroxide affects the rate, or how changing the quantity of catalyst affects it.

### BE PRECISE ABOUT WHAT YOU ARE TALKING ABOUT

Don't use vague expressions like 'I would change the amount of hydrogen peroxide.' Avoid the word 'amount' entirely. If you are talking about a solution, say whether you are intending to change the volume of the solution or its concentration. If you are talking about a solid, say whether you are talking about changing its mass, or how many moles of it you have got, or its surface area. You must be precise in order to get the mark.

Always remember that, as a general rule, one mark equals one piece of information, so three marks equals three pieces of information, and so on.

## MAKE SURE YOU ARE DESCRIBING A VALID (FAIR) TEST

Whatever it is that you are changing, *that must be the only thing which changes from one experiment to the next*. If you are changing the concentration of one solution, the concentration of every other solution involved must stay the same as before. The total volume must stay the same. The temperature must stay the same. The mass of any solid must stay the same, and so must its size (powder, small lumps, big lumps and so on) so that its surface area stays the same.

# MAXIMISING YOUR SUCCESS

The very best way of revising for the exam at the end of your course is to work through past papers set by your examiners. You should look especially at the most recent ones. You need to check your answers against the published mark schemes and against the comments in the Examiners' Reports. The Examiners' Reports are particularly useful for these practical-based questions, where what the examiners are looking for isn't always obvious.

In question papers from June 2011 onwards, practical-based questions are mixed up with more theoretical ones in the same exam papers. If you use older test papers to practise, note that there was a separate paper in which all the practical questions were found.

On the website accompanying this book, you will find advice about how to obtain past papers, mark schemes and reports for the latest specification.

# GLOSSARY

**absorb** take in (a gas or a liquid)

**acid** a substance that acts as a source of hydrogen ions ($H^+$) in solution (Arrhenius theory) or as a proton donor (Brønsted–Lowry theory)

**acid rain** rain which has a pH of less than about 5.6 It is caused when water and oxygen in the atmosphere react with sulfur dioxide to produce sulfuric acid, or with various oxides of nitrogen, $NO_x$, to give nitric acid

**activation energy** the minimum amount of energy required for a collision to be successful, i.e. to result in a reaction

**addition** a chemical reaction in which one molecule adds to another without taking anything away, to form a single product. For example, when alkenes react with halogens, the halogen atoms add onto the alkene molecule

**addition polymerisation** polymerisation of monomers containing a carbon–carbon double bond. A large number of monomer molecules add onto each other without anything else being formed

**alcohols** a homologous series of compounds which all contain an –OH functional group attached to a hydrocarbon chain

**alkali metals** group 1 elements including lithium, sodium, potassium, rubidium, caesium and francium Note, hydrogen is not an alkali metal

**alkanes** a homologous series of similar hydrocarbons in which all the carbons are joined to each other with single covalent bonds. These are saturated compounds with the general formula $C_nH_{2n+2}$

**alkenes** a homologous series of hydrocarbons which contain a carbon–carbon double bond. These are unsaturated compounds with the general formula $C_nH_{2n}$

**allotropes** different forms of the same element, for example diamond, graphite and $C_{60}$ fullerene are three allotropes of carbon

**alloy** a mixture of a metal with, usually, other metals or carbon. For example, brass is an alloy of copper and zinc, and steel is an alloy of iron and carbon

**amphoteric** substances that can react with both acids and bases to form salts

**anaerobic** in the absence of air

**anhydrous** without water

**anion** a negative ion, formed by atoms gaining electrons

**anode** the positive electrode in electrolysis, which attracts negative anions

**anomalous result** a result that does not fit in with the pattern of the others

**aqueous** something dissolved in water

**array** an ordered arrangement of things

**ash** the soft powder that remains after something has been burned

**atom** the smallest piece of an element that can still be recognised as that element

**atomic number** the number of protons in an atom

**Avogadro constant** the number of $^{12}C$ atoms in 12 g of $^{12}C$ ($6.02 \times 10^{23}$ mol$^{-1}$)

**balancing the equation** a process of putting coefficients in front of formulae so that the same number of atoms of each type is on both side of an equation

**barrier method** a method of rust prevention by coating iron with paint, oil, grease or plastic, so that oxygen/water cannot reach the iron/steel

**base** a substance that neutralises acids by combining with the hydrogen ions in them. They are usually metal oxides, hydroxides or ammonia. A soluble base is called an alkali, and it is a substance that acts as a source of hydroxide ions ($OH^-$) in solution (Arrhenius theory) or as a proton acceptor (Brønsted–Lowry theory)

**batch** a group of things that are produced or are dealt with together

**biofuel** a fuel that is made from biological sources, such as sugar cane or corn

**biopolyester** a polyester which is biodegradable

**bleach** a chemical used to make something paler or whiter, or to sterilise something

**boiling** the change of state from a liquid to a gas. It occurs at the boiling point

**bond energy** the amount of energy required to break 1 mole of covalent bonds in gaseous molecules, or the amount of energy released when 1 mole of covalent bonds are formed in gaseous molecules

**branch** something that grows out from the main part of an object

**bubbling** producing bubbles

**bulb** a glass object that produces light when electricity is passed through it. Or, the wider part of some plastic pipettes, which is squeezed to allow the pipette to be filled

**calorimetry** measuring the heat given out or taken in by a chemical reaction

**carboxylic acids** a homologous series of compounds which all contain a –COOH functional group attached to a hydrocarbon chain

**catalyst** substance that speeds up a chemical reaction by providing an alternative pathway of lower activation energy. Catalysts are not used up and remain chemically unchanged at the end of the reaction

**catalytic converter** a device used in cars to convert oxides of nitrogen and carbon monoxide into harmless nitrogen gas and carbon dioxide. It uses platinum, palladium and rhodium as catalysts

**cathode** the negative electrode in electrolysis, which attracts positive cations

**cation** a positive ion, formed by atoms losing electrons

**chemical means** methods which involve chemical reactions

**chromatogram** the adsorbent paper from paper chromatography showing the separation of different coloured substances

**clump** a small group or cluster

**coefficient** number written in front of formulae in a balanced chemical equation

**coil** continuous series of circular rings into which something such as wire or rope has been wound or twisted

**collision theory** states that for a reaction to occur, the reactant particles must collide with each other and the collision needs to have sufficient energy and the correct orientation

**combustion** a chemical reaction in which a substance reacts with oxygen (burns) to form products and heat

**competition reaction** displacement reaction between a more reactive metal and the oxide of a less reactive metal

**complete combustion** occurs when a hydrocarbon burns in sufficient oxygen (burns) and forms carbon dioxide and water as products

**compound** a substance that forms when two or more elements chemically combine The elements cannot be separated by physical means

**condensation** the change of state from a gas to a liquid. It occurs at the condensation point

**condensation polymerisation** polymerisation of monomers in which each time two monomers combine, a small molecule such as water or hydrogen chloride is removed

**condensation reaction** a chemical reaction in which two molecules combine to form a larger molecule with elimination of a small molecule such as water

**corrosive** a substance which can damage living tissue

**covalent bonding** strong electrostatic force of attraction between the nuclei of the atoms making up the bond and the shared pair of electrons

**cracking** a process in which long-chain alkanes are converted to alkenes and shorter-chain alkanes. It is carried out using silica or alumina as catalyst at a temperature of 600–700 °C

**crude oil** formed from the remains of living organisms when their soft tissue was gradually changed by high temperatures and pressures into a thick, black oil. It is a mixture of hydrocarbons

**crystallisation** a process in which a solute (soluble solid) is obtained from its solvent

**damp** slightly wet

**decomposition reaction** a chemical reaction in which a compound is broken down into its elements or simpler compounds

**dehydration** removal of water

**delocalised electrons** electrons that are no longer attached to particular atoms or pairs of atoms, but are free to move throughout the whole structure

**denatured** the loss of function of enzymes due to structural changes, usually caused by changes in temperature or pH

**dependent variable** what you measure in an experiment

**deposition** the change of state directly from a gas to a solid

**diatomic** a molecule that contains two atoms

**diffusion** the random movement of particles from an area of high concentration to an area of low concentration

**dimerise** forming a dimer – two identical molecules react to form a larger molecule

**dip** put something into a liquid and lift out again

**discharged** when an ion loses its charge by losing or gaining electrons

**displacement reaction** a chemical reaction in which a more reactive element replaces a less reactive one in its compound

**displayed formula** a formula that shows all the bonds in a molecule as individual lines. Each line represents a pair of shared electrons in a covalent bond

**dot-and-cross diagram** a representation of how electrons are arranged in ions or molecules

**double bond** atoms sharing two pairs of electrons in a covalent bond

**drive off** force something to go away

**dry ice** solid carbon dioxide

**ductile** a property of metal that allows it to be drawn out into wires

**dynamic equilibrium** equilibrium means the concentrations of the reactants and the products in a reversible reaction remain constant. Dynamic means the reactions are still continuing, but the rate of the forward reaction is equal to the rate of the reverse reaction

**electrode** a conductor through which electricity is passed into and out of an electrolyte

**electrolysis** the chemical change caused by passing an electric current through a compound that is either molten or in solution

**electrolyte** a liquid that undergoes electrolysis. Electrolytes are molten ionic compounds or solutions containing ions

**electrolytic cell** a compartment in which electrolysis occurs

**electron** a subatomic particle found in shells (energy levels) outside the nucleus of an atom. It has a relative mass of 1/1836 and a relative charge of 1−. For a neutral atom, the number of electrons = the number of protons = the atomic number

**electronic configuration** how electrons are arranged in the shells (energy levels) in an atom

**electrostatic attraction** the force of attraction between something that is positive and something that is negative

**element** a substance that cannot be split into anything simpler by chemical means. An element contains atoms of the same atomic number

**empirical formula** gives the simplest whole-number ratio of the atoms present in a compound. It can be worked out from experimental data

**end point** the point at which the indicator changes colour in a titration

**endothermic** reactions in which heat energy is taken in from the surroundings

**energetics** a study of energy change in chemical reactions

**enthalpy change** the amount of heat energy taken in or given out in a chemical reaction. It has the symbol $\Delta H$

**enzymes** biological catalysts, usually consisting of proteins

**ester** esters are organic compounds formed by the reaction of an alcohol with a carboxylic acid. They have the functional group –COO–

**esterification** a chemical reaction in which an alcohol and a carboxylic acid react together to form an ester

**evaporation** the change of state from a liquid to a vapour. It occurs at a temperature below the boiling point and only at the surface of the liquid

**excess** having more than enough of a reactant to react with all of something else

**exothermic** reactions in which heat energy is given out to the surroundings

**fade** lose colour and brightness

**fermentation** converting sugar such as glucose into ethanol and carbon dioxide using enzymes in yeasts

**filtrate** the liquid that comes through the filter paper during filtration

**filtration** a process to separate an insoluble solid from a liquid

**fizz** make many bubbles, producing a continuous sound

**flake off** come off in small pieces

**flash** shine suddenly and brightly for a short period of time

**flush out** force something out

**formula** a representation of a chemical showing the elements present

**forward reaction** from reactants to products (the left-to-right reaction)

**fossil fuels** these include coal, gas and fuels derived from crude oil, which all come from things that were once alive

**fractional distillation** a process to separate two liquids of similar boiling points, for example ethanol and water or the components of crude oil

**fractionating column** a piece of equipment used for separating vapours in fractional distillation

**freezing** the change of state from a liquid to a solid. It occurs at the freezing point

**freezing point** the temperature at which a substance changes its state from a liquid to a solid

**fuel** a substance that when burned in oxygen releases heat energy

**fume** strong-smelling gas or smoke that is unpleasant to breathe in

**functional group** an atom or a group of atoms that determine the chemical properties of a compound

**galvanisation** a method of preventing rusting by coating iron with a layer of zinc

**general formula** a formula applicable to all members of a homologous series, for example $C_nH_{2n+2}$ for alkanes and $C_nH_{2n}$ for alkenes

**giant** a structure in which there are no individual molecules and the bonding extends in all directions. There is no limit to the number of particles present

**giant ionic lattice** the arrangement of ions in an ionic compound in its solid state

**girder** a strong beam, made of iron or steel, that supports a floor, roof or bridge

**global warming** greenhouse gases, including carbon dioxide, trap the heat radiated from the Earth's surface (originally from the Sun) and lead to an increase in the temperature of the Earth and its atmosphere

**glow** produce a steady light

**greenhouse gas** gases such as carbon dioxide which can trap heat radiated from the Earth's surface (originally from the Sun)

**group** a vertical column in the Periodic Table. All elements in the same group have the same number of outer shell electrons

**half-equation** a balanced symbol equation to describe either oxidation or reduction

**halogen** group 7 element, including fluorine, chlorine, bromine, iodine and astatine

**hammered** been hit with a hammer to be forced into a particular shape

**homologous series** a series of compounds with similar chemical properties because they have the same functional group. Each member differs from the next by $-CH_2-$. The members show a gradation in physical properties

**hydrated** containing water

**hydration** the addition of water molecules to an unsaturated molecule, for example converting ethene to ethanol

**hydrocarbon** molecules containing carbon and hydrogen only

**hydrogen bond** a special type of intermolecular force found between water molecules and in some other substances

**hydrogen halide** compound formed between hydrogen and a halogen, with the formula HX where X stands for a halogen atom

**hydrogenation** the addition of hydrogen molecules to an unsaturated molecule, for example converting ethene to ethane

**hydroxonium ion** an ion formed when water accepts a proton from an acid. It has the formula $H_3O^+$

**incomplete combustion** occurs when a hydrocarbon burns in insufficient oxygen. Water is still formed as a product, but carbon monoxide or carbon are formed instead of carbon dioxide

**independent variable** what you vary in an experiment.

**indicator** a substance that has different colours depending on the pH

**inert** unreactive

**intermolecular force** force of attraction between covalent molecules, much weaker than covalent bonds within the molecules

**ion** charged particle formed when an atom (or group of atoms) loses or gains electrons

**ionic bonding** strong electrostatic force of attraction between oppositely charged ions, formed by the transfer of electrons from one atom to another

**ionise** become an ion. Usually refers to losing electrons to become a positive ion

**iron filings** small pieces of iron that are used in some chemical experiments

**isoelectronic** having the same number of electrons

**isotopes** different atoms of the same element, with the same number of protons but a different number of neutrons. Isotopes have the same chemical properties

**kinetic energy** a form of energy a substance possesses due to motion

**lattice** a regular array of particles

**Liebig condenser** a piece of equipment used to condense gas to liquid in distillation

**litre** a unit for volume. One litre = one $dm^3$

**lump** a mass of solid without a particular shape

**malleable** a property of metal which allows it to be hammered into different shapes

**mass number** the total number of protons and neutrons in the nucleus of an atom

**melting** the change of state from a solid to a liquid. It occurs at the melting point

**melting point** the temperature at which a substance changes its state from a solid to a liquid. A pure substance has a fixed melting point but a mixture may melt over a range of temperatures

**metallic bonding** electrostatic force of attraction between a lattice of positive ions and the sea of delocalised electrons

**microbial oxidation** oxidation by oxygen in the air in the presence of microorganisms such as bacteria and yeast

**minerals** naturally occurring, crystalline compounds in the Earth's crust

**miscible** liquids (or gases) which can mix with each other and form a homogenous mixture

**mixture** two or more substances not chemically combined that can be separated by physical means

**molar enthalpy change** the change in enthalpy when 1 mole of a particular reactant reacts

**molar mass** the mass of 1 mole of a substance

**molar volume** the volume occupied by 1 mole of a gas

**mole** a unit of the amount of a substance. A mole of anything contains the same number of particles as there are carbon atoms in 12 g of $^{12}C$ ($6.02 \times 10^{23}$ particles)

**molecular formula** shows the actual number of each type of atom present in a molecule (covalent compound) or formula unit (ionic compound)

**molecule** two or more atoms covalently bonded together. Molecules contain a certain fixed number of atoms

**molten** liquid state when a solid has melted

**monomer** molecules which can join up to form a polymer

**mono-substitution** a substitution reaction in which only one hydrogen atom in an alkane is replaced by a halogen atom

**native (metals)** metals which exist naturally as the uncombined element

**neutralisation reaction** a chemical reaction in which acids react with bases or alkalis to produce salts

**neutron** a subatomic particle in the nucleus of an atom. It has a relative mass of 1 and no charge. Number of neutrons in an atom = mass number – atomic number

**noble gas** group 0 elements, including helium, neon, argon, krypton, xenon and radon

**non-biodegradable** cannot be broken down by bacteria or fungi in the environment

**non-continuous variable** an independent variable which can only take certain values

**non-renewable resource** a finite resource that cannot be replaced, at least not for millions of years

**nucleon number** same as mass number

**nucleus** contains protons and neutrons in an atom

**octet** the octet rule states that atoms generally lose, gain or share electrons in order to have eight electrons in their outer shell

**ore** rocks that contain enough of the mineral for it to be worthwhile to extract the metal

**organic chemistry** branch of chemistry that studies the compounds of carbon except for the very simplest (like carbon dioxide, carbon monoxide, the carbonates and the hydrogencarbonates)

**oxidation** gain of oxygen or loss of electrons

**oxidising agent** something that oxidises something else by taking electrons away from it. An oxidising agent is reduced in a chemical reaction

**paper chromatography** a process to separate a mixture of coloured substances using adsorbent paper

**parent acid** the acid from which a salt is formed, for example sulfuric acid is the parent acid for sulfates

**particle** a small object. In chemistry, particle can be used to refer to atoms, molecules, ions or the subatomic particles, including protons, neutrons and electrons

**period** a horizontal row in the Periodic Table. All elements in the same period have the same number of occupied shells

**Periodic Table** a table in which elements are arranged in the order of increasing atomic number and in terms of chemical and physical properties

**pH scale** a scale to measure how acidic or how alkaline a solution is

**physical means** methods which involve changing temperature, dissolving etc., for example distillation, filtration, use of magnet, chromatography, crystallisation, for separating components of a mixture

**plot** draw a graph to represent information

**plunger** a part of a syringe that moves in and out

**polar** a polar covalent bond is one in which electrons are not equally distributed between the two atoms

**polymer** a large molecule made when many small molecules (monomers) join together. It consists of many repeat units

**polymerisation** joining up of lots of small molecules (monomers) to make one big molecule (polymer)

**pop** a sudden short sound, often loud

**position of equilibrium** a reference to the proportion of the various things in an equilibrium mixture. If the position of equilibrium of a reaction lies towards to the right, the equilibrium mixture contains a higher proportion of products than reactants

**pour** make something flow out of a container by holding it at an angle

**precipitate** a fine solid that is formed by a chemical reaction involving liquids or gases

**precipitation reaction** a chemical reaction in which a fine solid is formed from liquids or gases

**pressure** the force acting on something per unit area. Pressure has the SI unit of Pa ($N/m^2$) and is caused by molecules hitting the walls of their container

**proton** a subatomic particle in the nucleus of an atom. It has a relative mass of 1 and a relative charge of 1+. Number of protons in an atom = atomic number

**proton number** same as atomic number

**quicklime** calcium oxide

**radioactive** having an unstable nucleus that will emit particles and waves to achieve a more stable nucleus

**range** a number of things that are all of the same general type, or the extent of a variable or series

**rate** the speed at which the amount of reactants decreases or the amount of products increases. It is measured as a change in the concentration (or amount) of reactants or products per unit time

**reactivity series** a list of elements (mainly metals) in order of decreasing reactivity

**redox** reduction and oxidation. A redox reaction involves both reduction and oxidation occurring together

**reducing agent** something that reduces something else by giving electrons to it. A reducing agent is oxidised in a chemical reaction

**reduction** loss of oxygen or gain of electrons

**refinery gases** a mixture of methane, ethane, propane and butane. This mixture is commonly used as liquefied petroleum gas (LPG) for domestic heating and cooking

**relative atomic mass** the weighted average mass of the isotopes of the element, relative to the mass of 1/12th of a $^{12}C$ atom

**relative formula mass** the weighted average mass of a formula unit of a compound, relative to the mass of 1/12th of a $^{12}C$ atom. It is sometimes called relative molecular mass, when it refers to covalent molecules

**relative isotopic mass** the mass of one particular isotope of the element, relative to the mass of 1/12th of a $^{12}C$ atom

**repeat unit** the basic unit which repeats itself in a chain to form a polymer

**residue** the solid left on the filter paper during filtration

**reverse reaction** from products to reactants (the right-to-left reaction)

**reversible** a reversible reaction is one that can go both ways, in other words, reactants react to form products and products can react to form reactants

$R_f$ **value** the retardation factor, which is calculated as the distance moved by a spot (from the pencil line) divided by the distance moved by the solvent front (from the pencil line) on a chromatogram

**roasting** a process in which metal ores are heated in air to convert them into metal oxides

**rust** the corrosion of iron in the presence of water and oxygen. The rust formed has the formula $Fe_2O_3 \cdot xH_2O$, where $x$ is a variable number, and can be called hydrated iron(III) oxide

**sacrificial protection** a method of preventing rusting by attaching a block of a more reactive metal to iron. The more reactive metals undergo oxidation in preference to the iron

**salt** a compound formed when a hydrogen is replaced by a metal or ammonium in an acid

**saturated compound** a compound containing only carbon–carbon single bonds with no carbon–carbon double or triple bonds

**saturated solution** a solution that contains as much dissolved solid as possible at a particular temperature

**scale up** make larger

**scrubbing** removing pollutant gases from the gases produced in a combustion reaction

**set up** the way that something is organised or arranged. This can refer particularly to chemical apparatus

**shells** a series of levels where electrons are found outside the nucleus. Also called energy levels. Each shell can hold a certain number of electrons

**shielded/screened** a term used to describe how the electrostatic force of attraction on an electron from the nucleus is offset by inner shell electrons. The inner electrons shield the outer electrons from the full attraction of the nucleus

**shift** move from one place to another

**simple distillation** a process used to separate two liquids of different boiling points, or to separate the solvent and solid solute from a solution

**slaked lime** calcium hydroxide

**soak** make very wet

**solubility** the mass of solute which must dissolve in 100 g of solvent to form a saturated solution at a particular temperature

**solubility curve** a graph showing how the solubility of a solute in a particular solvent changes with temperature

**solute** the substance that dissolves in a solvent

**solution** the mixture formed when a solute dissolves in a solvent

**solvent** the liquid that a solute dissolves in

**soot** black powder consisting largely of carbon

**spark** very small pieces of burning material produced in a fire or by hitting and rubbing two hard objects together that leads to a sudden discharge of light

**specific heat capacity** the amount of heat needed to raise the temperature of 1 g of a substance by 1 °C

**spectator ion** an ion that is not changed in a chemical reaction. It is omitted from both the reactant and the product sides when writing ionic equations

**spit** send out small pieces of something, for example from a fire

**stability** a term used to describe the relative energies of the reactants and the products in a chemical reaction. The more energy a chemical has, the less stable it is

**straight chain** a molecule with no branches in its molecular structure

**structural formula** a formula that shows how the atoms are joined together in a molecule. It is often written in a condensed form by omitting all the carbon–carbon and carbon–hydrogen single bonds

**structural isomerism** the existence of two or more different structures with the same molecular formula

**structural isomers** molecules with the same molecular formula, but different structural formulae

**sublimation** the change of state from a solid to a gas

**subscript** a number or symbol written below the line

**substitution** a chemical reaction in which an atom or group of atoms in a molecule is replaced by a different atom or group of atoms. For example, when alkanes react with halogens in the presence of ultraviolet light, the hydrogen atoms in the alkanes are replaced by halogen atoms

**successful collisions** collisions with energy greater than or equal to the activation energy. These collisions result in reactions

**tangent** a straight line that touches a curve, but does not cut across it

**tarnish** become dull and lose colour

**thermal decomposition** decomposition reaction that requires heating to occur

**titration** a technique used to follow the course of a neutralisation reaction between an acid and an alkali. It can be used to find out how much of an acid/alkali reacts with a certain volume of an alkali/acid

**trace** a very small quantity

**trickle** flow slowly in drops or in a thin stream

**triple bond** atoms sharing three pairs of electrons in a covalent bond

**ultraviolet radiation (UV light)** the part of the electromagnetic radiation spectrum that has wavelengths between those of visible light and X-rays. It is invisible to the human eye

**undergo** experience something

**unsaturated** compounds that contain one or more carbon–carbon double or triple bonds

**vapours** gases that are produced when a liquid is heated

**vigourously** if something reacts vigorously, it does so rapidly and with energy

**vinegar** a dilute solution of ethanoic acid in water

**viscous** a liquid that is resistant to flow

**volatile** a substance that evaporates easily

**waft** move gently through the air

**water of crystallisation** water molecules that are part of a crystal structure in which they are chemically bound up with a salt. They are represented by $\cdot n H_2O$ in the formulae of compounds, for example $CuSO_4 \cdot 5H_2O$ means there are five water molecules associated with each $CuSO_4$ unit

**weighted average** an average which takes into account the percentage abundance of the components

**wick** the piece of thread/string in a candle, or alcohol burner, from which the flame comes when you light it

**word equation** a representation of a chemical reaction using the names of the chemicals only

**yeast** a group of single-cell fungi that contain enzymes to convert sugar into ethanol

**yield** the amount of something that is produced in a chemical reaction

# INDEX